DATE			

Traceable
Temperatures

WILEY SERIES IN MEASUREMENT SCIENCE AND TECHNOLOGY

Chief Editor

Peter H. Sydenham
University of
South Australia

Traceable Temperatures

An Introduction to Temperature Measurement and Calibration

J. V. Nicholas
Measurement Standards Laboratory of New Zealand

D. R. White
Measurement Standards Laboratory of New Zealand

JOHN WILEY & SONS
Chichester • New York • Brisbane • Toronto • Singapore

Copyright © 1994 by John Wiley & Sons Ltd,
Baffins Lane, Chichester,
West Sussex PO19 IUD, England
Telephone (+44) (243) 779777

Other Wiley Editorial Offices

John Wiley & Sons, Inc., 605 Third Avenue,
New York, NY 10158-0012, USA

Jacaranda Wiley Ltd, 33 Park Road, Milton,
Queensland 4064, Australia

John Wiley & Sons (Canada) Ltd, 22 Worcester Road,
Rexdale, Ontario M9W 1L1, Canada

John Wiley & Sons (SEA) Pte Ltd, 37 Jalan Pemimpin #05-04,
Block B, Union Industrial Building, Singapore 2057

Library of Congress Cataloging-in-Publication Data

Nicholas, J. V.
    Traceable temperatures / J. V. Nicholas, D. R. White.
        p.   cm.   — (Wiley series in measurement science and
    technology)
    Includes bibliographical references and index.
    ISBN 0 471 93803 3
    1. Temperature measurements.   2. Temperature measuring
instruments — Calibration.   I. White, D. R.   II. Title.
III. Series.
QC271.N48   1994   93-5247
536.5′0287 — dc20   CIP

British Library Cataloguing in Publication Data

A catalogue record for this book is available from the British Library

ISBN 0 471 93803 3

Typeset by Laser Words, Madras
Printed and bound in Great Britain by Bookcraft (Bath) Ltd

Contents

8   RADIATION THERMOMETRY   283

Series Editor's Preface

The Series provides authorative books, written by internationally acclaimed experts, on the topics that constitute measurement science (today also known as sensing) and its engineering.

One line of books published in the Series is those related to the measurement of key physical variables. The technology and underlying scientific principles of measurement of the variable *temperature* have been published in this Series under the title "Temperature Measurement".

One important aspect of applying any measurement device is that of traceable calibration to ensure the determinations made are accurate and legally valid.

This Series was fortunate in obtaining the rights and authors' devotion to carry forward an outstanding publishing project well known to a selected few in physical standards laboratories. "Traceable Temperatures" is here presented as a completely rewritten and updated version containing the latest knowledge on how to obtain accurate operation from temperature sensors.

Peter Sydenham
Editor in Chief

Preface

We expect this book to shake your faith in temperature measurement because, unlike most other books on measurement, it emphasises the things that go wrong. We have done this because we believe that only by knowing what can go wrong can you be confident that your measurement is sound. This book then provides you with the means to develop your measurement expertise and increase your confidence in your temperature measurements.

Traceable Temperatures is an introduction to temperature measurement and calibration. We have put an emphasis on calibration not because we want to train everybody to do calibrations, but because calibration is a simple example of sound measurement design. We have tried to cater particularly for the beginner with modest experience who wishes to acquire expertise or knowledge quickly. It is therefore more of a self-teaching text rather than a handbook, although we have included some reference material. We have, however, written it for a wide readership, ranging from the beginner seeking help to experienced scientists and engineers; in particular readers who find that temperature is only one of their measurement responsibilities. We do not expect the book to be read and digested at one sitting; we hope you will grow into it as you become more proficient.

This book began in 1981 as a set of notes for a series of one-day workshops on temperature measurement, designed primarily to assist those seeking laboratory accreditation. The notes formed the basis for a bulletin, also entitled *Traceable Temperatures*, which was published in 1982 by the New Zealand Department of Scientific and Industrial Research (DSIR). We used the bulletin as the text for on-going workshops over the next ten years.

Over that period the concept of traceability has gained almost overwhelming importance, with many nations investing heavily in systems to ensure that traceability can be readily achieved. Traceability now clearly links all the people, organisations, documents, techniques, and measurements within a large and diverse measurement community. If we are to communicate and interact easily and constructively with each other and our clients, we must also be systematic and talk the same 'language'. Unfortunately there are still too many areas where this ideal has yet to be achieved.

In preparing this edition of Traceable Temperatures we have completely rewritten the text and restricted some of the scope. This was necessary to present a systematic approach and include our approach to calibration. We have also attempted to capture the trends of recent developments in international standards relating to measurement. The most important trends relate to the harmonisation of treatments of uncertainty in measurement and an emphasis on quality assurance (QA) systems and procedures. The simple procedures that we outline should simplify the task of those having to prepare detailed procedures of their own. In many cases the information can be used directly.

The approach should also assist in the interpretation and implementation of documentary standards — as much as is practicable.

We have tried to make the book, and each chapter to a lesser extent, as self-contained as possible. For this reason we have not provided extensive references. Those who require more information are reaching beyond the scope of the book or asking difficult questions. If you require more information, the references at the end of each chapter are a good point to start. We have also listed some good general references below, which complement the treatment of thermometry given here. Your National Standards Laboratory is also a good source of advice.

We recommend that you read all of the book to gain a broad view of thermometry practice. If you require a rapid introduction we recommend as a minimum the first half of each of Chapters 2 and 3, all of Chapter 4, and all of the chapter covering the thermometer of your choice. If you are involved in QA systems or have an interest in how the measurement system works you will also find Chapter 1 useful. If some of the terms are new to you, you will note that we have *italicised* terms which have a specific meaning to thermometrists and metrologists when they are first defined or encountered. The corresponding entry is placed in **bold** type in the index.

If there is a single message that we wish to convey it is this:

> For a measurement to be successful, traceability must be addressed at the planning stage.

That is, measurement and calibration are not separable and traceability is not something we can sort out after the measurement.

General Reading

Temperature (2nd edition), T. J. Quinn, Academic Press, London (1990).

Thermometry, J. F. Schooley, Chemical Rubber Press, Boca Raton, Florida, (1986).

Temperature Measurement, L. Michalski, K. Eckersdorf and J. McGhee. Wiley, Chichester (1991).

Principles and Methods of Temperature Measurement, T. D. McGee. Wiley, New York, (1988).

Industrial Temperature Measurement, T. W. Kerlin and R. L. Shepard. Instrument Society of America (1982).

The first two books concentrate on the science behind temperature measurement and are recommended reading for researchers and those establishing the ITS-90 scale directly. The third and fourth provide a very broad outline of the theory and operation of almost all types of thermometers and are suited for readers requiring more general information. The last book is one of the few texts that treats thermocouples correctly. It has a strong industrial flavour with information on response times of thermometers.

PROCEEDINGS OF SYMPOSIA ON TEMPERATURE

Six symposia have been held under the general title of *Temperature Measurement and Control in Science and Industry*. The proceedings of the first held in 1919 were not published; those of the second were published in 1941 and are now known as *Temperature, its Measurement and Control in Science and Industry, Volume 1, 1941* (Reinhold Publishing Co.).

The third symposium was held in 1954 and its proceedings were published as *Temperature, its Measurement and Control in Science and Industry, Volume 2, 1955*, published by Reinhold (New York) and Chapman and Hall (London), edited by H. C. Wolfe.

The fourth symposium was held in 1961 and its proceedings were published as *Temperature, its Measurement and Control in Science and Industry, Volume 3, Parts 1, 2 and 3, 1962*, published by Reinhold (New York) and Chapman and Hall (London), edited by C. M. Herzfeld.

The fifth symposium was held in 1972 and its proceedings were published as: *Temperature, its Measurement and Control in Science and Industry, Volume 4, Parts 1, 2 and 3, 1972*, published by the Instrument Society of America, edited by H. H. Plumb.

The sixth symposium was held in 1982 and its proceedings were published as *Temperature, its Measurement and Control in Science and Industry, Volume 5, Parts 1 and 2, 1982*, published by the American Institute of Physics, edited by J. F. Schooley.

The seventh symposium was held in 1992 and its proceedings published as *Temperature, its Measurement and Control in Science and Industry, Volume 6, Parts 1 and 2, 1992*, published by the American Institute of Physics, edited by J. F. Schooley.

Most of the symposia have been sponsored by the American Institute of Physics, the Instrument Society of America, and the National Institute of Standards and Technology. They have brought together many scientists and engineers involved in all aspects of temperature. The resulting volumes form a most important reference for thermometry as they cover all its aspects from theory to everyday industrial practice.

Acknowledgements

In preparing this book we have been very conscious of the thousands of man-years of research that lie behind thermometry, and we are aware that there is very little in this book which we can call ours — except of course the mistakes which may have crept in, and for which we apologise. We owe a debt to our many colleagues around the world who have given very generously of their results, time and thought to help us refine our idea of how thermometry works. In particular we thank Ron Bedford, John Anscin and Ken Hill (NRC, Ottawa); Maurice Chattle, Richard Rusby (NPL); Billy Mangum, Jim Schooley, Greg Strouse, Jacqueline Wise, George Burns, Robert Saunders, John Evans (NIST Washington); Luigi Crovini, Piero Marcarino, Francesco Righini, Franco Pavese (IMGC); Trebor Jones, John Connolly, Robin Bently, Corrinna Holligan, Tom Morgan (CSIRO Division of Applied Physics, Sydney); Piet Bloembergen, Martin de Groot (Van Swinden Laboratory); John Tavener, Henry Sostman (Isothermal Technology); Murray Brown, Ralph Payne (Land Infra-red); Heinz Brixy (IAW Jülich); Ray Reed (Sandia National Laboratories); Alan Glover and Malcolm Bell (Telarc NZ).

We would also like to thank our colleagues and the staff at the Measurement Standards Laboratory for their support and constructive criticism: in particular Sheila Coburn who tells us she still enjoys typing despite our best efforts; John Breen for preparing all the computer-graphic line drawings, Terry Dransfield, Pene Grant-Taylor, Hamish Anderson and John Bellamy for preparing the other figures and photographs; and Barbara Bibby for editing the manuscript.

Acknowledgements for Figures and Tables

With the exception of some of the photographs, all of the figures have been prepared by the authors (and helpers). However, we acknowledge with thanks the following people and organisations for providing photographs or information on which some of the figures are based.

Figures 3.1, 3.2, 7.10
Photographs and drawings supplied by Isothermal Technology Ltd, United Kingdom.
Figures 3.10, 3.11, 3.12, 3.15
From the BIPM booklets: *Techniques for approximating the international temperature scale*, and *Supplementary information for the international temperature scale of 1990*.
Figure 4.5
NBS monograph 126, *Platinum resistance thermometry*, US Department of Commerce, 1973.
Figure 3.14
Dr T. P. Jones, CSIRO Division of Applied Physics, Australia.
Figure 5.2(a)
Based on drawing from information bulletin: Minco Products Ltd, United States.
Figure 5.2(b)
Based on drawing from information bulletin: Sensing Devices Ltd, United Kingdom.
Figure 5.9
D. J. Curtis, *Temperature, Its Measurement and Control in Science and Industry*, Vol. 5, pp. 803-12 (1982).
Figures 6.3 and 6.14
Measurements carried out by Dr C. M. Sutton, Measurement Standards Laboratory of New Zealand.
Figure 7.5
N. A. Burley *et al.*, *Temperature, Its Measurement and Control in Science and Industry*, Vol. 5, pp. 1159-66 (1982).
Figure 7.15
K. R. Carr, *Temperature, Its Measurement and Control in Science and Industry*, Vol. 4, pp. 1855-66 (1972).
Figures 8.4, 8.17
Reproduced by permission of Land Instruments International Ltd.
Figures 8.5, 8.6, 8.16, 8.18, Table 8.3
D. P. DeWitt and G. D. Nutter, *Theory and Practice of Radiation Thermometry*, Wiley Interscience, New York (1988). Copyright © 1988 John Wiley & Sons Inc. Reprinted by permission.
Figure 8.9
Data supplied by Dr J. E. Butler, Naval Research Laboratory, Washington DC, USA.

1
Traceability

1.1 INTRODUCTION

An experimenter reports the result of an experiment as $20°C \pm 1°C$. At first sight we would conclude that the experiment was well conducted. But can we be sure? What if the experimenter had used one of the thermometers in Figure 1.1? Anyone who has experience in measurements has seen such horrors far too often. How then can we accept and understand the experimenter's temperature measurement?

As long as we treat the experimenter and experiment in isolation, very little meaning can be attached to the result. Not until we realise that any measurement is the end product of a long chain of events involving many people can we start to consider the quality of the measurement. The chain involves the manufacturer who made the instrument, the calibration laboratory who calibrated it, the international body who defined the units, various groups who set the specifications and procedures for the manufacturer and calibrator, and finally the experimenter and research team who designed the experiment. With so many people involved, the problem appears to be compounded!

The concept of *traceability* has developed to solve the problem of ensuring that the many links of the chain underlying any measurement are secure. Basically we need to know if all those involved have carried out their jobs correctly and we need to affirm this in a simple manner. Obviously a community effort is needed.

This chapter explores the wider aspects of traceability as it affects temperature measurements. Ensuring traceability will have a greater impact on your temperature measurements than any technological advance. Therefore an understanding of all its implications will make your task easier.

This chapter is split into two parts with a bridging section:

- The first part of the chapter is a selected history of thermometry. The history of thermometry is a fascinating subject in itself and the references given at the end of the chapter should be read for different perspectives. However, a different emphasis is presented here, stressing points that will have to be addressed by any system of traceability for thermometers. One way to view the history is to see it as presenting the concerns of individuals about the thermometers they have invented, which then spread outwards to the whole scientific community for a solution and finally involve the international community. The history will give you insights regarding the meaning of temperature and its measurement units, and will give you some background regarding other concerns which still need to be met.

Figure 1.1
Two identical spirit thermometers immersed in a well-stirred bath of chilled water at
4°C. One indicated temperature is clearly physically impossible, and the accuracy of
the other thermometer is also outside the tolerance for this type of thermometer.
Both thermometers showed no other physical problems and were as supplied by the
manufacturer to a standard specification. Such gross faults can be found in all types
of thermometers.

- The bridging section takes up the problem of defining terms, especially 'calibration' and 'traceability'. These words mean different things to different people with
  resultant misunderstandings. The meanings have developed historically and a clear
  understanding is needed to follow the modern implications of the words.

- In the second part of the chapter national systems are examined. The thrust of the
  history is effectively reversed. Now it is the international community which has a
  concern, namely that the individual who carries out measurements should not only do
  so reliably but also be able to prove it. Nations have to have measurement systems
  which enable the user to readily meet these requirements. We examine the main

features of the organisations involved and the role they play in ensuring traceability. However, the onus is clearly on the individual to carry out the measurement correctly. We therefore end up considering how to enable the individual to design temperature measurements, not only how to measure the physical temperature of a process but also how to link the measurement with the measurement systems. The rest of the text takes up this theme in more technical detail.

We should like to emphasise that our main concern is with the measurement of temperature, in that the practices outlined are those which have evolved for thermometry. While much of what is covered can be applied generally to other physical measurements and units, significant modifications may be necessary.

1.2 THE DEVELOPMENT OF THE CONCEPT OF TEMPERATURE

The history of thermometry can provide an insight into the problems associated with any temperature measurement. Four areas of concern are identified here and are coded P, S, M and T so that the related historical developments can be highlighted. Some of the problems raised have not yet been solved. However, most of the issues have been addressed, and these form the core of the subject matter of this text, namely how to make acceptable temperature measurements.

The four areas of concern are as follows:

- *Physical (P)* What is the meaning of temperature? It took nearly 300 years from the invention of the first thermometer until a good physical theory of temperature was developed.
- *Scale (S)* How should the measurement units for temperature be defined? Not until 1960 was a suitable unit defined and even now its practical realisation as a temperature scale is an on-going saga.
- *Measurement (M)* Can the measurement of temperature be approached scientifically rather than empirically, i.e. based on physical theory rather than experimental observation only? Temperature measurements are influenced by many physical variables and the evaluation of the resulting errors requires a systematic approach.
- *Traceability (T)* What is required for a measurement to be understood and accepted by others? Traceability as a concept has evolved to meet this challenge and is still undergoing evolution. Traceability requires that the three concerns above are adequately dealt with.

This brief history of the development of thermometry is of necessity selective and simplified so that the above features can be highlighted, and an emphasis is placed on important steps which bear on traceability. Some of the ideas encountered in the history will be met again later in the chapter.

1.2.1 Early thermometry — the measurement of heat

'Hot' and 'cold' were considered two separate properties of matter by the ancient Greeks, and this point of view prevailed at the time the first thermometers were invented. (At the time matter was considered to have only four properties, the other two being 'wet' and 'dry'.) Intuitive concepts of temperature developed because personal comfort is often determined by the feeling of hot and cold. The invention of the thermometer meant that concepts of temperature and heat needed alteration.

While the origin of the first thermometer is for historians to argue over, temperature-measuring devices had appeared in Italy by 1600 AD. The earliest devices were called *thermoscopes* and were based on the expansion and contraction of air with temperature (see Figure 1.2). Increasing temperature would cause the liquid level to fall in the stem. String appears to be tied to the stem to mark the liquid level and temperature changes could be measured off with a compass. Applications were found for the investigation of atmospheric temperatures and the monitoring of temperatures of medical patients.

In spite of the obvious instrumental difficulties with thermoscopes they were used for some time. One of their faults was that they also responded to a change in atmospheric pressure. The effect was not known originally, as there was no concept of an atmospheric pressure. The pressure effect was later removed by making corrections. Two important measurement concepts can be seen to have developed at this stage, namely:

Thermometers respond to more than one physical parameter   (M1)

and

Errors can be corrected for if they are known.   (M2)

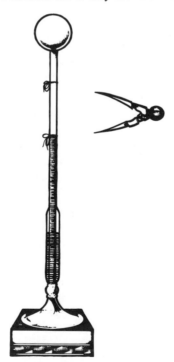

Figure 1.2
Early air thermometers or thermoscopes required a compass to measure the change in levels.

Two other problems — the lack of portability and the evaporation of water — led to the development of sealed thermometers. By the 1650s liquid-in-glass thermometers, similar to today's thermometers, were in production (see Figure 1.3). Great care and consistency were needed in producing what was later to be known as pure alcohol: at the time there was no concept of a pure substance as we use it today. Skilled Florentine glassblowers were able to construct the thermometers with sufficient dimensional consistency that different thermometers could give similar readings. The consistency was sufficient to enable the meaningful comparison of temperature measurements. There were, of course, no calibration services available or even known reference temperatures to check the thermometers. The measurement feature to be noted here is:

Consistent thermometers are often made in accordance with empirical rules.   (M3)

Our interest now moves on to the development of calibration methods and methods of making scales. A significant development in liquid-in-glass thermometry in this regard was the realisation that mercury gives a more linear scale than does spirit. The practical way of engraving a scale, by marking two points on the stem and dividing the interval into steps, also served to define the temperature scale, and this practice persisted in thermometry up to 1960. Even the modern scales are not entirely free of it. Thus two essential features of thermometry scales are introduced:

Linearity of temperature scales is important   (S1)

Figure 1.3
An early Florentine spirit-in-glass thermometer. By the mid-seventeenth century the glass-blower's art had reached a stage where consistent thermometers could be made.

and

Temperature scales are defined by their practical construction methods.   (S2)

Calibration was not possible until confidence had been established in physical situations which have fixed temperatures. Some of the early choices, such as the temperature of the first frost of the season, may seem quaint to us now but they were the first benchmarks for temperature scales. Accepted modern reference points, such as the melting point of ice, were considered variable because, among other factors, there was no concept of a pure substance on which to base a distinction between water obtained from various sources. Instead advances were based on the erroneous belief that all underground caverns were at the same temperature. However, since the temperature inside an individual cavern hardly varied during the course of a year, they were suitable for drift studies and other intercomparison work. Confidence slowly grew in various physical situations which allowed them to be used as reference temperatures. Another essential feature of thermometry scales is established:

Naturally stable temperatures are useful as references.   (S3)

After 1710 Fahrenheit was able to make reliable mercury-in-glass thermometers by using fixed points to calibrate them. While many of his procedures for producing quality thermometers appear to have been trade secrets it is fairly certain that he used the melting point of ice, the armpit of a healthy man and an ice/salt mixture as reference temperatures. The use of mercury allowed the scale to be divided into linear steps between the fixed points.

The development of Fahrenheit thermometers was a watershed point in the development of thermometry. They represented quality precision instruments, based on empirical arguments, that were used to measure a yet to be understood physical quantity. Thermometers were used to uncover the physical nature of heat and hence the physical basis of temperature itself. Investigators vouched for the quality of a temperature measurement and its traceability by declaring that it was Mr Fahrenheit's thermometers that had been used and thus we conclude that:

Reputation is an important factor in the assessment of measurement quality.   (T1)

The scale used by Fahrenheit became very popular in English-speaking countries, but not all early thermometers had the same scale. Figure 1.4 shows a thermometer which measures the degrees of cold with the boiling point of water at $0°$ and the freezing-point of water at $152°$. Which way up the scale went seemed to depend on whether the country of origin had a hot climate or a cold climate, e.g. in Russia people were more interested in how cold it was. When Celsius proposed his centigrade scale he had the boiling point of water at $0°$ and the freezing-point of water at $100°$. It appears to have been Linnaeus (see Figure 1.5) who turned it the other way up to give the scale we use today.

We can draw two further conclusions from this early thermometer development that still affect the quality or traceability claims made by many thermometer users today:

Temperature readings can be useful even if not fully understood   (T2)

Figure 1.4
A thermometer from the mid-eighteenth century.
It still has a large bulb and an inverted scale.

and

> Measurement and calibration methods are often highly empirical.   (T3)

1.2.2 Meteorological temperatures — an empirical approach

The various attempts to determine air temperature for meteorological purposes exemplify
the highly empirical approach taken by the early thermometrists. One of the earliest sets

Figure 1.5
Part of the frontispiece of *Hortus Cliffortianus* by Linnaeus 1737. The scale is similar to the present-day Celsius scale except for the abnormal position of 1 and the lack of negative signs.

of measurements made with a thermometer noted the change in temperature during the day. This was considered to correspond to how hot or cold a person felt. The next step was to find the temperature variations at other places. To do this a network of similar thermometers (such as the Florentine thermometer in Figure 1.3) were required, or alternatively thermometers were exchanged with those of other observers.

Problems arose when observers compared geographical sites that they considered similar. An observer at one site, where the thermometer was kept in the living room heated by a fire, would find less temperature variation than an observer who kept the thermometer in a spare bedroom. Another observer, who thought that thermometers should be located outside the window on the sunny side of the house, found an even wider temperature variation.

It took some time before everybody became convinced that meteorological readings should be taken outdoors, even though temperature variations could be just as great as indoors. Readings in the shade were different from those in direct sunshine; very windy sites gave different values to those from calm sites; isolated sites contrasted with sites near buildings.

What then is the correct air temperature? The answer for meteorological purposes is to define a method for taking the measurements — a method which gives a temperature reading as unaffected as possible by the surroundings, air movement, and direct sunlight. This requires the thermometer to be screened. A variety of screens have been developed, the most successful being a double screen developed by Stevenson. Today his design is the basis used for nearly all meteorological observations world-wide (see Figure 1.6).

The use of such a screen raises two important issues regarding traceability. The first is the fact that an empirical definition of a measurement device has had to be used in order to give meaning to a temperature measurement. Meteorological day temperature is basically the air temperature measured with a long time-constant which is not defined but depends on the screen and thermometer. In reality, the air temperature is known to fluctuate very rapidly. The screen also provides an unknown reduction in the measurement errors due to air movement and solar radiation. The World Meteorological Organisation gives no performance figures to control these quantities. Instead they are empirically fixed by specification of a loose guideline for the screen. However, there is a requirement that the maximum difference between screens should be less than $3°C$. Guidelines are also given about the sites on which the screens are located, to further reduce variability.

The second issue is the calibration of the screen. From an instrumental viewpoint we should consider both the screen and its internal thermometer as an air-temperature sensor and both should be calibrated together for their ability to determine the air temperature. For a variety of reasons it is not possible to calibrate the complete screen. Only the internal thermometer is calibrated, normally to an uncertainty of $±0.1°C$. Intercomparisons of screens at the same site are sometimes made for evaluation purposes, but seldom to calibrate them. Thus the comparison of data taken from different sites or even at different times is problematical, because there is very little information on which to base an uncertainty measurement.

1.2.3 Thermodynamic temperature — a physical approach

The development of reliable thermometers led to the development of scientific theories about heat and temperature. The thermometers showed that there must be an underlying physical principle which is independent of people's feelings of hot and cold. For example, when early thermometers were used to take temperature measurements of lake water which felt cool in summer and warm in winter, the thermometer showed that the lake water was cool in winter and warm in summer. Thermometers were clearly measuring something quite different from what people felt. Indeed, even hot and cold were

Figure 1.6
A large double-louvred thermometer screen of the Stevenson type. The screen contains a thermograph, wet bulb and dry bulb thermometers, and maximum and minimum thermometers. Note that the screen is located clear of buildings and well off the ground.

eventually found to be related to the same property called heat and were not separate attributes.

Temperature was originally considered a measure of the amount of heat in a body. Even today many people take this as an intuitive meaning for temperature. This changed with the development of the theory of heat which then gave rise to the new science of *thermodynamics*: the study of heat and work.

Around 1850, Kelvin was able to use thermodynamics to define temperature in terms of thermal efficiency. Kelvin had a choice of the functional form for the temperature, that is, the scale linearity has a certain amount of arbitrariness to it. A logarithmic function had certain advantages except that it complicated many formulae. In the end his choice followed as closely as possible the existing concept of linearity, based largely on the expansion of mercury (see highlight (S1) above). This choice meant that when a substance has zero heat it also has a zero temperature. Temperatures which follow this are sometimes called *thermodynamic temperatures* or *absolute temperatures*.

In 1909, Caratheodory further extended the theory of thermodynamics and developed a more fundamental concept of temperature. This concept showed that any choice of temperature scale was arbitrary.

These thermodynamic studies showed clearly that temperature is not the amount of heat but a heat density or heat potential, i.e. temperature is an intensive property of matter, not an extensive one. Denoting temperature as a heat potential means that temperature determines the direction of heat flow, which by convention is from high to low temperatures.

We have now reached the point in history after which there is very little further development of temperature as a scientific concept, and so we turn our attention to considering the nature of temperature. The concept of temperature arises from thermodynamics and therefore temperatures can exist only for systems that can be described by thermodynamics.

Thermodynamics is basically a statistical theory and applies only to systems containing a sufficiently large number of particles. Not all situations can have a temperature ascribed to them, e.g. ascribing a temperature to a single atom is meaningless. A meaning for temperature also implies the existence of *thermodynamic equilibrium*, i.e. a state where the thermodynamic variables do not change with time. In practice equilibrium can be achieved only approximately, either as a quasi-equilibrium where variations are slower than our ability to measure them, or over a local region where changes are small. In situations where there is no equilibrium we can still record a temperature because the thermometer sensor itself is usually a region of local equilibrium, even though it may not be in thermal equilibrium with its surroundings; meteorological air temperatures are such an example. These situations are not the exception and many empirical temperature measurements rely on the temperature sensor to define the temperature recorded. Thus we arrive at a very important principle in the interpretation of temperature measurements:

A thermometer reads its own temperature.   (M4)

Unexpected results can arise from the temperature concept because of the arbitrary way in which the scale is chosen. It is therefore essential to follow definitions. Consider a closed quantum mechanical system which can have a well-defined temperature. It is possible to put such a system into a stable state which has a negative temperature. The negative values are not simply due to a linear shift in the origin of the scale. Negative temperatures are not colder than the zero absolute temperature: instead they are hotter than an infinite temperature! This strange discontinuity can be removed by choosing an equally valid and arbitrary temperature scale T_1 where $T_1 = -1/T$. Even the convention that heat flows from high to low temperatures is arbitrary.

We can now identify three essential physical aspects of temperature:

Temperature is an intensive property of a body,   (P1)

Temperature indicates the direction of heat flow,   (P2)

and

Some systems do not have a temperature.   (P3)

Having an adequate concept of the physical meaning of temperature is of no use to us unless there is a way of measuring the thermodynamic temperature unambiguously. A rigorous relationship between temperature and some physical variable is needed in order to make a thermodynamic thermometer.

Unfortunately there is no rigorous thermodynamic treatment for the two most common types of thermometers, namely those based on the thermal expansion of liquids and those based on the electrical resistance of metals. The thermal expansion of gases is, however, well understood and provides a practical means of obtaining thermodynamic temperatures. *Gas thermometry* is based on the ideal gas law,

$$PV = NkT \qquad (1.1)$$

where P is the pressure of the gas,
    V is the volume of the gas,
    N is the number of molecules of the gas,
    k is Boltzmann's constant, 1.3807×10^{-23} J/K,
    T is the thermodynamic temperature.

Because absolute measurements of the four quantities P, V, N and k cannot be made with sufficient accuracy, the gas law is usually used in its ratio form:

$$\frac{PV}{P_0 V_0} = \frac{T}{T_0} \qquad (1.2)$$

where T_0 is the defined temperature of a fixed point, usually the triple point of water (see Chapter 3).

However, the inherent simplicity of this equation is lost when making real measurements. Because molecules have a finite size, no real gas follows the ideal gas behaviour, although hydrogen and helium give close approximations. The container to hold the gas changes size with temperature and pressure and also absorbs gas on its surfaces. Temperature measurement with gas thermometry is therefore a lengthy process but it has been, until recently, virtually the only way to measure thermodynamic temperatures.

Gas thermometry has been used extensively to develop temperature scales. Figure 1.7 shows an early type of gas thermometer.

The thermodynamic properties of ideal gases have also been used in acoustic thermometry to determine thermodynamic temperatures. Otherwise only two other systems are used for thermodynamic thermometry, namely *radiation thermometry* and *noise thermometry*. Both of these use the ratio technique to reference the temperature to a fixed point.

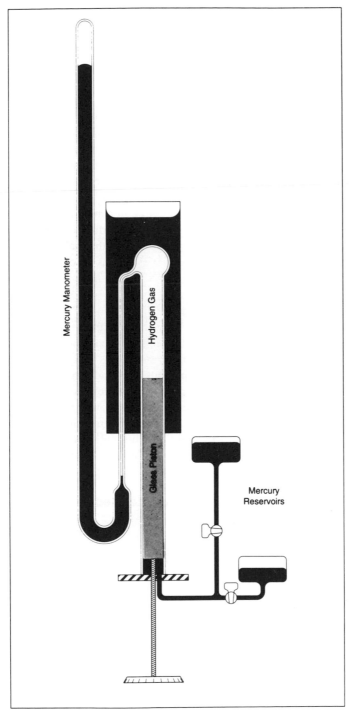

Figure 1.7
An early constant-pressure hydrogen gas thermometer designed by Kelvin. The volume of gas in the thermostatted chamber can be altered by adjusting the glass piston so that the pressure from the mercury manometer is constant. The other mercury containers provide sealing for the piston.

The radiation thermometer is based on the Stefan–Boltzmann law (see Chapter 8):

$$E = \varepsilon \sigma T^4 \tag{1.3}$$

where E is the energy radiated per unit area of a surface,

ε is the emissivity where $\varepsilon \le 1$,

σ is the Stefan–Boltzmann constant, 5.6705×10^{-8} W m^{-2}K^{-4}.

Because of the strong temperature dependence, the method has been more applicable at high than at low-temperatures. However, with the advent of cryogenic radiometers the useful range extends below 0°C and has provided a valuable check on gas thermometry.

Noise thermometry is based on the noise voltage generated due to thermal fluctuations in a resistor. The noise is usually called Johnson noise and can be represented to a good approximation by the equation

$$\overline{V^2} = 4kTR\Delta f \tag{1.4}$$

where V is the noise voltage (averaged),

R is the resistance,

Δf is the bandwidth of the measurement.

The full version of the equation should be used for extreme frequencies or temperatures. As the noise voltage is very minute this technique has only been used at high and low-temperatures where other noise sources are smaller.

Good agreement amongst the different methods is an indication that the main sources of errors have been found and the new temperature scale (see Chapter 3) is based on the best modern evaluations. Even so there still remains about 3 mK uncertainty in the temperature in the room temperature range. Temperature measurements for science and industry require an uncertainty of 1 mK in their definition and as yet this is not possible thermodynamically.

1.2.4 The Metric Treaty — temperature defined

The Metric Treaty, signed by several countries in 1875, is now adhered to by about 50 countries. The importance of the Treaty is not so much that it establishes a metric system of units but rather that nations agree on the meaning of measurement units. Unambiguous meanings for units are essential in order to avoid confusion in commerce, science and engineering. The Treaty agreement concerns the practical way in which units are applied to the measurement of physical quantities. The theoretical meaning attached to the physical quantity itself comes from the relevant field of science or engineering.

The treaty sets up a general conference which passes formal resolutions to which the member nations are expected to adhere. The conference, CGPM, has an organisational structure as given in Figure 1.8. Nearly all resolutions come from recommendations from the other parts of the organisational structure, to which member nations have ample opportunity to contribute. The CGPM authorises the International System of Units (SI) which defines seven base units, of which temperature is one, and relates all other units to the base units. The CGPM also recommends research and development for the maintenance of the SI. New technologies often need new measurement methods and higher

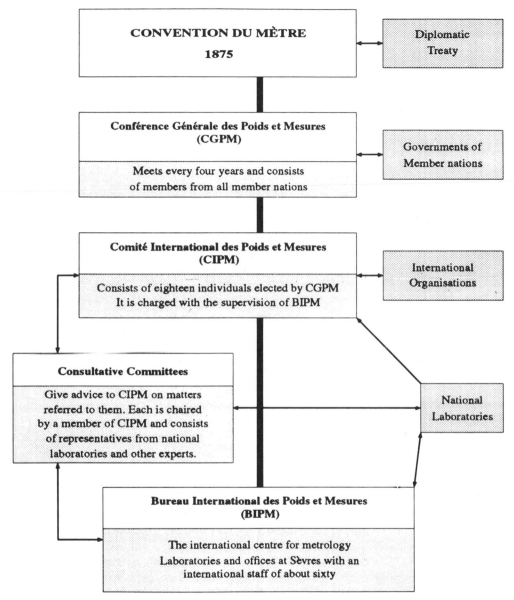

Figure 1.8
Organisational structure under the Convention du M ètre.

accuracies, and thus can expose weaknesses in existing definitions of the units. Coordination of the research and development is through the CIPM, the BIPM, the Consultative Committees and their interaction with national laboratories and universities. We emphasise that the range of concern is with the way the definitions and their dissemination can be carried out practically, not how they could be done in principle.

With the Treaty there has been a profound change in the way in which the quality of a measurement is assessed. Previously, as in highlight (T1), credibility depended on reputation. Thus, claiming that Mr Fahrenheit's thermometers had been used in your measurements would indicate the degree of trust others could put in your measured temperatures, because of the general recognition of the quality of Fahrenheit thermometers. Clearly this situation becomes untenable when the number of thermometer manufacturers and users become large: even more so if new thermometer makers decide to use their own proprietary temperature scale to identify their thermometers. Internationally agreed units allow measurement quality to be claimed on the basis that the international methodology has been followed.

Thermodynamic methods of temperature measurement, as noted in the previous section, give neither the accuracy required nor the convenience of use. Therefore the more recent history of thermometry has been dominated by the need for a defined *temperature scale* which, while empirically based, gives a close approximation to the thermodynamic temperature. The scale must be practically defined: highlight (S2) regarding scales is thus still relevant today.

Once the scale has been defined, temperature becomes traceable, since the chain of traceability leads back to the defined temperature scale. A scale temperature is commonly referred to as 'temperature' and the term 'thermodynamic temperature' is used if a distinction is to be made even though the thermodynamic temperature is the 'true temperature'.

The first official temperature scale was the Normal Hydrogen Scale adopted by the CGPM in 1889 and was based on the work of Chappius. Hydrogen was used in a gas thermometer but without all the corrections needed to be truly thermodynamic (see Figure 1.7). Two fixed points, the ice point, 0°C, and the boiling point of water, 100°C, were used to fix the scale making it a centigrade scale. This scale allowed very accurate thermometry techniques to be developed.

However, the gas thermometer was very difficult to use and a more practical scale was required. The result was the International Temperature Scale of 1927, ITS-27, which was also a centigrade scale. The scale covered a limited temperature range from −190°C upwards and the scale consisted of a set of defined fixed points together with specified instruments and equations to interpolate between fixed points. The instruments selected were resistance, thermocouple and radiation thermometers of a special type. The thermocouple was recognised as unsatisfactory but there was no practical alternative.

The International Temperature Scale of 1948, ITS-48, essentially tidied up some of the unsatisfactory features of ITS-27, and with the lowest temperature raised to −183°C. Temperature scales of necessity were compromises between existing techniques and newly emerging techniques.

With the adoption of the International System of Units (SI) the CGPM finally adopted a definition for a unit of temperature in 1960. The definition fixed the value for the triple point of water and thus did away with the centigrade scale then in use. This puts the unit of temperature on the same footing as all the other base units of the SI. As a consequence of this change the temperature is to be expressed in kelvin, K, *not* as degrees Kelvin or °K.

Defining the meaning of the unit does not get around the problem of providing a practical scale of high-resolution. Revision was also made of ITS-48 as the International Practical Temperature Scale, IPTS-48, with the word 'practical' in its title to emphasise the point that it did not necessarily follow thermodynamic temperatures.

In 1968 an International Practical Temperature Scale, IPTS-68, was adopted. IPTS-68 was not, in principle, a centigrade scale, even though the boiling point of water was taken as 100°C. IPTS-68 was an upgrade on IPTS-48, but with considerable numerical changes due to better thermodynamic determinations of temperatures. The lowest temperature was 13.8 K. IPTS-68 was revised in 1975.

The lowest temperature of each scale was determined by the research activity of the time. IPTS-68 very soon needed extension due to rapidly developing research in the area. In 1975 a Provisional Temperature Scale, EPT-76, was introduced for the range 0.5 K to 30 K.

The current scale is the International Temperature Scale of 1990, ITS-90, and is covered in more detail in Chapter 3. A major change from previous scales is the dropping of reference thermocouples by extending the range of platinum resistance thermometers and radiation thermometers. A variety of ranges are introduced that overlap with each other to allow flexibility in the choice of fixed points and temperature changes.

Since 1889 these defined scales have provided the official way for traceability to be claimed. There can be marked differences between the scales and care is needed when comparing temperature results from different years (or even the same year!) to ensure they were made with the same scale. BIPM provides tables to convert one scale to another and Appendix B gives the table to convert IPTS-68 to ITS-90.

1.2.5 An overview — the way ahead

The historical survey has come to two definite end points:

- a physical meaning for temperature; and
- a scale to measure temperature.

The survey has also highlighted features of direct applicability in modern thermometry practice. The features (P, S, M and T) are re-stated below and their relevance to topics covered in this text is outlined.

Two of the major concerns of the early inventors of thermometers have now been essentially answered. The highlighted features connected with the physical meaning (P) and the scale (S) are:

(P1)	Temperature is an intensive property of a body;
(P2)	Temperature indicates the direction of heat flow;
(P3)	Some systems do not have a temperature;

and

(S1)	Linearity of temperature scale is important;
(S2)	Temperature scales are defined by their practical construction methods;
(S3)	Naturally stable temperatures are useful as references.

In contrast with the above, the two other concerns, connected with measurement (M) and traceability (T) are not completely met. They are still on-going concerns for today's thermometer users and much of this text deals with solutions to them. The highlighted features are:

(M1)   Thermometers respond to more than one physical
        parameter;
(M2)   Errors can be corrected for if they are known;
(M3)   Consistent thermometers are often made in accor-
        dance with empirical rules;
(M4)   A thermometer reads its own temperature;

and

(T1)   Reputation is an important factor in the assessment of mea-
        surement quality;
(T2)   Temperature readings can be useful even if not fully un-
        derstood;
(T3)   Measurement and calibration methods are often highly em-
        pirical.

As we shall see later (Section 1.3.2), traceability (T) is about a chain of events linking the measurement (M) of a physical temperature (P) to its value expressed accurately in terms of the temperature scale (S). Thus all the highlighted features are pointers to essential information that the thermometry user needs to take into account. However, not all features can be adequately dealt with in this text.

The temperature properties of a physical system are the subject matter of thermodynamics (P1) and heat transfer (P2). Both subjects are beyond the scope of this text which assumes only a basic knowledge of heat in order to study common calibration requirements. On the other hand, the temperature scale is fundamental to calibration. The current scale is examined in Chapter 3 and it will be seen that the three features above ((S1), (S2) and (S3)) form the foundation of the scale. As noted in Section 1.2.3 a significant modification is that the concept of linearity is firmly tied to thermodynamics.

Most of this text is concerned with the physical problem of making a temperature measurement (M) and establishing its traceability (T). The first two measurement features ((M1) and (M2)), identifying what affects a measurement and how to assess errors, are the basis of the main subject matter of this text. A systematic measurement-science approach is outlined in Section 1.4.4. The basic mathematical tools for handling errors are covered in Chapter 2 while Chapter 4, on calibration, discusses how to obtain the data on which to assess the dominant errors for a thermometer. Chapters 5–8 then provide the details for treating the four main types of thermometers: resistance, liquid-in-glass, thermocouple and radiation thermometers.

The issue of making suitable thermometers (M3) can also be tackled systematically in a scientific manner using the principles outlined in this text rather than using an empirical approach. However, the detail more properly belongs to instrumentation science, which is outside the scope of this text.

The introduction of an empirical theme picks up on the last two highlighted features on traceability (T2) and (T3). Empiricism is especially acute in thermometry because a thermometer always returns a temperature reading of some kind (M4). Even if there is no defined temperature (P3) or thermal equilibrium has not been established, an apparent temperature can be assigned to the system. Because it is often very difficult to relate the temperature of the thermometer to the temperature of the physical system under study, methods are needed to cope with this. For traceability the approach is to document what is done using quality assurance procedures (see Sections 1.3.2 and 1.4.1) and develop a consensus through documentary standards (see Section 1.4.3). In Chapter 2 an approach is given for making uncertainty assessments in these situations. Chapters 5–8 show examples of empirical approaches to calibration.

Finally, we look at the first traceability feature (T1) regarding reputation. Reputation is still an important criterion on which to base a claim that traceable procedures have been followed. The big difference is that nowadays the reputation is not that of the experimenter or the manufacturer but that of an independent accreditation body which is familiar with the measurement practices of the user (see Section 1.4.1).

1.3 METROLOGICAL TERMS

In the historical outline we have seen the development of a definition for temperature. Definitions for the terms used in a subject are very important for its orderly development. In turn, as the subject develops, this development causes the definitions to change. *Metrological terms*, which are the definitions, jargon and notation used in the science of measurement, are no exception.

A consequence of changing definitions is that at any one time there may be no universal meaning for a metrological term. Also, because metrology often has legal implications some meanings become frozen but vary from country to country. International agreements which involve any metrology will therefore require clearer meanings for metrological terms. The BIPM in conjunction with other international agencies has published an official glossary of terms in an attempt to more closely define meanings for international acceptance.

The two important terms, *calibration* and *traceability*, do not have universal meanings. Both terms appear to be evolving in meaning and are strongly interrelated. The evolution seems to be occurring as international requirements for measurement practices become more specific. The words are moving from general descriptive meanings to strong technical meanings. We examine the meaning of these two terms here because they are crucial to understanding the main thrust of this text.

The meaning of other metrological terms used in this text should be clear from the definitions given here and from their context. The references given at the end of the chapters should be consulted if you are unsure.

1.3.1 Defining calibration

The word 'calibration' has changed its meaning with time and three distinct meanings can be distinguished. The more modern meaning, which we wish to use here, does not appear in dictionaries before 1940 and indeed even one current modern dictionary does not give it. Dictionaries, of course, follow the general use of a term and not necessarily the technical usage.

The word 'calibration' appears to derive from an Arab word for a mould for casting metal. This is presumably because calibration originally referred to the making of guns by casting metal. Calibration referred either to the means used to determine the calibre (or bore) of the gun or to the determination of the range of the gun. Instrument users are not likely to be confused by this usage but the same cannot be said for the other two developed meanings.

The second and more common meaning of the word 'calibration' is the checking and adjusting of the bore of an instrument and the application of a scale by the manufacturer. Extension of the concept of a bore can be readily applied to other devices such as electrical meters or even alloys which are adjusted for desired resistance or thermo-electric properties. That is, a calibration refers to the set of operations carried out by an instrument manufacturer in order to ensure that the equipment has a useful measurement scale. A slight extension of this meaning is the facility most manufacturers provide to enable the user to make adjustments to an instrument after repair or to compensate for drift. This process of adjustment is often what users expect when they send their instruments in for calibration.

The third meaning for calibration, as used by metrologists, is defined by the ISO as:

Calibration

The set of operations which establish, under specified conditions, the relationship between values indicated by a measuring instrument or measuring system, or values, represented by a material measure, and the corresponding known values of a measurand.

Notes

1   The result of a calibration permits the estimation of errors of indication of the measuring instrument, measuring system or material measure, or the assignment of values to marks on arbitrary scales.

2   A calibration may also determine other metrological properties.

3   The result of a calibration may be recorded in a document, sometimes called a calibration certificate or a calibration report.

4   The result of a calibration is sometimes expressed as a calibration factor, or as a series of calibration factors in the form of a calibration curve.

This definition is best seen as a pointer to a meaning of calibration acceptable to modern quality assurance (QA) and accreditation practices (see Section 1.4.1). It represents the current practice and if the emphasis is put on the Notes the definition gives rise to a valuable concept of calibration. The Notes can be restated in a stronger form:

• An assessment of the uncertainties in the instrument readings is required.

• Other important properties must be evaluated, such as pressure effects, the influence of vibrations and immunity to electrical noise.

• A calibration certificate must be produced.

• The calibration provides the means for interpreting the readings. Further, any correction method for the errors should apply unambiguously to all readings.

Clearly this definition is different from the previous meaning and in practice many calibrations do not meet all the criteria. We will generally reserve the word 'calibration' for this third meaning, which corresponds to a formal assessment of the accuracy of an instrument. The second meaning will be referred to as *adjusting* or *checking* the instrument.

Normally no adjustment of the instrument is expected during the calibration so that the instrument is physically unchanged after calibration. In practice the calibration laboratory may have made some adjustment before the calibration proper, but usually it assumes that the instrument is a 'black box' in a good state of repair and adjustment. At the end of the calibration the instrument is not better, but the certificate may allow a better accuracy to be achieved by processing the readings, as well as increasing the confidence in the readings from an evaluation of the uncertainties. Besides the technical benefit there is a legal benefit for traceability, since the certificate provides a formal declaration to demonstrate to all that the instrument performed to the specification required of it.

The distinction and confusion of these last two meanings for calibration can be seen by considering so-called 'self-calibration'. Modern instruments usually convert an analogue signal to a digital signal and use a microprocessor to process the signal. A degree of 'intelligence' is sometimes added to allow the instrument to check itself against an internal reference and make adjustments to its scale accordingly. Thus this can be considered a self-calibration in terms of the second meaning but not in terms of the third meaning. Self-calibration is contradictory because an independent reference standard is required for the calibration of an instrument to be traceable. Further, calibration philosophy is mostly built around passive instruments whose properties change slowly with time; not instruments which make rapid unscheduled and unrecorded changes. Self-calibration does not therefore do away with the need for calibration but properly implemented may extend the time between calibrations. We add the observation that measurement and calibration procedures try to minimise intelligent interference because it is so unpredictable!

One purpose of calibration is do something that is basically impossible, namely predict the future behaviour of the instrument. This is essential if the user is to be able to claim traceability during 'reasonable' use of the thermometer. Reasonable use will involve among other things the regular checking at, say, an ice point, to ensure that the certificate values are still appropriate. The prediction will require the use of previous calibration records, adjustment and repair records, the specification of the instrument, and the known behaviour of similar instruments (the *generic history*) in order to establish the past instrument stability and infer the future stability and the period the instrument can be used with confidence. We would like to emphasise that many calibrations rely heavily on the generic history of instruments and therefore may appear not to cover all the points raised here. The apparent simplicity of such calibrations cannot be extended to instruments with no generic history.

Instrument manufacturers do not necessarily calibrate all their instruments before sale and instead they rely on their quality assurance systems to produce good instruments. Failure of new equipment to meet specification is unfortunately common, with some studies showing up to 50% failure. Our current experience of a 10% to 15% failure rate for thermometers is more typical. Therefore a calibration obtained through an accredited calibration laboratory gives an assurance of performance. In general there appears to be no way to determine the quality of the instrument from the reputation of the manufacturer. For example our observations are that reputable high-precision instrument makers can produce near faultless performance for the measurement capability but the data processing and computer interface may not be so satisfactory.

1.3.2 Defining traceability

The word 'traceability' has a wide range of related meanings, the most appropriate dictionary meaning being 'the ability to follow a path exactly'. Compare this with the ISO definition:

Traceability
The property of a result of a measurement whereby it can be related to appropriate standards, generally international or national standards, through an unbroken chain of comparisons.

This definition really only adds one piece of information, but an important piece. It tells us where the path starts and ends, otherwise the definition merely replaces the metaphor of a path with the metaphor of a chain. Note that the thickness of the chain links or the width of the path is not specified; compare this with the dictionary use of 'exactly'. Also, the starting point of the chain is the measurement and not the measuring instrument which is often the case in other definitions. Using the measurement as an end point is practical but, at least in the case of temperature, this procedure needs extension.

Let us first consider the wide range of meanings that can be attached to traceability. As an example, take the contract requirement that the thermometers should be 'traceable to NIST' (NIST is the National Institute for Standards and Technology, the national standards laboratory for the USA, formerly known as NBS, the National Bureau of Standards). By taking the general meaning of traceability it is easy to show that the requirement means nothing. NIST has been very active in temperature measurements, making many major advances, and so it would be very difficult to find a thermometer anywhere in the world that could not be traced back to NIST. We may need a large team of investigators to do so, as some of the paths may be very convoluted and long and some paths may be tenuous, but in principle it could be done. In fact if it was not possible then it represents a failure of the national standards laboratories because one of their functions is to provide for uniform measures, nationally and internationally.

Clearly this meaning of traceability is not useful and some paths need to be excluded, especially those that are too hard to trace and those that are so broad that there is no confidence in the accuracy. Perhaps we should have started at the other end of the path and have asked NIST what paths to follow. Now we find that NIST, while it can pass opinions, has no legal authority to say which are approved paths. Thus we have ended up in an 'all or nothing' situation with respect to traceability.

We thus need another way of determining what traceability means and, as with other words with variable meaning, we have to rely on the context in which they are used to decide on the meaning. Thus if a contract requires traceability then hopefully the context in which it is asked will determine the meaning; certainly the dictionary will not be helpful!

We examine two distinct ways in which traceability appears to be used. The first model, the filing cabinet model, goes like this: for each measuring instrument used there shall be a calibration certificate kept in the filing cabinet, which allows the location to be traced of another filing cabinet, and so on until in a filing cabinet is found a certificate bearing the NIST logo (or the NBS logo, since no timeframe was defined!). When the filing cabinet model is seen as the minimum extra needed to be done by a competent laboratory in order to obtain traceability, then the model is reasonable. However, when it becomes all that any laboratory needs to do for traceability, then the filing cabinet model is meaningless. Possession of a certificate is no assurance that measurements will be carried out competently.

The issue of laboratory competence brings us to the second model, the laboratory accreditation model, which relies on the reputation of the accreditation authority to assert the quality of measurements or calibrations. Not all accreditation authorities are

Table 1.1.   International requirements related to the quality of tests and measurements.

(a) *ISO Standards*

ISO 8402 : 1966	Quality — Vocabulary
ISO 9000 : 1987	Quality management and quality assurance standards — Guidelines for selection and use
ISO 9001 : 1987	Quality systems — Model for quality assurance in design/development, production installation and servicing
ISO 9002 : 1987	Quality systems — Model for quality assurance in production and installation
ISO 9003 : 1987	Quality systems — Model for quality assurance in final inspection and test
ISO 9004 : 1987	Quality management and quality system elements — Guidelines
ISO 10012-1 : 1991	Quality assurance requirements for measuring equipment — Part 1: Management of measuring equipment

(b) *ISO/IEC Guides*

Guide 2	Definitions (1986)
Guide 25	Requirements for technical competence of calibration and testing laboratory (1990)
Guide 38	Requirements for the acceptance of testing laboratories (1983)
Guide 40	Requirements for the acceptance of certification bodies (1983)
Guide 43	Proficiency testing (1984)
Guide 45	Presentation of test results (1985)
Guide 49	Quality manual for a testing laboratory (1986)
Guide 54	Recommendations for the acceptance of accreditation bodies (1988)
Guide 55	Recommendations for the operation of testing laboratory accreditation systems (1988)

(c) *ISO/IEC/BIPM/OIML Documents*

International vocabulary of basic and general terms in metrology (VIM): 1984	Issued by BIPM, IEC, ISO and OIML

equal and we specifically refer to those that follow ISO Guide 25 (see Table 1.1) and thus are integrated into the ISO 9000 quality assurance systems. Under such a scheme the traceability requirement is almost inseparable from the quality assurance requirements.

Detailed management requirements are expected for the measurement equipment such as:

- selection and purchase procedures,
- calibration by approved laboratories,
- transportation safeguards,
- appropriate storage conditions,
- use by qualified personnel,
- re-calibration periods established,
- checks and repairs scheduled and recorded,
- data handling checked independently,

- mistakes detected and dealt with,
- records held for a specified period.

Most requirements are plain commonsense. A record needs to be kept so that we know what happened to the equipment 'from the cradle to the grave'. Thus traceability implies not that there is a path but that there is a path which is so well signposted that we will not be led astray. That is, the accuracy of the instrument is demonstrable.

All this emphasis on keeping records should not obscure the fact that traceability is fundamentally about having the technical competence to make measurements to the required accuracy. By careful study of appropriate texts, such as this one, you should gain sufficient knowledge to convince yourself that you can measure temperatures accurately. Once you can convince yourself then you will have sufficient evidence to convince any knowledgeable technical person. Legal obligations may also have to be met and these may have their own peculiar requirements. Hopefully they should follow the technical requirements. The laboratory accreditation model can fall into the same trap as the filing cabinet model if all the emphasis is on the paperwork.

Having covered a range of meanings for traceability we now examine what important properties traceability should have if it is going to be of use. Clearly the context for traceability will be within the laboratory accreditation model. Two further aspects of traceability need to be considered. They affect the end point of the path traced; namely *accuracy* and *feasibility*.

Accuracy does not imply that a measurement is carried to as many decimal places as possible. Accuracy is about having the confidence that a measurement result gives the value of the temperature within a known range. Chapter 2 discusses the issue of accuracy further. One way to define traceability would have been as 'the ability to demonstrate the accuracy of a measurement'. Then all the traceability requirements are a checklist to ensure all relevant factors affecting the accuracy are taken into account. There is a remaining difficulty in that the end point is not clearly specified in this description.

The other concern is feasibility. Traceability must be practical in that any requirement is achievable. Therefore consensus agreement on what is acceptable is required when the ideal is not possible. For example to insist on all temperature measurements being made to the current temperature scale would be wrong in that temperatures lower than 0.65 K are not defined. Another area where consensus compromise may be required is when no suitable calibration service is readily available due to the lack of infrastructure in the national measurement system.

To conclude we give a definition of traceability in order to highlight the points raised.

A definition of traceability

Traceability is the ability to demonstrate the accuracy of a measurement in terms of the SI units.

Notes

- demonstrate — the evidence should be clear, simple as possible and readily available. Proof that such evidence exists will usually be through independent laboratory accreditation schemes.

Continued on page 25

Continued from page 24

- accuracy – this will usually be expressed as the uncertainty in the measurement, but subjective appraisals may need to be allowed for where appropriate.
- SI units – in practice the scientific and legal meanings of all units derive from the SI definitions and hence they are the logical end for the path traced but this requirement may not be feasible for all quantities.

1.4 NATIONAL MEASUREMENT SYSTEM

Having a knowledge of what is necessary for good measurements (i.e. achieving traceability) is of no use unless the theory can be put into practice. Much of what is required is beyond the direct influence of an individual. Fortunately many governments take an interest in their national ability to measure accurately because this ability affects the country's wealth and welfare. Good measurement practice requires several services to be readily available:

- training of staff in measurement techniques;
- supply of measuring instruments;
- repair and servicing of instruments;
- calibration of instruments;
- specifications and procedures for measurements; and
- certification of measurement results.

The total of all these services inside a country can be considered together as the National Measurement System (NMS), a concept which has grown in importance since 1970. Planning of the NMS by government ensures that the necessary services are available. This enables regulatory and contractual requirements to be met by industry.

Only four of the main components of the NMS, namely training, calibration, specification and certification, are considered further here. Requirements on the user with respect to supply and repair of instruments are covered by these four. Note that the system requirements for calibration are quite different from those for repair and servicing, even if carried out by the same organisation. The traditional view of calibration would have seen it as part of the service check.

Training is essential throughout the whole NMS to ensure that those involved are technically competent to make measurements. Unlike the other three components, there is no recognised organisational structure to achieve this. Indeed the subject matter often lacks a formal basis for an educational curriculum.

Figure 1.9 outlines the main external components relevant to traceability, that is calibration, specification and certification. The figure gives only the more visible organisations in a well-organised NMS. In practice there are many more interactions. An important role of the organisations is to ensure that there is a basis for agreement on what is to be measured, at what accuracy, and how to carry out the measurement.

In Figure 1.9 the main traceability path is shown by the heavy line. Documentary input which affects the path is also shown by the other lines. Accreditation bodies for testing and calibration can be quite separate from each other. The International Labora-

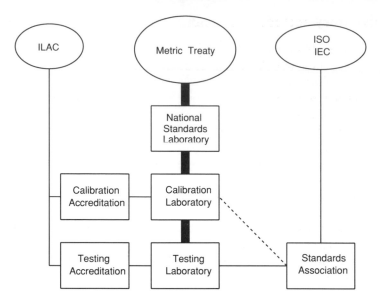

Figure 1.9
Traceability links in a national measurement system.

tory Accreditation Conference (ILAC) provides a forum for cooperation between bodies. Tests are usually made to specifications. The specifications may be based on those recommended by the National Standards Association or adopted from international ones. General documentary standards come under the International Standards Organisation (ISO). Electrical standards come under the International Electrotechnical Commission (IEC).

Science research laboratories need the calibration services of the NMS. They may not require the formal specification standards and accreditation services. Science is largely international in scope and relies on peer evaluation to provide the equivalent of specification standards and accreditation. Discerning peers will accept results only from other researchers known for their careful measurement methods as are well documented in their published papers.

Exercise 1.1

- Draw a diagram similar to Figure 1.9 for your temperature measurements. Then note how the procedures differ for any other measurements you make.

- If possible obtain the names and addresses of the organisations involved in these procedures and the name of a contact person.

- Include any linkages to organisations outside your NMS.

- Indicate in the diagram if you are subject to more than one accreditation body or standards association and show their international linkages.

- Scientific users can give the relevant scientific references instead of the organisations.

- Are your measurements traceable according to the definitions given in this chapter?

1.4.1 Laboratory accreditation bodies

To be useful traceability must be clearly demonstrable to others. While you may be happy that your path is well-defined, other people may not have the time or resources to check it out. They may, however, accept the word of a third party who has followed the path. Accreditation schemes provide the means to have the path checked to internationally accepted requirements. The accreditation body empowers an organisation to issue certificates for product conformity and the certificates are recognised as if they had the authority of the accreditation body. That is, an organisation does not have to rely on its own reputation to certify the quality of the product. The independent accreditation body needs to have its own reputation and hence acceptability. International acceptance of accreditation bodies does not come through ILAC. Rather, it relies on mutual agreements between bodies in different countries.

A useful concept is to consider the measurements as a transaction involving a supplier and a customer. The supplier is the laboratory or person carrying out the measurements, and can be identified as a single entity. The customer, the user of the measurement results, may be an individual or a group or the public in general. The important point about customers is that if they are not satisfied you may find yourself out of business. Customers for measurements are often not the purchasers of the product, e.g., the funding agency for a science research laboratory may have no use for the measurements made, but expects that the real customers, other scientists and engineers, will. Requirement for accreditation will usually come from the customer who wants to be assured of quality.

First look at a transaction. In a supplier–customer transaction the customer may choose on reputation or require the supplier to be inspected, assessed or audited by the customer. With two parties involved, determining the customer's requirements and meeting them should be straightforward, but unfortunately very few transactions involve only two parties. For example an aircraft manufacturer needs to make aircraft safe for passengers, to enable an airline to sell tickets. Neither the passengers nor the airline are the direct customers for the measurement made, say, to ensure the correct heat treatment of the aircraft parts. The state agency for airworthiness is the customer and their requirements have to be met by the measurement supplier. If the aircraft is to fly to other countries then there are more customers' requirements to meet and the supplier hopes that none are incompatible.

Two problems for the measurement supplier arise. The first can be stated as 'many customers, many inspections', and the overheads of the measurement laboratory could be high to meet all the inspection requirements. The other problem is more subtle. The state agency, for example, appears as a third party looking after the interests of the ultimate user of the product, the general public. To the measurement laboratory, however, the state agency is the main customer and hence the state agency is an interested party both making the rules and enforcing them. A legal analogy is that the legislature and judiciary are the same, which is not a desirable state of affairs.

An international solution to both of these problems is the development of disinterested laboratory accreditation schemes as per ISO Guide 25 (see Table 1.1 for a list of relevant documents). The reputation of the accreditation authority assures all customers of the quality of the measurements. The schemes are run on a national basis and obtain international recognition through mutual recognition agreements. Since 1980 such schemes have developed rapidly to meet the growing international requirements for several interrelated concerns: trade, safety, quality assurance and product liability. Laboratory personnel need to be aware of the difference in approach when seeking accreditation under these schemes. The accreditation authority is not a customer but instead it is a supplier of a service, and should be treated as such and not as if it were a regulatory authority. The service to be provided is the acceptance of your test and measurement results by a wide range of international customers.

Measurement quality is almost impossible to determine from the result alone and only by knowing what went into the measurement can we really judge its reliability. *Quality assurance* (QA) systems are ideally suited for this, hence the ISO Guide 25 on laboratory accreditation is integrated into the ISO 9000 series on quality assurance. The multitude of factors in the operation of a laboratory must be well under control. The main concerns are outlined below:

- The laboratory's management must be committed to a quality scheme by ensuring that policies and objectives are communicated to, understood, and implemented by all laboratory personnel. Of special concern is safety.

- Independence and financial stability of the laboratory are desirable. Where the laboratory is part of a larger organisation it is particularly important for the laboratory to act independently.

- Quality systems must to be properly documented. Procedures need to be written to cover not only how the test equipment is to be used and looked after but also the responsibilities of the staff.

- Documentation control is needed to ensure staff are using the latest procedures.

- Staff should be properly qualified and a regular training programme in place.

- New work must be reviewed to understand its requirements and determine if the laboratory can carry it out.

- There must be procedures for the identification of items for test, and safe handling and storage to ensure the integrity of the items.

- Records should be made and kept not only for test results but also other quality actions. Regulatory or contractual requirements may determine the length of time to hold records.

- Test reports and certificates need to be well-specified in terms of content and format.

- Participation in proficiency testing programmes and other statistical techniques can enhance the confidence in procedures.

- Inspection of the laboratory should be carried out by the accreditation body on site at regular intervals. In-house audits, both scheduled and random, are required.

- Accommodation and environment provisions need to be adequate for staff, test equipment and test samples.

- Procurement of equipment, consumables and services requires procedures to see that they are of appropriate quality, e.g. calibration certificates supplied with test equipment.

- There must be adequate management and control of test equipment. This is an essential feature for traceability which translates into the whole life of a piece of test equipment being properly documented, e.g. for any given time it should be possible to find where it was, who was using it and in what state of repair or calibration it was.

- Calibration of test equipment needs to be carried out in a timely and proficient manner. This is another essential feature for traceability and the laboratory needs to decide which calibrations it needs to carry out itself and which can be contracted out.

- Complaints and corrective action procedures are essential, especially the control of sub-standard testing work.

Many of the above features will be covered in more detail in subsequent chapters. Others will clearly depend on local circumstances and details of these will arise from discussions between the laboratory and the accreditation authority.

For a scientist engaged in academic research the formal accreditation schemes may seem of only commercial relevance. However, a careful study of the above features will reveal that they can be translated into good scientific practices for a research laboratory and in particular provide good traceability for measurements. Experiments often require new measurement techniques with which the experimenter is unfamiliar and the pressure to publish and obtain funding does not permit the time to gain the experience and understanding needed, whereas many commercial measurements are repeated and repay the investment of ensuring their quality. Peer review of scientific papers, which is the closest the scientific community comes to formal quality control, seldom considers the measurement quality.

1.4.2 National standards and calibration services

For a calibration to have legal status even a member nation of the Metric Treaty will need laws concerning measurement units. For the laws to be effective a national standards laboratory needs to hold the primary physical measurement standards. An important function for the laboratory is to provide for uniform measures nationally and internationally. National requirements can be met by providing calibration services to disseminate the measurement standards. Enforcement is usually not required for temperature measurements. For important trade measures, a separate weights and measures inspectorate is often established. The national standards laboratory will be the normal end point of the traceability chain inside a country. The laboratory in turn ensures that its standards match the SI units.

Where best accuracies are required the national laboratory will realise the physical definitions of the units as recommended in the SI and maintain these as the primary standards. For temperature the relevant ranges of the defined scale can be established. International obligations can also be met by using reference standards calibrated by the BIPM or other national standards laboratories. National laboratories have been very active in developing suitable techniques for realising the temperature scale and have established effective and relatively simple methods to do so. Many organisations requiring best accuracies are therefore capable of establishing the relevant parts of ITS-90 independently of the national laboratory. Provided that proper procedures are followed this will give good technical accuracy. To satisfy legal requirements demonstration of traceability to the national laboratory is needed.

Calibration services of a national laboratory transfer the primary standard to other reference standards. Lower level calibration laboratories use these standards to provide calibration services closer to the needs of the industry they serve. Because any calibration is an important link in the chain to SI, independent calibration accreditation organisations are often established to control the quality of service. More stringent requirements over that of test laboratories is required, especially for the care of instruments and in the assessment of accuracy. Future ISO guides or standards may cover these points. For the maximum acceptance of your measurements choose a calibration service which has international acceptance through its accreditation body.

A calibration chain for temperature should not be long, about four links or less from a realisation of ITS-90. Each calibration link lowers the accuracy by a factor of 3 or more and temperature is only known to a relatively low accuracy. The practice of labelling links in the calibration chain as a primary standard, secondary standard and so on, is not particularly useful in thermometry except as a general description. It is better to label the standard by the accuracy it can achieve and its intended use. A *working standard* is one used frequently to calibrate or check other thermometers. A *reference standard* may have the same quality as the working standard but is used less frequently. Most thermometer users will have a thermometer as a working standard. The reference standard will be an ice point in order to check if a re-calibration is needed.

1.4.3 Standards associations and specifications

The advantages of calibration are lost if a measurement using the equipment is not performed properly. Most measurements are part of a test to duplicate a standard test condition. Procedures or specifications should be written by those involved to cover these tests. While the specification may be unique for a particular contract, it is more likely to be of a general nature. Specifications which arise from consensus agreement can become standard specifications or documentary standards. Industrial user groups, professional bodies or national standards associations usually approve these.

Ideally the specification should be on the performance required and not the method to achieve it. Unfortunately there is often a poor understanding of the physical basis of tests. An empirical approach is therefore used. For such tests the resulting method has to be followed exactly to obtain consistent results. Because the consensus approach is very slow there is often a reluctance to update test specifications even though they may be clearly inadequate.

National standards associations have an important role to ensure that specification standards used nationally do not clash with each other. For example, the fire protection requirements of a local building code should be consistent with those of the insurance company. In turn the national standards association ensures consistency with the international bodies, ISO and IEC.

International standards are becoming increasingly important for trade. The General Agreement on Tariffs and Trade (GATT) has established rules to reduce non-tariff trade barriers by using international documentary standards. Conformance with these requirements will also be through the ISO standards relating to quality, test and measurement.

1.4.4 Measurement science — acquiring the skill

Measurement is a fundamental task in science and engineering. Most people learn measurement by experience rather than by formal training. Ideally traceability calls for a good physical understanding of every step of the measurement process. Often, as we have noted several times, an empirical approach is taken instead. The wide diversity of measuring equipment has meant that training is often very specific to the instrumentation used. This can leave the beginner (and even the experienced) with the impression that there is no underlying principle.

The real difficulty lies with the wide range of subjects which measurement science covers. In order to see this more clearly consider Figure 1.10, which shows the information flow involved in a typical temperature measurement. Like most flow diagrams the choice of blocks and their arrangement depends on the application. A measurement can be considered as gaining information about the physical world in a form we can use. Thus a measuring instrument can be considered as an information machine. Traceability involves assessing the influence each stage has on the information to help guarantee that the information remains useful.

The assessment of the effect of each block on the information involves considerable knowledge and effort by the user. Most blocks represent major areas of study (see the measurement science references given at the end of this chapter). For example, if the signal is in electrical form then the two signals blocks represent telemetry and electronic instrumentation. Fortunately these subjects are sufficiently advanced that the user may not need to know all the internal details. The performance of the equipment can often be specified adequately so that the user can treat it as a black box, i.e. only the inputs and outputs need be considered. In a less advanced subject the user may need quite detailed knowledge of the equipment in order to operate it. The user often becomes an active part of the signal processor. For example, compare a voltage measurement made on an old potentiometer with that on a modern digital voltmeter.

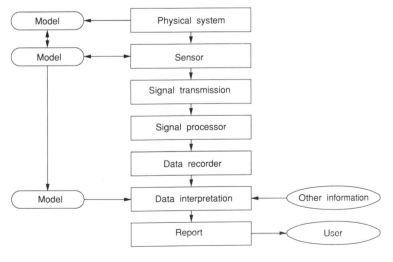

Figure 1.10
Information flow for a temperature measurement system.

The task can be simplified where there are well-established techniques or methods. For example, this book details the most important issues for several commonly used thermometers. This does not imply that all the problems using the thermometers are solved. The users still have the responsibility to ensure that the circumstances apply to their measurements. In particular users need to learn to distinguish the unusual problem requiring more investigation. Careful analysis will be needed for users to decide how to obtain the information required, either from experimentation, consultation or education.

A diagram similar to Figure 1.10 should be drawn up for the temperature measurement. Two examples are given later, Figure 2.10 for a general immersion thermometer and Figure 7.4 for a thermocouple. We need to consider how the information at each stage might be degraded and become untrustworthy. A convenient way to do this systematically is to consider all possible energy interactions. The interactions may be between the environment and parts of the measurement system, between the various blocks, or between components inside each block. The most common forms of energy can be classified as:

- electrical;
- mechanical;
- chemical;
- thermal;
- radiant;
- acoustical;
- nuclear;
- gravitational; and
- magnetic.

Other forms of classification can be used. Where a temperature sensor is particularly prone to interference then a further breakdown can be made, e.g. the influence of chemical energy may be broken into oxidation, corrosive attack, metallurgical change, and metal migration. The list is to aid evaluation and a rigid classification is not wanted, e.g. vibrations are considered under acoustical energy but they could be considered separately or under mechanical energy.

The first six energy interactions are the most commonly met and are dealt with in the text. Nuclear and magnetic energy are usually only of concern in the presence of a major source. However, for very sensitive measurements, very low-intensity sources, such as cosmic rays or the Earth's magnetic field, can cause interference. The specialised literature should be consulted for advice. In contrast most measurements are made in the presence of a large gravitational field and some equipment will not operate correctly if not properly aligned with it, e.g. held horizontally or vertically.

Obviously evaluating the effects of each of these forms of energy on each part of the measurement chain can take a long time and may involve considerable research. Indeed, this is the case for state-of-the-art measurements such as determining the thermodynamic temperature with a gas thermometer. Users need to develop a sense of what is appropriate for their case. Where the measurement uses variations on known methods or equipment a more limited assessment can be made. This is the normal case with a calibration where the known generic history of a sensor and its instrumentation determines a suitable calibration method.

The most difficult part to evaluate or find help for is the sensor's interaction with the physical system of interest. In the case of a calibration the system can be modelled simply; the temperature only varies slowly with time or distance. In a more complex case, such as a reformer furnace, the temperature profile depends on a large number of chemical reactions at different temperatures as well as various heat-transfer mechanisms. A model is now much more difficult, requiring several specialities and a team effort to develop. Note that a model does not have to be 'correct' to be useful. The model should determine the most appropriate measurements to make and the accuracy expected. Refinement of a new model will be based on the measurements by a process of iteration.

The sensor is the heart of the temperature measurement. Using the sensor (or thermometer) to ensure that the traceability remains clear is the main concern of thermometry. Very many temperature sensors are available and they need to be chosen and perhaps modified to suit the user's purpose. A general-purpose thermometer may serve no purpose at all! Often compromises are needed, e.g. accuracy is sacrificed to achieve a robust thermometer. The user needs to clearly establish the expected accuracy and not rely on suppliers and manufacturers who have little experience in thermometry.

Three aspects of the sensor can be recognised to assist in its evaluation:

- a physical interface;
- a transducer; and
- a signal interface.

The transducer converts one form of energy to another; for a thermometer the thermal conversion is usually to electrical energy. The terms 'sensor' and 'transducer' are often used with the same meaning because the transducer is often physically inseparable from its interfaces. A model, e.g. an equation, will relate the input temperature to the signal output. This thermometer equation, besides being of sufficient accuracy, should be useful in the assessment of errors and interactions. For example, we shall see that a commonly used model for thermocouples, while accurate by itself, is not adequate to deal with the main source of error (Chapter 7).

Interfacing the transducer to the physical system requires a good knowledge of thermal transfer mechanisms. Experimentation is often needed to determine the thermal parameters. For example, the adequacy of an immersion depth for a thermometer probe can be checked simply by increasing or decreasing the depth. As we have seen, the temperature measured will be that of the transducer itself. It is therefore important that this temperature correctly reflects the temperature of interest. For example, meteorological air temperature measurement requires adequate radiation screens and a controlled air flow for the transducer. Chapter 4 on Calibration also covers the essential thermal parameters for good thermometry.

The signal interface needs to be such that the inevitable heat flow does not cause any difficulties. Also the signal transmission must not load the transducer, nor must the transducer load the subsequent instrumentation. This is a link in the traceability chain which is often compromised for instrumental simplicity, e.g. a three-lead resistance measurement rather than the ideal four-lead measurement. Appropriate methods of interfacing are given in the relevant chapters.

An important feature for the overall sensor model is the ability to describe the influence the sensor has on the physical system being measured. It is desirable that the perturbation is small. For many natural systems this is the best we can hope for. In a manufactured system the design should ensure that the perturbation is small and optimal placement of the sensor should achieve the desired accuracy. Good examples of such design are the temperature fixed points covered in Chapter 3.

For a temperature measurement the signal transmission is shown in Figure 1.10 as a separate item. For other sensors the transmission could be considered a part of the sensor or instrumentation. The separation is to emphasise that the signal has to pass through a region of unknown temperature gradient — a situation usually avoided in other measurements. A robust strategy is needed to avoid errors.

The signal processor is the instrument package for the sensor and often it is a complete measuring instrument in its own right, e.g. a potentiometer. However, a calibration of the instrument without its sensor does not constitute a calibrated thermometer, e.g. the use of uncalibrated thermocouple wires with a calibrated potentiometer does not give a traceable temperature measurement. Modern instrumentation practice should be capable of delivering a well-specified instrument and the user will need to interpret the specification for their application. Besides the input/output characteristics, the main environmental influences should be given, e.g. the temperature coefficient, vibrational tolerance, immunity to electrical noise. Where any of these are critical to the accuracy of the temperature measurement an independent assessment is needed. Often this can be included in the calibration. The instrumental accuracy will generally be higher than that of the sensor and it is important not to confuse these two.

Modern digital instruments require little skill from the operator apart from the ability to accurately record the result. Thus an error assessment of the instrument can be made on an objective basis. Where instruments still require human intervention to make adjustments and convert an analogue output to digits, a quantitative error assessment is more difficult. QA procedures can help reduce the incidence of errors, but even for the most skilled operators it is difficult to judge if they are consistently accurate.

The other three steps in the measurement chain, the data recorder, the data interpretation and the report, will also be under the control of the laboratory's QA system. They are usually carried out by human agency or by computer and hence error avoidance is more relevant than error assessment. Physical interactions with energy sources still need consideration, e.g. errors in reading a digital display can occur if the room lighting is too bright. Because these parts of the measurement are common to all measurements they are well covered by laboratory procedures and hence only the temperature-related concerns will be covered in this text.

Thermometric principles will be involved in the model used for the data interpretation. The model will combine the model of the physical system with that of the temperature sensor in order to interpret the result and carry out an error analysis. Chapter 2 on Uncertainties gives appropriate methods for treating errors. Additional information is often needed to complete the analysis, such as calibration data or the results of the measurement of other physical quantities.

Good thermometry which is useful to you and others requires careful and painstaking work. The systematic approach to traceability outlined in this chapter should serve as a guideline to that goal. However, the ideal is not always attainable. There are many gaps in knowledge, availability of devices, and infrastructural services. With more informed thermometer users, hopefully the gaps will become the exception rather than the rule.

FURTHER READING

A History of the Thermometer and its Use in Meteorology, W. E. Knowles-Middleton, Johns Hopkins, Baltimore (1966).
A very readable history with a good discussion of the developing concept of temperature.

Handbook of Measurement Science, Wiley, Chichester.
Vol 1 *Theoretical Fundamentals*, Ed. P. H. Sydenham (1982).
Vol 2 *Practical Fundamentals*, Ed. P. H. Sydenham (1983).
Vol 3 *Elements of Change*, Ed. P. H. Sydenham and R. Thorn (1992).
An extensive coverage of all aspects of measurements.

Quality Assurance in Research and Development, G. W. Roberts, Marcel Dekker, New York (1983).
A quality approach to research before ISO 9000. Shows that lack of calibration is one of the most frequent errors. Gives a discussion on the meaning of calibration and traceability.

Introduction to Measurement Science and Engineering, P. H. Sydenham, N. H. Hancock, and R. Thorn, Wiley, Chichester (1989).
A good introductory text on measurement with most of the terms defined and explained.

2

Uncertainties in Measurement

2.1 INTRODUCTION

All measurements, even the best planned and best carried out, are subject to error. The confidence we place in the result of any measurement depends on our perception of the magnitude of the error. An estimate of the accuracy of a measurement is known technically as the uncertainty of measurement. In this chapter we introduce error analysis, the means by which we make objective estimates of the uncertainty. We begin by developing the concept of an error distribution and then develop the simple statistical tools for assessing and combining uncertainties. Following a section on the presentation of results we cover more advanced topics. The advanced sections may be read by all, but are most appropriate for scientists and engineers who may be designing measurements or calibrations.

While the treatment we give here is based on the BIPM's recommendations and the current practice of calibration laboratories, we expect that the treatment will also provide sufficient background information to enable the reader to understand any of the current codes of practice, in particular the new ISO guide (see references at end of the chapter). However, the uniformity of practice between client and supplier is the most important factor in determining the code of practice. If your science, trade or industry has its own code of practice, then you must adopt that code.

Traditional approaches to error analysis tend to view the measurement in hindsight and to analyse the uncertainties only after the results have been obtained. However, a more robust approach is to carry out an uncertainty analysis before the measurement is assembled. In this way the most significant sources of error are recognised early, and the measurement planned to ensure that the errors are under control.

The two keys to successful uncertainty analysis are: firstly, a good model of the system which includes the physical variables affecting the measurement, and secondly, a set of tools — mathematical equations — which tell us how these variables affect the measurement. Such analysis is especially important for novel or non-repeatable measurements. In these cases experimenters do not have the luxuries of a history of proven methods or an opportunity to repeat the measurement if they fail to measure a critical variable.

The early sections of this chapter concentrate on the relatively simple techniques for uncertainty analysis which are applicable generally to most measurements; the last three sections deal with the mathematical tools for analysing more complex measurements. The chapter may be read in full without the equations as we have provided sufficient discussion in the text to convey the ideas and principles. The use of the most important equations is also demonstrated several times in the following chapters; and they are often available as functions on pocket calculators.

2.2 UNCERTAINTIES — WHAT THEY ARE AND HOW THEY ARE USED

Consider:

- a news report describing a group of 1000 people;
- a contents label on a box of 1000 wood screws;
- and a shipping docket for 1000 television sets.

Although none of these reported measurements has an explicit statement of uncertainty, we know from experience and context what magnitude of error to expect in each count of 1000. The news report may be in error by several hundred, the box will probably contain a few extra screws, and there will be exactly 1000 television sets in the shipment. Thus the concept of uncertainty in measurement is not new to us: builders, traders, and navigators, for example, have been acutely aware of errors and uncertainties for centuries.

Nowadays it is relatively rare for measurements to bother us because of large errors. The errors in our measuring instruments — our watches, thermostats, meters, rulers, etc. — are for most practical purposes negligible. This happens by design: for most of the measurements we encounter there is a well-defined mechanism for controlling the uncertainty. Here are some examples.

- There are documentary standards describing nuts and bolts in terms of their thread size, pitch, length, with a well-defined tolerance for each property. Manufacturers conform to the standards so that nuts from one manufacturer fit bolts from another.

- Most countries have 'fair trading' regulations governing the meaning of nett weight on contents labels. To meet the regulations, manufacturers must adjust their production line to overfill the containers sufficiently to ensure that the incidence of underfilling is infrequent. Manufacturers therefore have considerable financial interest in knowing and reducing the uncertainty in the weight of the product placed in the containers.

- Similarly, for health reasons there are regulations governing the cooking and sterilisation of foodstuffs: too low a temperature may not kill all micro-organisms, too high a temperature will degrade the taste and texture of the foodstuffs.

In all of these examples there is a strong incentive for manufacturers to understand the uncertainties in their measuring equipment. Indeed, product liability laws and the desire to minimise overfilling are two of the major driving forces for improved measurements. Because intuition and experience alone are not able to define uncertainties sufficiently accurately to satisfy trade regulations and standards, scientific techniques have evolved which give objective measures of the departures of measurements from ideal.

All three of these examples are at the working end of the measurement chain. Less obvious are the calibrated instruments: the masses, micrometers, thermometers that trace these measurements back to the international definitions of the units. And without traceability to some common standard all the quality control effort is pointless.

For the metrologists who must maintain traceability, an understanding of uncertainties is vital, as is uniformity of practice. Unfortunately there is as yet no robust theory of errors or universally accepted practice for determining the uncertainty in measurements. This situation is well recognised, as the ISO definition shows:

> **Uncertainty in measurement:**
> An estimate characterising the range of values within which the true value of a measurand lies.

The key words in this definition are 'estimate' and 'characterising'. There is considerable scope for flexibility as to how the uncertainty is determined. This places a burden on both supplier and client in any measurement transaction: each must know what the other expects or provides. Clearly a single uniform practice is desirable, otherwise it is difficult to deal with clients or suppliers with a variety of practices, it is difficult to compare measurements and instruments because uncertainties are characterised differently and it may even be impossible to conform to documentary standards. It is hoped that more uniform practice will develop with the rapid adoption of quality assurance (QA) systems and the publication of the ISO guide. One of the important features of recent QA systems is the drive to harmonise practices.

2.3 MEASUREMENTS AND DISTRIBUTIONS

A measurement is usually made in the belief that there is a *true value* for the quantity being measured, for example, for the length of a rule or the temperature of a calibration bath. A little thought shows that this is not necessarily the case: the rule will not have perfectly parallel scale markings and the bath will not be perfectly uniform in temperature.

Not being able to define exactly what is meant by variables such as the length of the rule or the temperature of the bath is known generally as the *problem of definition*: for almost all measurements there is no single true value for the measurand. We will get back to this problem in the discussion on the realisation of the temperature scale, but for the moment we will assume that for all measurements there is a quantity which can be thought of as the true value.

Whenever we carry out a measurement we know the measured value will not equal the true value. The difference between the two is the *error*:

$$\text{error} = \text{measured value} - \text{true value}. \tag{2.1}$$

When this measurement is repeated we usually get a number of different values. By making enough measurements we can then build up a picture which enables us to make predictions about the likelihood of making a measurement and obtaining a particular result. This picture is called a *distribution*. Usually the distribution cannot be known exactly, but the more measurements are made, the greater will be the confidence in the measured distribution.

A distribution of measurements is shown in Figure 2.1. This particular distribution would appear to have a *bias* of some kind since all the measured values are higher than the true value: this problem is discussed later under Systematic errors (Section 2.5).

For the purposes of studying errors we characterise a distribution by two parameters: the *mean* and the *variance*. The mean is an average of measured values, while the

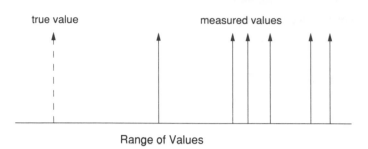

Figure 2.1
A distribution of measured values around a true value.

variance is a measure of the spread of the distribution. A third essential definition is the *standard deviation*, which is the square root of the variance. The standard deviation is proportional to the width of the distribution.

It is not possible to know the true values of the mean μ and variance σ^2; they can only be estimated from the measurements. The best estimate of the mean of the distribution is the arithmetic mean, m, of the measurements:

$$m = \frac{1}{N} \sum_{i=1}^{N} X_i \tag{2.2}$$

where X_i are the N measurements of x. The mean or the average estimates the centre of the distribution.

The best estimate for the true variance, σ^2, is the variance s^2

$$s^2 = \frac{1}{N-1} \sum_{i=1}^{N} (X_i - m)^2 \tag{2.3}$$

where s is the standard deviation.

The symbols m and s^2 are used to describe the mean and variance in situations like the above, which are based on measured, experimental values of the distribution. The symbols μ and σ^2 are used in situations where the values can be calculated exactly from theoretical considerations. (This convention is not universal. The symbols s and σ are often used for statistical functions on pocket calculators but may follow a different convention for their meaning. Consult your calculator handbook.)

Example 2.1   Calculating the mean and variance

Calculate the mean and variance of the temperature (in degrees Celsius) of a chilled water-bath for which the following 20 measurements have been made: 6.6, 6.5, 7.0, 6.4, 6.5, 6.3, 6.6, 7.0, 6.5, 6.5, 6.3, 6.0, 6.8, 6.5, 5.7, 5.8, 6.6, 6.5, 6.7, 6.9.

The measurements constitute the readings X_i. They are arranged in ascending order and tabulated using f, the frequency of occurrence for a given reading, as seen in the first three columns below. As a check, the sum of the frequencies should equal the number of measurements.

Continued on page 41

Continued from page 40

Results X_i	Frequency f	fX_i	Deviation $(X_i - m)$	$(X_i - m)^2$	$f(X_i - m)^2$
5.7	1	5.7	−0.785	0.616	0.616
5.8	1	5.8	−0.685	0.469	0.469
6.0	1	6.0	−0.485	0.235	0.235
6.3	2	12.6	−0.185	0.034	0.068
6.4	1	6.4	−0.085	0.007	0.007
6.5	6	39.0	+0.015	0	0
6.6	3	19.8	+0.115	0.013	0.039
6.7	1	6.7	+0.215	0.046	0.046
6.8	1	6.8	+0.315	0.099	0.099
6.9	1	6.9	+0.415	0.172	0.172
7.0	2	14.0	+0.515	0.265	0.530
Totals	20	129.7			2.281

The mean m is then determined:

$$m = \frac{1}{N} \sum fX_i = \frac{129.7}{20} = 6.485°C$$

Note that the mean is written here with three decimal places while the original readings have only 1 decimal place. Guidelines on rounding and presentation of results are described in Section 2.9.

Once the mean has been calculated, the last three columns of the table can be filled in and the variance calculated as

$$s^2 = \frac{1}{N-1} \sum f(X_i - m)^2 = \frac{2.281}{19} = 0.120.$$

Hence the standard deviation is the square root of the variance:

$$s = 0.346°C.$$

Figure 2.2 shows a histogram of the 20 data points used in Example 2.1. The vertical axis on the left-hand side is the sample *frequency*, namely the frequency with which the results occur within the ranges indicated by the vertical bars, while the right-hand axis is the *probability* of obtaining a result. The probability is calculated as the frequency divided by the total number of measurements. Given enough measurements, the probability information allows us to make predictions about the likelihood of the results. For example, we can expect about 3 out of every 10 measurements to yield a result in the range 6.45 to 6.55.

As the number of measurements is increased the shape of the distribution becomes better determined and smoother. The curve obtained for an infinite number of measurements and an infinite number of sections is known as the limiting distribution for the measurements.

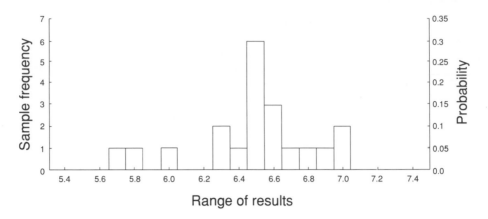

Figure 2.2
Histogram of the 20 measurements of Example 2.1.

Exercise 2.1

Calculate the mean and standard deviation for the following 13 measurements of the freezing-point of indium:

156.5994	156.5988	156.5989
156.5991	156.5995	156.5990
156.5989	156.5989	156.5986
156.5987	156.5989	156.5984
156.5986		

(Hint: To simplify the averaging calculation, consider only the last two digits of each number, 94, 91, etc. The final mean is then the calculated mean plus 156.590.)

2.4 TWO USEFUL DISTRIBUTIONS

The limiting distributions for many physical processes are well enough known to be determined from theoretical considerations. In this section we look at two theoretical distributions that are useful in determining uncertainties.

2.4.1 The rectangular distribution

Whenever we read the scale on a rule or meter we are unable to determine the reading to better than, say, one-tenth of a scale division. Rounding the readings in this way introduces an error known as *quantisation error*. Quantisation is the term describing the process of converting any continuous reading into a set of discrete numbers. For example, a $3\frac{1}{2}$ digit thermometer converts a continuous temperature scale into 1 of 3999 decimal numbers which we use to represent the temperature.

Figure 2.3(a) shows the input/output characteristic for a digital thermometer with a resolution of 1°C. When the digital thermometer reads 100°C, the true temperature could

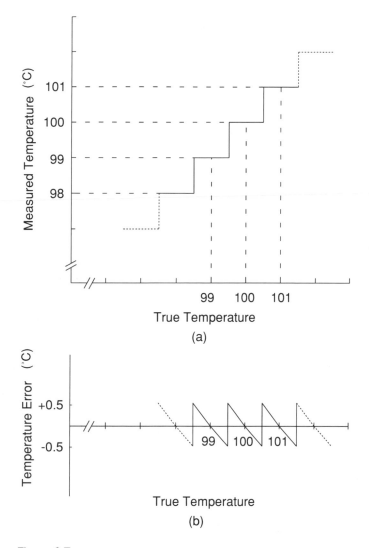

Figure 2.3
Quantisation error: (a) the input/output characteristic of an ideal digital thermometer; (b) meter error versus true temperature.

be anywhere in the range 99.5°C to 100.5°C. Figure 2.3(b) shows the error for the meter as a function of true temperature. This graph shows two things. Firstly, the maximum error is ±0.5°C, and secondly, the error is uniformly distributed over this range, any error in the range −0.5 to +0.5°C being equally likely.

The distribution of the quantisation error is one example of the uniform or rectangular distribution shown in Figure 2.4. For the rectangular distribution all results between the lower and higher limits X_L and X_H are equally likely. For the quantisation error the limits are $X_L = -0.5$°C and $X_H = +0.5$°C.

The probability of a single result lying within a certain range is given by the area under the distribution curve enclosed within the range. For example, the probability of a

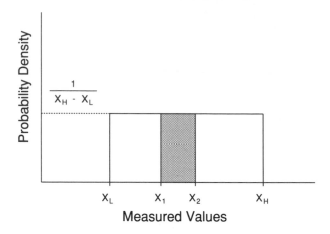

Figure 2.4
The rectangular probability distribution. The probability of obtaining a result between X_1 and X_2 is given by the shaded area.

result X lying between X_1 and X_2 in Figure 2.4 is given by

$$P(X_1 < X < X_2) = \frac{X_2 - X_1}{X_H - X_L}. \tag{2.4}$$

Note that the total area under the graph is unity; that is, the probability of a result between X_L and X_H is 1.

For a rectangular distribution the mean and variance are known exactly and are given by

$$\mu = \frac{X_L + X_H}{2} \tag{2.5}$$

and

$$\sigma^2 = \frac{(X_H - X_L)^2}{12}. \tag{2.6}$$

If we use Δ to represent the resolution of a digital instrument $(X_H - X_L)$, then the variance of the quantisation error is

$$\sigma^2 = \frac{\Delta^2}{12} \tag{2.7}$$

and the range of results for a rectangular distribution is

$$\mu \pm \Delta/2 \quad \text{or} \quad \mu \pm \sqrt{3}\sigma. \tag{2.8}$$

The rectangular distribution is a useful tool for characterising some uncertainties. Simply by assigning upper and lower limits to an error, we obtain a value for the mean which may be applied as a correction, and a variance which characterises the uncertainty. We will demonstrate this later.

As observed at the beginning of this section, quantisation occurs for both analogue instruments and digital instruments. It also occurs when we report the average of a set of measurements, since the result is not reported to an infinite number of decimal places. Quantisation error is usually introduced at least twice into most measurements.

2.4.2 The normal distribution

Within a large group of measurements we expect on the average to find most of the results close to the mean with progressively fewer further from the mean. One distribution which describes this type of behaviour is the *normal* or *Gaussian distribution*. Mathematically the distribution is described by the probability density

$$p(x) = \frac{1}{\sigma\sqrt{2\pi}} \exp(-(x-\mu)^2/2\sigma^2) \qquad (2.9)$$

where μ is the mean and σ^2 is the variance of the distribution. Figure 2.5 shows the classic bell shape of the Gaussian distribution. As with the rectangular distribution the probability of obtaining a result between two values X_1 and X_2 is the area under the curve between these two values:

$$P(X_1 < X < X_2) = \int_{X_1}^{X_2} p(x)\mathrm{d}x. \qquad (2.10)$$

This function is usually tabulated because it is difficult to evaluate numerically.

Example 2.2 Using Gaussian probability tables

By using Table 2.1, which tabulates the area under the Gaussian distribution, determine the percentage of measurements likely to fall within $\pm 1\sigma$, $\pm 2\sigma$ and $\pm 3\sigma$ of the mean.

_____ *Continued on page 46*_____

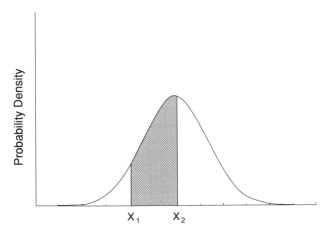

Figure 2.5
The Gaussian or Normal probability distribution.

Table 2.1. Area under the Gaussian probability distribution.

The percentage probability of finding x
within $\mu \pm k\sigma$

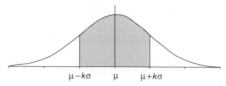

k	0.00	0.01	0.02	0.03	0.04	0.05	0.06	0.07	0.08	0.09
0.0	0.00	0.80	1.60	2.39	3.19	3.99	4.78	5.58	6.38	7.17
0.1	7.97	8.76	9.55	10.34	11.13	11.92	12.71	13.50	14.28	15.07
0.2	15.85	16.63	17.41	18.19	18.97	19.74	20.51	21.28	22.05	22.82
0.3	23.58	24.34	25.10	25.86	26.61	27.37	28.12	28.86	29.61	30.35
0.4	31.08	31.82	32.55	33.28	34.01	34.73	35.45	36.16	36.88	37.59
0.5	38.29	38.09	39.69	40.39	41.08	41.77	42.45	43.13	43.81	44.48
0.6	45.15	45.81	46.47	47.13	47.78	48.43	49.07	49.71	50.35	50.98
0.7	51.61	52.23	52.85	53.46	54.07	54.67	55.27	55.87	56.46	57.05
0.8	57.63	58.21	58.78	59.35	59.91	60.47	61.02	61.57	62.11	62.65
0.9	63.19	63.72	64.24	64.76	65.28	65.79	66.29	66.80	67.29	67.78
1.0	68.27	68.75	69.23	69.70	70.17	70.63	71.09	71.54	71.99	72.43
1.1	72.87	73.30	73.73	74.15	74.57	74.99	75.40	75.80	76.20	76.60
1.2	76.99	77.37	77.75	78.13	78.50	78.87	79.23	79.59	79.95	80.29
1.3	80.64	80.98	81.32	81.65	81.98	82.30	82.62	82.93	83.24	83.55
1.4	83.85	84.15	84.44	84.73	85.01	85.29	85.57	85.84	86.11	86.38
1.5	86.64	86.90	87.15	87.40	87.64	87.89	88.12	88.36	88.59	88.82
1.6	89.04	89.26	89.48	89.69	89.90	90.11	90.31	90.51	90.70	90.90
1.7	91.09	91.27	91.46	91.64	91.81	91.99	92.16	92.33	92.49	92.65
1.8	92.81	92.97	93.12	93.28	93.42	93.57	93.71	93.85	93.99	94.12
1.9	94.26	94.39	94.51	94.64	94.76	94.88	95.00	95.12	95.23	95.34
2.0	95.45	95.56	95.66	95.76	95.86	95.96	96.06	96.15	96.25	96.34
2.1	96.43	96.51	96.60	96.68	96.76	96.84	96.92	97.00	97.07	97.15
2.2	97.22	97.29	97.36	97.43	97.49	97.56	97.62	97.68	97.74	97.80
2.3	97.86	97.91	97.97	98.02	98.07	98.12	98.17	98.22	98.27	98.32
2.4	98.36	98.40	98.45	98.49	98.53	98.57	98.61	98.65	98.69	98.72
2.5	98.76	98.79	98.83	98.86	98.89	98.92	98.95	98.98	99.01	99.04
2.6	99.07	99.09	99.12	99.15	99.17	99.20	99.22	99.24	99.26	99.29
2.7	99.31	99.33	99.35	99.37	99.39	99.40	99.42	99.44	99.46	99.47
2.8	99.49	99.50	99.52	99.53	99.55	99.56	99.58	99.59	99.60	99.61
2.9	99.63	99.64	99.65	99.66	99.67	99.68	99.69	99.70	99.71	99.72
3.0	99.73	—	—	—	—	—	—	—	—	—
3.5	99.95	—	—	—	—	—	—	—	—	—
4.0	99.994	—	—	—	—	—	—	—	—	—
4.5	99.9993	—	—	—	—	—	—	—	—	—
5.0	99.99994	—	—	—	—	—	—	—	—	—

Continued from page 45

Table 2.1 lists the probability that the result lies within k standard deviations of the mean. Using the values for $k = 1, 2,$ and 3 we find that:

> 68.27% of measurements lie within 1σ of the mean;
> 95.45% of measurements lie within 2σ of the mean;
> 99.73% of measurements lie within 3σ of the mean.

Example 2.2 showed how to develop three important rules for the Gaussian distribution. These rules help us to characterise and interpret uncertainties. With a little approximation and rewording they are also easy to remember:

- 1 in 3 measurements lie outside $\mu \pm 1\sigma$;
- 1 in 20 measurements lie outside $\mu \pm 2\sigma$;
- almost no measurements lie outside $\mu \pm 3\sigma$.

Figure 2.6 shows the histogram of Figure 2.2, upon which the corresponding Gaussian distribution has been overlaid. Although the histogram is very different from the Gaussian curve in appearance, the distribution of results obeys the three distribution rules rather closely. Both of these features are typical of real measurements.

Exercise 2.2

Using the Gaussian probability Table (Table 2.1) characterise the ranges containing 50%, 95% and 99% of measurements.

When a number of random variables are added together the resulting distribution tends to be Gaussian. This is shown in Figure 2.7 for sums of variables with a rectangular distribution. The mathematical proof of this tendency for all distributions to become Gaussian is known as the *central limit theorem*. It is this property that makes the Gaussian distribution useful: whenever we collect and sum uncertainties we can, with some justification, assume that the resulting distribution is Gaussian. Also, because many natural processes tend to be Gaussian we have a good intuitive feeling for the distribution.

Exercise 2.3

When tossing a die, each of the six faces are equally likely to lie face up. Thus the probability distribution for the throw of the die is rectangular.

(a) Record the results of 24 throws of a single die and plot the histogram.

(b) Record the results of 48 throws of two dice. The sum of the two is a number between 2 and 12. Plot the histogram of the results. This should have a triangular distribution (see Figure 2.7(b)).

(c) Calculate the mean and variance for the results in (a) and (b). You should find that the mean and variance for (b) are both twice that for (a).

(d) If you have plenty of time, you might like to record the results for three dice and compare the distribution with Figure (2.7c).

(Note: A useful guide for plotting histograms is to have about 4 results per section, e.g. 4 throws for each of 6 faces of the die gives 24 throws total.)

2.5 RANDOM AND SYSTEMATIC ERRORS

When carrying out a measurement we generally recognise two types of error. The most obvious is the so-called *random error*, which causes a sequence of readings to be scat-

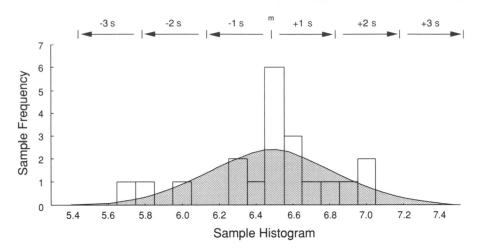

Figure 2.6
Histogram of Figure 2.2 with corresponding Gaussian curve overlaid.

tered unpredictably. The second type of error, the *systematic error*, causes all the readings
to be biased away from the true value of the measurand. Systematic errors are usually
associated with uncalibrated equipment, experimental oversights, errors in theory, and
imperfect observations. While the term 'systematic' has a strong intuitive implication
suggesting that the error is in some sense predictable, this meaning cannot be translated
into an unambiguous technical definition. Indeed, traditional treatments of errors that
have attempted such a definition have resulted in controversy.

Instead, the modern definitions of random and systematic error are based only on the
premise that a systematic error causes bias in the results, whereas a random error does
not. Hence, according to the ISO definitions:

Systematic error:
The mean of a large number of repeated measurements of the same measurand minus
the true value of the measurand.
Random error:
The result of a measurement minus the mean result of a large number of repeated
measurements.

The modern treatment of errors is based on methods for determining and then describing the distributions of errors. Once a distribution for an error has been described in
terms of its mean and standard deviation it is then a relatively simple task to determine
the correction and the uncertainty. According to the ISO definition:

Correction:
The value that, added to the uncorrected result of a measurement, compensates for an
assumed systematic error.

Thus the correction applied to a measurement is the negative of the mean error.

Similarly the standard deviation is used to characterise the random error in the measurement. We discuss procedures for determining the uncertainty in detail in Sections 2.6
and 2.7.

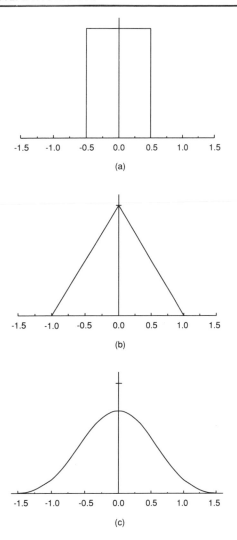

Figure 2.7
The distributions of results obtained by summing measurements with a rectangular distribution: (a) the rectangular distribution; (b) the sum of two rectangular variables gives rise to a triangular distribution; (c) the sum of three rectangular variables yields a 'nearly' Gaussian distribution.

Quantisation error gives a good illustration of the difference between the traditional and modern views of random and systematic errors. Referring to Figure 2.3 we can see that for any value of the true temperature the reading of the digital thermometer, and hence the error, can be predicted exactly. This suggests that the error is systematic. However, the converse is not true. Given any reading there is an infinite number of temperatures within ±0.5C of the reading, all of which are equally likely to have been the true temperature. The error is then uncorrectable, and on the average does not bias the measurement — is it best described as a random error? This is one of many difficulties that have led metrologists away from predictability as the basis for classifying systematic errors. The modern view considers only the distribution of the error in the measurements; since the mean of the error distribution is zero, quantisation error is a purely random error.

While the distribution-based definitions overcome many of the problems affecting traditional treatments of uncertainties, some of the difficulties that were part of the motivation for the older classification remain.

One of the main motivating factors was to distinguish between the random error, which can be assessed objectively from the sequence of measurements, and the systematic error, which often cannot. In situations where the systematic error cannot be measured by carrying out auxiliary experiments, the experimenter may have to rely on intuition and experience to guess the uncertainty. Now the idea of doing something so highly subjective as guessing is distasteful to most metrologists: the potential for scientists either to exaggerate their abilities by claiming excessively low uncertainties, or to claim very conservative uncertainties, just in case they have missed something, is highly undesirable.

The older classification of systematic error sought to single out the subjective estimates of uncertainty for special attention, in effect stating that 'These numbers are not especially trustworthy.' The statement of uncertainty that accompanies the measurement would have two components: firstly, an overall estimate of the uncertainty due to random errors obtained by summing individual random uncertainties in *quadrature* (summed as the root-sum-square); and secondly, an estimate of the systematic uncertainty, which is the linear sum of the contributing systematic uncertainties. The use of the linear summation was in part based on the high degree of correlation between systematic errors in some experiments, and in part due to the lack of confidence in the estimate.

The fundamental problem with the subjective estimates of uncertainty was the lack of traceability. There was no assurance that an appropriately competent person repeating or scrutinising the measurement would make the same assessment of the errors. The solution to maintaining the traceability for such assessments (indeed for any assessment) has two facets. Firstly, we must provide the experimenter, who is usually in the best position to judge the uncertainty, with the best tools available for making the most honest and consistent estimates as is practical. Secondly, the experimenter must record the calculations and arguments in support of the assessment in sufficient detail to allow the unambiguous retracing of all the steps leading to the assessment. This allows confirmation or improvement of the assessment at a later date.

The BIPM recognises two categories of uncertainty:

Type A: Those uncertainties that are evaluated by statistical methods.
Type B: Those uncertainties that are evaluated by other methods.

Type A assessments of uncertainty are based on the variance or standard deviation of a measured distribution, and are therefore assessed objectively. Type B assessments rely on information derived from other sources: theory, other people's work, and occasionally guesswork. While it is tempting to associate Type A with random error and Type B with systematic error, it is misleading to do so. For example, a calibration is a measurement that is designed to assess objectively the systematic error in an instrument, and is a Type A assessment. A theoretical assessment of the uncertainty caused by random electrical noise is a Type B assessment of a random error.

2.6 CHARACTERISING 'TYPE A' UNCERTAINTIES – CONFIDENCE LIMITS

Example 2.2 shows that the variance is a good way of characterising the width of a distribution. For most distributions the square root of the variance, namely the standard deviation, is directly proportional to the width. Thus the standard deviation can be used to characterise the uncertainty, and the result of a measurement reported as

$$\text{value} = \text{mean} \pm \text{standard deviation.} \tag{2.11}$$

An uncertainty expressed as the standard deviation is known as the *standard uncertainty*. However, the range characterised by the standard deviation includes only 68% of all measurements. While this is an acceptable characterisation for scientific measurements, there are many other measurements, including calibrations, for which the users of a result like wider uncertainty characterisations that include almost all the measurements. In cases where higher confidence is required the results should be reported as

$$\text{value} = \text{mean} \pm k \times \text{standard deviation.} \tag{2.12}$$

where k is the multiplying factor which increases the quoted range to include a greater proportion of the measurements. The k factor, also known as the *coverage factor*, can be determined from the Gaussian probability table (Table 2.1). For example, using a value of $k = 2.6$ would characterise the uncertainty by a range (the *confidence interval*) which is expected to include 99% of all measurements.

The probability statement we make, 'expected to include P% of the measurements' states the *confidence level* (CL) for the report. Note that k has to be very large to include all measurements. In practice there is a compromise: $k = 2$ (95% CL) and $k = 3$ ($\sim$ 99% CL) are common choices. This is one area where codes of practice for evaluating uncertainties differ markedly. It is important when reporting a measurement that the confidence level is reported along with the result and the uncertainty.

Note the convention that the symbol U is used for any uncertainty while the standard deviation, σ or s, may be used for the standard uncertainty.

Because the mean μ and the variance σ^2 cannot be known exactly, we must consider the decrease in confidence level due to the lack of exactness in the estimates m and s^2 of the mean and variance. It can be shown (see Exercise 2.8 below) that the mean of a set of N independent measurements is also random and is distributed with a variance

$$\sigma_m^2 = \sigma^2/N \tag{2.13}$$

where σ^2 is the variance in the measurements.

Similarly, the variance is distributed with a variance (for large N) of

$$\sigma_{s^2}^2 = 2\sigma^4/N. \tag{2.14}$$

Equations (2.13) and (2.14) show that the picture of the distribution based on m and s is fuzzy and this fuzziness reduces the confidence placed in the uncertainties.

There are two ways to remedy this situation. Firstly, take more measurements to increase the confidence in the estimates of the mean and variance. Secondly, increase the

k factor to account for the increased uncertainty. But how many extra measurements are required, and by how much must the k factor be increased?

The answer to both of these questions can be found from a special distribution known as *Student's t-distribution*. The tables (Table 2.2) for this distribution are similar to the Gaussian probability table except that they depend also on the number of measurements. Actually the third parameter is v, the number of degrees of freedom. This is the number of spare pieces of information available. Where N measurements are made and a mean calculated, there are $N - 1$ degrees of freedom. Strictly the k values for characterising the confidence intervals $m \pm ks$ are $(1 + 1/N)^{1/2}$ times the standard t-distribution values listed in Table 2.2; however for useful values of N the error is not significant.

Figure 2.8 illustrates Student's t-distribution for several values of v. The most important feature of the curves is the very long tails on the distributions for low values of v. In order to establish a given confidence limit, the k factors for the longer-tailed distributions must be larger in order to enclose the same area and represent the same confidence level. The distribution becomes more and more like the Gaussian distribution as the number of degrees of freedom increases.

Table 2.2. Student's t-distribution.

Values of k for specified confidence limits, P, as a function of the number of degrees of freedom v. Where N measurements are used to determine ρ parameters, the number of degrees of freedom is $v = N - \rho$. When calculating a mean $v = N - 1$.

P is the percentage probability of finding X within $m \pm ks$

$P(\%)$ v	50	68.3	95.0	95.5	99.0	99.7
1	1.000	1.84	12.7	14.0	63.7	236
2	0.817	1.32	4.30	4.53	9.92	19.2
3	0.765	1.20	3.18	3.31	5.84	9.22
4	0.741	1.14	2.78	2.87	4.60	6.62
5	0.727	1.11	2.57	2.65	4.03	5.51
6	0.718	1.09	2.45	2.52	3.71	4.90
7	0.711	1.08	2.36	2.43	3.50	4.53
8	0.706	1.07	2.31	2.37	3.36	4.28
9	0.703	1.06	2.26	2.32	3.25	4.09
10	0.700	1.05	2.23	2.28	3.17	3.96
11	0.697	1.05	2.20	2.25	3.11	3.85
12	0.695	1.04	2.18	2.23	3.05	3.76
13	0.694	1.04	2.16	2.21	3.01	3.69
14	0.692	1.04	2.14	2.20	2.98	3.64
15	0.691	1.03	2.13	2.18	2.95	3.59
16	0.690	1.03	2.12	2.17	2.92	3.54
17	0.689	1.03	2.11	2.16	2.90	3.51
18	0.688	1.03	2.10	2.15	2.88	3.48
19	0.688	1.03	2.09	2.14	2.86	3.45
∞	0.675	1.00	1.96	2.00	2.58	3.00

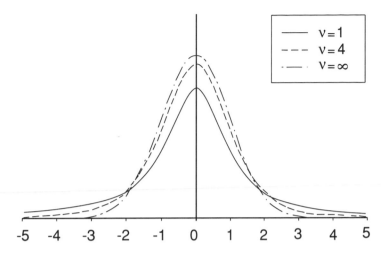

Figure 2.8
Student's t-distribution for different values of v, the number of degrees
of freedom. Note the long tails on the distributions for small values of v.

Example 2.3   Determining confidence limits with Student's t-tables

Using Table 2.2 which tabulates the area under Student's t-distribution, calculate the k factor for 95%
confidence limits for a result determined from 6 measurements.

Looking up the entry for $P = 95.0$ and $N = 6(v = 5)$ we find that $k = 2.57$. That is, we expect
95% of measurements to lie within $m \pm 2.57s$.

Exercise 2.4

Find the minimum number of measurements required to achieve 99% confidence limits corresponding to
± 3 times the standard deviation.

A close look at Table 2.2 shows that the largest values of k occur at the top right-
hand corner of the table, i.e. the uncertainty is largest for small numbers of measurements
and large confidence limits. These are situations to be avoided in practice if relatively
small uncertainties are required. A reasonable compromise must be reached between
the desire for large confidence levels and the need for the number of measurements
to be practical. A 95% confidence level is considered acceptable. The 95% confidence
level requires 5 or more measurements to keep k to values less than 3.0. This choice
of confidence level also ensures that the k values are very similar for all distributions.
With larger confidence levels the k values are strongly dependent on the shape of the
tails of a distribution, and therefore dependent on the least likely measurements. The
95% confidence level is becoming the preferred option for characterising uncertainties
in non-scientific reporting.

Equation (2.13) for the variance in the mean raises an important issue, since many
texts advocate the use of the so-called 'one-upon-N' rule, namely $s_m^2 = s^2/N$, as the

means of estimating the uncertainty in measurements. As a general rule s_m is rarely an appropriate measure of uncertainty. It is appropriate only if the errors are purely random — this is an aspect we discuss in detail in Section 2.10.3; and secondly, only if the measurand has a single true value — this is discussed now.

Consider the 20 measurements of temperature in Example 2.1. Since the bath does not have a single true temperature the most accurate report of the bath temperature is

$$\text{bath temperature} = m \pm ks. \tag{2.15}$$

The confidence interval $\pm ks$ then truly characterises the range of temperatures within the bath. As the number of bath temperature measurements increases, the picture of the distribution becomes clearer but the distribution of bath temperatures does not change. Thus the uncertainty in the bath temperature is almost independent of the number of measurements we make.

If, however, we wished to measure the average bath temperature (perhaps to estimate the average power required to heat it) then the measurement is best reported as

$$\text{average bath temperature} = m \pm ks_m$$

$$= m \pm ks/\sqrt{N}. \tag{2.16}$$

For measurements of averages the confidence interval does decrease as the number of measurements is increased.

For calibrations this distinction between the variance itself and the variance in the mean is important. One characterises the errors in the use of the thermometer, while the other characterises the errors in the calibration. A calibration measures the distribution of errors in a thermometer's readings. Such a distribution may be represented by the wider curve in Figure 2.9. The mean of the distribution is then used to determine the correction for the temperature readings, eliminating the systematic error. The standard deviation s characterises the remaining (random) error in the thermometer's readings. It is an estimate characterising the range of values around the corrected readings within which the true temperature lies. It is used to calculate the uncertainty reported on the certificate.

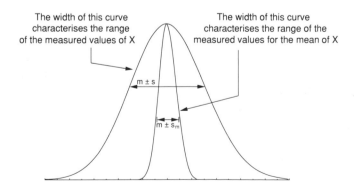

Figure 2.9
Gaussian distributions highlighting the difference between the standard deviation and the standard deviation of the mean.

On the other hand, the standard deviation s_m, the narrow curve in Figure 2.9, characterises the accuracy with which we know the correction. It determines how precisely the correction is reported.

2.7 ESTIMATING AND CHARACTERISING 'TYPE B' UNCERTAINTIES

Type B uncertainties are those determined by other than statistical means. These other means include, for example, theoretical models of the measurement, the work of other experimenters, intuition, experience, and even guesswork. The need to estimate Type B uncertainties arises, for example, when single measurements are made or when corrections are applied to eliminate known errors. Indeed, one of the simplest examples is the uncertainty in the readings of a reference thermometer. In this case the uncertainty is reported on the calibration certificate, so is very easily assessed. In most cases, however, Type B uncertainties are not so easily assessed. As with Type A uncertainties the key is to determine the error distribution.

The assessment process has three main stages:

(1)   Identify the error.
(2)   Describe the error.
(3)   Determine a mean and variance for the error distribution.

All three of these stages must be documented and recorded. This ensures that the measurement is traceable, and therefore open to scrutiny, and can if necessary be improved at a later date as new information or expertise becomes available.

Stage 1, the identification of the error, is often made easier with the construction of a simple model of the measurement. We have shown a very general model of a temperature measurement in Figure 2.10. Before measurements are made, a little time should be spent thinking about the physical processes occurring in and between each block of the model. Imperfections in a process, or external influences on a process, usually give rise to errors.

For the thermometers which are used to maintain the temperature scale, all of the major interactions have been thoroughly investigated so that any errors that arise are well known. The use of less established thermometers carries a risk of large unidentified errors, i.e. the first stage in the error assessment process is incomplete. For all the temperature sensors discussed in this book we describe the most significant errors affecting that particular thermometer. However, there is no guaranteed method for identifying all sources of error. At best one can explore various models of the measurement and research other workers' approach to the measurement.

Once the sources of error have been identified, the second and third stages of the assessment proceed as follows:

•   Collect as much information and advice on the error as is available. Subsidiary measurements which vary the experimental conditions can be very useful.

•   Based on this information, develop a picture of the likely distribution of the error.

•   Decide whether a Gaussian or rectangular distribution best describes the distribution of the error. In some cases there may be sufficient information to identify the real

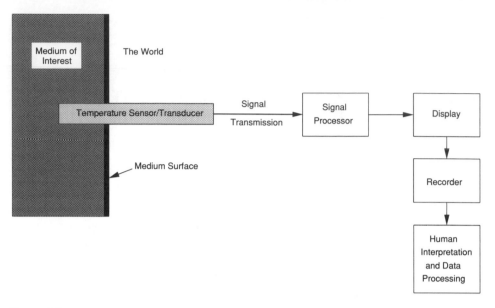

Figure 2.10
A general model of a temperature measurement. Consideration of the processes in the various blocks of the model often exposes potential for errors.

distribution which may be of another kind, for example Poisson, binomial or chi-square (see the references at the end of the chapter).

- Estimate the limits for the distribution and the confidence level associated with the limits. When a distribution is known to be non-symmetric, the limits may also be non-symmetric.

- Derive the mean and variance; the mean may be used to make a correction and the variance to characterise the uncertainty in the corrected measurements.

- Record all of the assumptions and the reasoning leading to the estimates, so that the rationale is clear and unambiguous.

To quote one metrologist, 'The experimenter must recognise that he is quoting betting odds. . . If he has formed his error estimate honestly, avoiding both undue optimism and undue conservatism, he should be willing to take both sides of the bet.'

Once the values for the correction and uncertainty have been determined, the result is corrected and the uncertainty may be reduced to a standard deviation or variance for listing and summing with all the other uncertainties.

Example 2.4   Assessment of the self-heating error for a resistance thermometer

When resistance thermometers are used a current is passed through them to sense t he resistance. The resulting power dissipation causes the thermometer to be at a slightly higher temperature than its surrounds. This effect is known as self-heating (see Section 5.4.4).

Continued on page 57

Continued from page 56

The magnitude of the temperature rise is proportional to the power dissipated and to the thermal resistance between the sensing element and the surrounds. The thermal resistance depends both on the internal construction of the probe and the immediate environment around the probe. Resistance thermometers are calibrated in a well-stirred water bath which keeps the thermal resistance low, so that the self-heating is typically only a few millikelvin. In most applications the self-heating is similar to that in calibration so that negligible error occurs.

However, for some measurements, notably air-temperature measurements, the self-heating effect can be as high as several tenths of a degree. The effect is therefore an important source of error in an air-temperature measurement. How can it be assessed?

For the best accuracy the effect should be measured by varying the sensing current (see Section 5.4.4). This is the subsidiary experiment approach. However, when this is not possible, the estimation of the correction and uncertainty becomes difficult. We find ourselves in a situation where we know that there is an error somewhere between a few millikelvin and perhaps 500 mK, and that it definitely biases the measurement. This is where experience and guesswork may come into the picture. Let us look at two cases.

An inexperienced thermometrist may not have access to any other information. Then the likely error may be best described by a rectangular distribution with limits of 0.05°C and 0.45°C. The correction is therefore estimated to be −0.25°C, and the standard uncertainty (based on equation (2.8)) is ±0.12°C.

An experienced thermometrist will seek, from the manufacturer of the thermometer, information on the self-heating properties under various conditions; in stirred or still water, moving or still air, and as a function of the sensing current.

Armed with this information and knowledge of the measurement conditions the experienced thermometrist may decide that the likely error is best characterised by a Gaussian distribution with a mean of 0.15°C and 95% confidence limits of ±0.1°C. In this case the correction is estimated to be −0.15°C and the standard uncertainty ±0.05°C.

The example above is typical of those uncertainty estimates that are highly subjective, and generally distasteful to metrologists. The most difficult aspect of this problem is assigning confidence limits. How can it be said with any confidence that there is a 95% likelihood of the error lying within 0.05 and 0.25°C? Is it reasonable to assign 68% or 95% confidence limits? The answer is: probably not. The best that can be done is to allow the experimenter, who is usually in the best position to judge the uncertainty, to make the most considered and honest estimate possible.

The assessment of the error is only one facet of the problem. If we place ourselves in the position of the reader of a report, we find that we cannot assess the quality of a Type B error assessment (and therefore the quality of the result) unless we also know the assumptions which have been made. For example, if an experimenter were to report only that 'a correction of −0.25 ± 0.12°C was made for self-heating', then it is not possible to verify or update the assessment in the light of new information. Thus all Type B error assessments must be accompanied by a statement of the assumptions made in the assessment.

The examples that follow have been chosen to illustrate other aspects of temperature measurement as well as a range of problems often encountered in the assessment of Type B uncertainties.

Example 2.5   Assessment of the uncertainty in the definition of the temperature scale

Currently, most of the ITS-90 temperature scale is realised (see Chapter 3) by platinum resistance thermometers calibrated at the defined fixed points. One obvious question about this type of scale is 'How can different thermometers, calibrated at different fixed points and using different interpolation equations, possibly realise exactly the same temperature scale?'. The answer is clearly that they do not! So how can we assess the uncertainty associated with such definitions?

As implied in the question, there are three contributing factors to the irreproducibility of the scale. Firstly, the different properties of the different platinum resistance thermometers give rise to a type of irreproducibility known as *non-uniqueness*. Secondly, on some parts of the ITS-90 scale several alternative interpolation equations are available, depending on the fixed points used. Variations in the realised scale caused by the use of the different equations are known as *subrange inconsistency*. These two contributing factors to the scale uncertainty affect all realisations similarly. The third factor, the different temperatures of the fixed points, depends on many factors including the purity of the metals, and is a matter which must be considered by the individual laboratories operating fixed points.

Figure 2.11 shows an example of non-uniqueness for the section of the scale between 14 K and 273 K. This is the result of experimental and theoretical work carried out by the metrologists who formulated the ITS-90 scale.

It is clear from Figure 2.11 that the uncertainty due to non-uniqueness is about 0.5 mK. Thus national standards laboratories may only quote total uncertainties to about 1 mK for platinum resistance thermometers.

This problem of definition limits the ultimate accuracy of all measurements. For many measurements other sources of uncertainty dominate so we rarely have to consider the uncertainty in the definition of the quantities we measure.

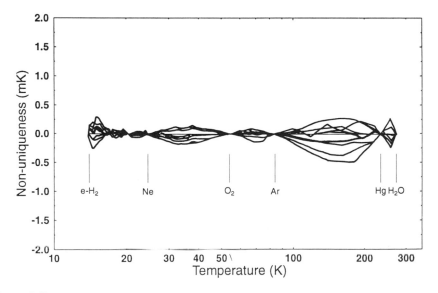

Figure 2.11
The non-uniqueness of the ITS-90 scale in the range 14 K to 273 K. Because every standard platinum resistance thermometer realises a slightly different ITS-90 there is an uncertainty in the definition of temperature.

— *Continued from page 60* —

From Chapter 8 of this book the experimenter finds that the uncertainty in the emissivity propagates according to

$$U_T = \lambda \left(\frac{T}{1200} \right)^2 \left(\frac{100 U_\varepsilon}{\varepsilon} \right) \qquad (2.17)$$

where λ in micrometres is the operating wavelength of the thermometer, T is the temperature in kelvins, and U_ε is the uncertainty in the emissivity.

For a measurement near 1200 K (930°C) and a thermometer operating at 4μm, the equivalent temperature uncertainty U_T is

$$U_T = 4 \times \left(\frac{1200}{1200} \right)^2 \times 100 \times \left(\frac{0.05}{0.85} \right) °C$$

$$= \pm 24°C \qquad (95\% \text{ CL}).$$

Exercise 2.5

(a) Without reference to any other clock, make an assessment of the error (correction and uncertainty) in the time indicated by a familiar watch or clock. This is a Type B assessment. Base your assessment on its past behaviour — is it normally slow or fast? How long since it was adjusted, etc.?

(b) Now locate a means of checking your clock against a local time standard. How good was your assessment?

Exercise 2.6

Repeat Exercise 2.5 for a measuring tape. Use a calibrated engineering rule to check the tape after the assessment.

Exercise 2.7

Repeat Exercise 2.5 for a low-cost mercury-in-glass thermometer. Use an ice point to check the thermometer after you have made your assessment.

2.8 THE UNCERTAINTY IN SUMS AND DIFFERENCES — COMBINING UNCERTAINTIES

In most measurements there is more than one source of uncertainty. In a calibration, for example, there are uncertainties in the reference thermometer readings, arising from the non-uniformity of the calibration bath, as well as in the readings of the thermometer under test. In order to determine the overall uncertainty, we need to know how to combine all the contributing uncertainties. There are standard statistical procedures for this.

Firstly, we assume that the errors are independent. The case where errors are not independent (i.e. the errors are *correlated*) is more difficult and will be discussed in Section 2.10.3. Secondly, we use a very powerful result from distribution theory, as follows:

> When independent variables are summed, the variance of the resulting distribution is the sum of the variances of the individual distributions.

This is true for all types of distributions and is the reason why we relate all uncertainties to the variance or standard deviation.

The formal result is: for any linear combination of independent variables $u, v, \ldots$,

$$z = au + bv + \cdots \tag{2.18}$$

where a and b are constants, then the means m_u, m_v sum to give

$$m_z = am_u + bm_v \cdots \tag{2.19}$$

and the variances sum to give:

$$\sigma_z^2 = a^2\sigma_u^2 + b^2\sigma_v^2 \cdots \tag{2.20}$$

Using this equation we can estimate the total variance from our estimates of the individual variances.

However, determining the uncertainty for 95% confidence limits from the total variance may not be so easy. For the case when the variances s_u and s_v are determined with different degrees of freedom, there is no easy formula for the uncertainty. An approximate and slightly pessimistic formula for the uncertainty is

$$U_z = (k_u^2 a^2 \sigma_u^2 + k_v^2 b^2 \sigma_v^2 \cdots)^{1/2} \tag{2.21}$$

where $k_u, k_v, \ldots$ all correspond to the same confidence level.

For the case when the number of degrees of freedom are the same for all variables this reduces to

$$U_z = k\sigma_z \tag{2.22}$$

where $k = k_u = k_v$. In many cases one contributing variance dominates the others, so that it is sufficient to use equation (2.22) as an approximation for equation (2.21) with the k value appropriate to the dominant contributor.

As you gain experience with equations (2.20) and (2.21) you will find that the contribution of the largest uncertainties to the total is made even more important because the quantities are squared in the equations. Often the smaller uncertainties can be ignored in the calculation.

Example 2.8   Calculation of total uncertainty

A total-immersion mercury-in-glass thermometer is used in partial immersion to determine the temperature of an oil bath. The average and standard deviation of the nine temperature measurements is

$$\text{measured temperature} = 120.68 \pm 0.04°C \quad (1 \text{ sigma})$$

Continued on page 63

Continued from page 62

The calibration certificate for the thermometer shows that a correction of $-0.07°C$ should be applied at $120°C$, and the uncertainty reported on the certificate is ± 0.02 (95% CL). To correct for the use of the thermometer in partial immersion, a stem correction of $+0.42°C$ is also applied: the uncertainty in the correction is estimated as $\pm 0.03°C$ (1 sigma).

Calculate the corrected bath temperature and its uncertainty.

The three contributing measurements and their uncertainties can be summarised as follows, where all measurements are in $°C$:

Temperature reading	120.68 ± 0.04 (1 sigma), $v = 8$	(Type A)
Certificate correction	-0.07 ± 0.02 (95% CL)	(Type B)
Stem correction	$+0.42 \pm 0.03$ (1 sigma)	(Type B)

The corrected bath temperature is given by the measured temperature plus the two corrections. Comparison with equation (2.18) shows that the constants a, b, and c are equal to 1. Hence:

Corrected bath temperature $= 121.03°C$

Uncertainty $= \pm\{(2.31 \times 0.04)^2 + (0.02)^2 + (2 \times 0.03)^2\}^{1/2}$

$\qquad\qquad\quad = \pm 0.11°C$ (95%CL)

(*Note*: The k factor of 2.31 is determined from Student's t-distribution for $v = 8$ and $P = 95\%$. The factor of 2 arises from the general rule cited in Example 2.2.)

Example 2.9   Determining the uncertainty in a temperature difference

Determine the uncertainty in the measurement of a temperature difference

$$\Delta T = T1 - T2, \qquad\qquad (2.23)$$

where the measured uncertainties in $T1$ and $T2$ are s_{T1} and s_{T2} respectively. As a first approximation it may be assumed that the errors in the measurement of the two temperatures are independent, although the errors are likely to be highly dependent if the same thermometer was used for both measurements. We investigate this example with correlated measurements later (see Example 2.13).

Comparing equation (2.23) with (2.18), we recognise that the total variance can be found from equation (2.20) with $a = 1$ and $b = -1$. Hence

$$\sigma_{\Delta T} = (\sigma_{T1}^2 + \sigma_{T2}^2)^{1/2}. \qquad\qquad (2.24)$$

In this case we wish to report the uncertainty as the standard uncertainty. Hence we can report the temperature difference as

$$\Delta T = T1 - T2 \pm (s_{T1}^2 + s_{T2}^2)^{1/2}. \qquad\qquad (2.25)$$

Uncertainties which are summed according to equation (2.24), as the root-sum-square, are said to be summed in *quadrature*.

Exercise 2.8

Using equation (2.20) for the sum of variances, prove that σ_m^2, the variance in the mean of a series of N measurements, is σ^2/N, where σ^2 is the variance of a single measurement of X.

$$\left(Hint : \text{The mean } m = \frac{X_1}{N} + \frac{X_2}{N} + \cdots + \frac{X_N}{N} \right)$$

(*Note*: This result is simply quoted without proof as equation (2.13), and some implications are discussed in Examples 2.14 and 2.15.)

2.9 PRESENTATION OF RESULTS

Throughout the analysis of numerical data one or two extra *guard digits* should be carried beyond the expected precision of the results (see Example 2.1, for instance). This is not because there is any meaningful information carried in the extra digits: they are there to prevent cumulative *rounding errors* from contributing additional uncertainty. This procedure effectively prevents the introduction of quantisation error (Section 2.4.1). Once the final results and uncertainties have been calculated the best precision for reporting the numbers can be determined. The final result is then rounded to the appropriate precision and rounding errors have minimal effect.

We have already observed that the calculated mean and variance are estimates and not known with precision, therefore they should not be reported too precisely. But how do we know how precisely to report them? Let us deal with the standard deviation first.

In Section 2.5, equation (2.14) shows that we can only know the estimated variance with a relative precision of $\sqrt{2/N}$. Through the square root relationship between variance and standard deviation it follows that the uncertainty is known with a relative precision of $1/\sqrt{2N}$. In cases where N is less than 50 (most measurements), it is therefore meaningless to report the uncertainty to any better than one significant digit. There are two exceptions to this rule. Firstly, when the most significant digit is a 1 or 2 it is quite acceptable to report to 5 or perhaps 2 in the next digit. Secondly, in high-precision work where N is large, it is acceptable to report to 2 significant digits. While this guide applies specifically to Type A uncertainties it is also a good guide for the presentation of Type B uncertainties.

The rule for reporting the result is then very simple: report the result to the same decimal place as the uncertainty. This ensures that extra meaningless digits are not reported while at the same time the additional quantisation error is negligible.

Example 2.10   Numerical presentation of result and uncertainty

In Example 2.1 we determined the mean and standard deviation of 20 measurements as

$$m = 6.485°C$$

and

$$s = 0.346°C.$$

Report these results to an appropriate level of precision.

Continued on page 65

Continued from page 64

The standard deviation is known to a relative precision of 1 part in 6, so we should not report the standard deviation to any greater precision than about ±0.06°C. A reasonable approximation is then

$$s = 0.4°C.$$

The mean should be reported to the same precision:

$$m = 6.5°C.$$

As a summary statement the result may then be presented as

$$\text{bath temperature} = 6.5 \pm 0.4°C \qquad (1 \text{ sigma}).$$

Or if 95% confidence limits are required, the same result could be presented as

$$\text{bath temperature} = 6.5 \pm 0.7°C \qquad (95\% \text{ CL}).$$

(*Note*: $k = 2.09$ for $v = 19$ (see Table 2.2).)

The above example illustrates the most important features which must be included in the statement of the result, namely:

- definition of the measurand (bath temperature);
- the estimate of the value;
- the estimate of the uncertainty;
- how the uncertainty is characterised (1 sigma or 95% CL);
- the units of measurement.

2.9.1 SI conventions

An important element in report presentation is the proper use of the SI conventions for the unit symbols and names. Some simple rules will help.

(1) When written in full, the name of all SI units start with a lower-case letter, except at the beginning of a sentence,
    e.g.   kelvin   not   Kelvin or degrees kelvin
            degrees Celsius   not   Degrees Celsius
    Note: degrees centigrade is obsolete.

(2) The symbols are lower case except when named after a person. Hence K is the symbol for kelvin.

(3) It is usual to use symbols for denoting units when reporting numerical results. The symbol may be separated from the last digit by a single space:
    e.g. 273.15K or 273.15 K
    Numbers less than 1 should have a single zero before the decimal point. A comma may be used as decimal point:
    e.g. 0.1°C or 0,1°C   not .1°C

(4)   When written in full, the names of the units may be made plural according to the rules of English grammar,
      e.g. "temperature difference in kelvins".

(5)   When joining a prefix and SI unit symbol, there is no space between the prefix symbol and the unit symbol.
      The most commonly used prefixes are:

Factor	Prefix	Symbol
10^9	giga	G
10^6	mega	M
10^3	kilo	k
10^{-3}	milli	m
10^{-6}	micro	μ
10^{-9}	nano	n

      e.g.   10 mK or 10 m°C   not   10m K   and not   10 milliK
             several millikelvin   not   several mK

(6)   When reporting quantities with compound units formed by the product of two or more units, the unit symbols should be separated by a dot or a space, e.g. for metre-kelvin: m.K or m K not mK, which implies millikelvin.
      When reporting quantities with compound units formed by ratios of two or more units, exponentiation or a single solidus may be used. Parentheses should be used to prevent ambiguities,
      e.g.   W/m^2 or W.m^{-2}
             J/(kg.°C) or J.kg^{-1}.°C^{-1} not   J/kg/°C.

(7)   There are currently no formal rules for the presentation of uncertainties. However, the following three styles are in common usage:
      100.315 ± 0.012 K
      100.315 K $\pm$ 12 mK
      100.315(12) K
      The uncertainty may also be stated in the text.

(8)   There are also no formal rules for the presentation of confidence limits.
          In scientific reporting only standard uncertainties should be reported, either as a variance or as a standard deviation, with a clear statement as to which is used.
          In reports requiring higher confidence limits, the meaning of the uncertainty should be stated in full for maximum clarity, e.g....'where the uncertainty of ±24 mK is assessed with 1 standard deviation of 12 mK and the confidence level of 95% is based on a Gaussian distribution with a k factor of 2.0'.

2.9.2 Scientific reporting

Scientists preparing papers should be aware that for the scientific community (a major client) the paper constitutes not only a report of the measurement but also a formal record of the measurement. It is after all papers and not laboratory notebooks that are referenced to establish credibility. Especially when reporting the results of novel or non-standard measurements, all assumptions made in the assessment of the uncertainties must be included.

When presenting novel or non-standard measurement techniques the presentation should follow the BIPM guide; all known uncertainties are reported and characterised as standard deviations or 1 sigma values and Type A uncertainties are additionally itemised with the number of degrees of freedom. Where correlations are known, the covariance or correlation coefficient should also be reported (see Section 2.10.3). This presentation serves the long-term interests of science in three ways:

(a) By itemising each source of uncertainty you make it possible for someone else to see directly what the most important uncertainty contributions are. This indicates the strengths and weaknesses of the particular technique, suggests where research should be directed to improve the technique and, if one or more sources of error look too large or are not identified, it will be clear that there are flaws or failings in the measurement. All these items serve to make the next measurements better.

(b) By reporting the standard deviation rather than confidence limits, you avoid the problems of deciding whether the distribution is truly Gaussian, and therefore whether the Student's t-distribution is appropriate for determining confidence limits.

(c) By reporting the number of degrees of freedom for each of the Type A uncertainties and the standard deviations, you make it possible for the total uncertainty to be computed according to any code of practice.

In cases where the measurement techniques are well-established the BIPM style is unnecessarily detailed. Therefore, according to the purpose of the paper and the needs of the client, the full itemised list can be shortened, for example by dropping the number of degrees of freedom, and not considering less important sources of uncertainty. However, even the most abbreviated form must state the result, the uncertainty, the confidence limits and the sources of error accounted.

2.10 ADVANCED TOPICS

In this section we introduce more advanced topics which require elementary knowledge of differential calculus for the best understanding. The section should be read by scientists and engineers who are involved in the designing and planning of measurements or calibrations.

2.10.1 Propagation of uncertainty — uncertainties in products

A problem that is commonly encountered is the determination of the uncertainty in temperature when the uncertainty is known in terms of some other quantity. For example, when using resistance thermometers we measure resistance, and the temperature is inferred from our knowledge of the resistance–temperature relationship for the thermometer. So if we know the uncertainty in the resistance measurement, what is the resulting uncertainty in the temperature?

This problem is closely related to the determination of uncertainties in sums and differences. Once again we give the general result first. We also assume again that all variables are independent, and that the uncertainties in any quantity are small relative to the quantity itself.

For any function of independent random variables:

$$z = f(u, v, \ldots), \tag{2.26}$$

the variance is given by

$$\sigma_z^2 = \left(\frac{\partial f}{\partial u}\right)^2 \sigma_u^2 + \left(\frac{\partial f}{\partial v}\right)^2 \sigma_v^2 + \cdots. \tag{2.27}$$

This equation is known as the *propagation-of-uncertainty* formula. Note that the formula for sums and differences (equation (2.20)) is a special case of this equation.

We now consider a special case where z is a product of three other variables:

$$z = u \cdot v \cdot w. \tag{2.28}$$

From equation (2.27), the variance is

$$\sigma_z^2 = u^2 v^2 \sigma_w^2 + u^2 w^2 \sigma_v^2 + v^2 w^2 \sigma_u^2. \tag{2.29}$$

When there are a large number of variables this is a cumbersome formula and it is more conveniently expressed in terms of the *relative variances*. By dividing by $z^2 = u^2 v^2 w^2$ and denoting the relative variance in z as $\rho_z^2 = \sigma_z^2/z^2$ the relative-variance formula is obtained:

$$\rho_z^2 = \rho_w^2 + \rho_v^2 + \rho_u^2, \tag{2.30}$$

where the relative variances ρ are usually expressed in percent.

This is a much more compact equation and its form is very similar to that for sums and differences (equation (2.20)) so it is reasonably easy to remember. This equation also has all the variables separated so that the relative importance of the contributing uncertainties is more obvious.

Example 2.11   Estimating the uncertainty in stem corrections

The stem-correction formula enables the reading on a liquid-in-glass thermometer to be corrected for the fact that some of the mercury in the column is not fully immersed (see Section 6.3.8 for details). The temperature correction is given by

$$\Delta T = N(t_2 - t_1)k, \tag{2.31}$$

where N is the length of the emergent column in degrees Celsius;
t_1 is the mean temperature of the emergent column in use;
t_2 is the mean temperature of the emergent column during calibration;
k is the expansion coefficient of mercury = $0.00016°C^{-1}$.
Now, given the uncertainties in N, $t_1 - t_2$ and k, what is the uncertainty in ΔT?

Continued on page 69

Continued from page 68

By applying equation (2.27) directly we get

$$\sigma_{\Delta T}^2 = (t_1 - t_2)^2 k^2 \sigma_N^2 + N^2 k^2 \sigma_{t_1 - t_2}^2 + N^2 (t_1 - t_2)^2 \sigma_k^2. \qquad (2.32)$$

By inserting the values for the known uncertainties we can now determine the uncertainty in the correction. But this is the cumbersome form of the formula. By dividing through by $(N(t_1 - t_2)k)^2$ we get the simpler equation, in the form of equation (2.30), namely

$$\frac{\sigma_{\Delta T}^2}{(\Delta T)^2} = \frac{\sigma_N^2}{N_2} + \frac{\sigma_{t_1 - t_2}^2}{(t_1 - t_2)^2} + \frac{\sigma_k^2}{k^2} \qquad (2.33)$$

or

$$\rho_{\Delta T}^2 = \rho_N^2 + \rho_{t_1 - t_2}^2 + \rho_k^2, \qquad (2.34)$$

where the ρ are the relative uncertainties which may be expressed in percent.

Typically the relative uncertainty in N, the length of the emergent column, is of the order of 2%, as is the uncertainty in k (which is not truly constant). The greatest source of uncertainty is in the temperature difference of the exposed column, $t_1 - t_2$. Typically the relative uncertainty may be 5% or more. Substituting these values into equation (2.34) we find that the total relative variance is

$$\rho_{\Delta T}^2 = 4 + 4 + 25 \qquad (2.35)$$

so that the relative uncertainty (1 sigma) in the correction is about 6%.

Exercises 2.9 and 2.10 develop other useful formulae for common functions.

Exercise 2.9

(a) Use the propagation-of-uncertainty formula (equation (2.27)) to show that for power laws of the form $z = T^n$ the uncertainty propagates as

$$\frac{\sigma_z}{z} = n \frac{\sigma_T}{T}$$

or equivalently $\rho_z = n\rho_T$.

(b) For the specific example of a total radiation thermometer which uses the Stefan–Boltzmann law

$$L = \varepsilon \frac{\sigma}{\pi} T^4$$

show that the uncertainty in the temperature inferred from measurement of total radiance L and an estimate of the emissivity ε is

$$\sigma_T = \frac{T}{4} \left[\left(\frac{\sigma_\varepsilon}{\varepsilon} \right)^2 + \left(\frac{\sigma_L}{L} \right)^2 \right]^{1/2}.$$

Exercise 2.10

(a) Use the propagation-of-uncertainty formula (equation (2.27)) to show that for exponential laws of the form $z = \exp(kT)$ the uncertainty propagates as

$$\frac{\sigma_z}{z} = k\sigma_T.$$

(b) A spectral-band radiation thermometer approximately obeys Wien's law

$$L_{m,\lambda} = \varepsilon \frac{c_1}{\lambda^5} \exp\left(\frac{-c_2}{\lambda T}\right).$$

Show that the uncertainty in measured temperature inferred from measurements of spectral radiance $L_{m,\lambda}$ and emissivity is

$$\sigma_T = \frac{\lambda T^2}{c_2} \left[\frac{\sigma_{L_{m,\lambda}}^2}{L_{m,\lambda}^2} + \frac{\sigma_\varepsilon^2}{\varepsilon^2}\right]^{1/2}.$$

2.10.2 Curve fitting by least squares

We have discussed how to determine the mean and variance for a set of measurements. However, in some instances we need to determine the values for parameters other than the mean. With resistance-thermometer calibrations, for example, we want to find good values for the coefficients in the quadratic equation relating resistance to temperature. In principle, the three quadratic coefficients could be determined from three measurements only. However, just as a single measurement is a poor estimate of a mean, the resulting estimates for the coefficients would not be good estimates. There would also be no indication of how well the thermometer's behaviour is described by a quadratic equation.

Thus we require a method of determining the best values for a small number of parameters from a larger number of measurements. Another way of looking at the problem is: we require a method which will fit the best curve to a set of points.

The best technique available for this purpose is known as the *method of least-squares* and naturally extends the definition of the mean. In addition to providing estimates of the values of the coefficients, the method of least-squares also provides estimates of the variance of residual errors, as well as uncertainties in the values obtained for the coefficients. With some restrictions on its use, the method of least-sqaures provides all that is required to fit calibration data to calibration equations.

We give here a very brief description of the method of least-sqaures, an introduction only. Readers requiring more information are referred to the books listed at the end of the chapter, which are reasonably tutorial and include example programs. We begin first with an outline of the technique and then follow with an example.

We assume firstly that we are trying to determine the coefficients $a_0, a_1, a_2, \ldots$ in a calibration equation of the form

$$y = f(x) = a_0 + a_1 x + a_2 x^2 + a_3 x^3 + \cdots \tag{2.36}$$

and that we have made N measurements (x_i, y_i) of the relationship between x and y. We shall see in other chapters how most calibration equations can be manipulated into the form of equation (2.36) if they are not already in this form.

The least-squares method requires us to find values of the coefficients that minimise the sum of the squares of deviations between the measured and fitted values of y. We thus need to minimise the function χ^2 given by:

$$\chi^2 = \sum_{i=1}^{N} \left(y_i - \left(a_0 + a_1 x_i + a_2 x_i^2 + \cdots \right) \right)^2. \tag{2.37}$$

The minimum value is found by setting to zero the derivatives of χ^2 with respect to each of the coefficients. This yields one equation for each of the unknown coefficients. For a fit to a quadratic, or second-order, curve there are three equations, one for each unknown parameter:

$$\frac{\partial \chi^2}{\partial a_0} = -2 \sum_{i=1}^{N} \left(y_i - \left(a_0 - a_i x_i - a_2 x_i^2 \right) \right) = 0$$

$$\frac{\partial \chi^2}{\partial a_1} = -2 \sum_{i=1}^{N} \left(y_i - \left(a_0 - a_i x_i - a_2 x_i^2 \right) \right) x_i = 0$$

$$\frac{\partial \chi^2}{\partial a_2} = -2 \sum_{i=1}^{N} \left(y_i - \left(a_0 - a_i x_i - a_2 x_i^2 \right) \right) x_i^2 = 0. \tag{2.38}$$

These are known as the *normal equations* of the method of least-squares. They are most succinctly written in matrix notation, which also shows the pattern of the equations more clearly. Appendix A lists all of the calibration equations recommended in this book and the corresponding normal equations. For a second-order fit the equations are

$$\begin{pmatrix} N & \Sigma x_i & \Sigma x_i^2 \\ \Sigma x_i & \Sigma x_i^2 & \Sigma x_i^3 \\ \Sigma x_i^2 & \Sigma x_i^3 & \Sigma x_i^4 \end{pmatrix} \begin{pmatrix} a_0 \\ a_1 \\ a_2 \end{pmatrix} = \begin{pmatrix} \Sigma y_i \\ \Sigma y_i x_i \\ \Sigma y_i x_i^2 \end{pmatrix} \tag{2.39}$$

or symbolically,

$$[A] \cdot a = b, \tag{2.40}$$

where $[A]$ is a matrix and a and b are vectors.

The unknown coefficients are then found by inverting the matrix $[A]$:

$$a = [A]^{-1} \cdot b \tag{2.41}$$

The overall standard error-of-fit s is found by substituting these values of the coefficients into equation (2.37), hence:

$$s^2 = \frac{\chi^2}{N - \rho} \tag{2.42}$$

where ρ is the number of coefficients. Note that s is the standard deviation of the residual errors in the fit. That is, the least squares technique finds values of the coefficients that minimise the variance of the residual errors. Note also the division by $N - \rho$. This is the number of degrees of freedom or the number of spare pieces of information we have: N measurements with ρ of them used to determine the coefficients.

The equivalent variances in a_0, a_1, a_2, propagated from the standard error-of-fit are estimated by

$$s^2_{a_{i-1}} = [A]^{-1}_{ii} s^2 \tag{2.43}$$

As with the variance in the mean (equation (2.13)), these uncertainties decrease as the number of measurements is increased. The off-diagonal elements of $[A]^{-1} s^2$ are the *covariances* (see Section 2.10.3).

Exercise 2.11

Apply the method of least-sqaures to the equation $y = m$, i.e. use least squares to fit a constant to a set of N data points, hence show that

(a) $m = \dfrac{1}{N} \Sigma y_i$

(b) $s^2 = \dfrac{1}{N-1} \Sigma (y_i - m)^2$

(c) $s^2_m = \dfrac{1}{N} s^2.$

These are the standard equations for the mean, variance, and variance in the mean (equations (2.2), (2.3) and (2.13)).

Exercise 2.12

Apply the method of least-squares to fitting the straight line $y = mx + c$ and hence show that

(a) $c = \dfrac{\Sigma x_i^2 \Sigma y_i - \Sigma x_i \Sigma x_i y_i}{N \Sigma x_i^2 - (\Sigma x_i)^2}$

   $m = \dfrac{N \Sigma y_i x_i - \Sigma x_i \Sigma y_i}{N \Sigma x_i^2 - (\Sigma x_i)^2}$

(b) $s^2 = \dfrac{1}{N-2} \Sigma (y_i - mx_i - c)^2$

(c) $s^2_c = \dfrac{s^2 \Sigma x_i^2}{N \Sigma x_i^2 - (\Sigma x_i)^2}$

   $s^2_m = \dfrac{N s^2}{N \Sigma x_i^2 - (\Sigma x_i)^2}$

There are two restrictions on the use of least-squares analysis for fitting calibration equations. Firstly, the calibration equation must interpolate correctly between calibration points. Linear interpolation between calibration points for a highly non-linear thermometer, for example, would introduce extra and unnecessary error. For all the thermometers discussed in this book we describe calibration equations which have been proved, by application to many thermometers, to be good interpolators. In some cases the equations have been derived from the theory of the thermometer.

The second restriction on the use of least-squares analysis arises from the need to demonstrate that the thermometer under test behaves as expected. This is accomplished by using a relatively large number of calibration points and demonstrating that the measurements consistently follow the fitted calibration equation. A strong pattern in the signs of the residual errors in a least-squares fit is a good indicator of a thermometer which is damaged and no longer well described by the established equation. From a purely statistical point of view the number of measurements should be such that the number of degrees of freedom is no less than 5. This ensures that the k factors from Student's t-distribution (Table 2.2) are reasonably small. However, a satisfactory demonstration of the validity of a calibration equation requires a few more measurements. We recommend a minimum of 3, preferably 4, data points per unknown coefficient. Thus when fitting a quadratic equation for a resistance thermometer, a minimum of 9 to 12 points is required.

Example 2.12   Fitting a quadratic equation for a platinum resistance thermometer

One form of the resistance temperature relationship (see Section 5.1.4) for a platinum resistance thermometer above 0°C is:

$$R(t°C) = R(0°C)(1 + At + Bt^2).\qquad(2.44)$$

The equation can be expanded to a form suitable for least squares fitting:

$$R(t°C) = R(0°C) + R(0°C) \cdot A \cdot t + R(0°C) \cdot B \cdot t^2.\qquad(2.45)$$

Describe a procedure for obtaining the coefficients $R(0°C)$, A and B.

By comparing equations (2.36) and (2.45) we can identify:

$$y = R(t°C),$$
$$x = t,$$
$$a_0 = R(0°C),$$
$$a_1 = R(0°C) \cdot A,$$
$$a_2 = R(0°C) \cdot B,$$

The equations were must solve are, from equation (2.39):

$$\begin{pmatrix} N & \Sigma t_i & \Sigma t_i^2 \\ \Sigma t_i & \Sigma t_i^2 & \Sigma t_i^3 \\ \Sigma t_i^2 & \Sigma t_i^3 & \Sigma t_i^4 \end{pmatrix} \begin{pmatrix} a_0 \\ a_1 \\ a_2 \end{pmatrix} = \begin{pmatrix} \Sigma R_i \\ \Sigma r_i t_i \\ \Sigma R_i t_i^2 \end{pmatrix}\qquad(2.46)$$

where R_i are the values of the resistance measured at temperatures t_i.

Continued on page 74

Continued from page 73

The solution to this example is presented in Appendix A as a Pascal program. The program is intended not only to illustrate the least-squares technique but also to provide a starting point for programming other least-squares routines. One crucial feature of the program is the pivot algorithm associated with the matrix inversion. This ensures that the full machine precision is preserved during computations. Without it the program may decide that the matrix $[A]$ is singular and halt, or worse, deliver results affected by round-off errors.

Figure 2.13 shows a sample output of the program. The data are taken from the DIN 43760 standard for platinum resistance thermometers. Such tables are very useful for proving and debugging fitting programs. Most of Figure 2.13 is self-explanatory. The least-squares problem set by equations (2.44) and (2.46) minimises the variance of the differences between the measured and fitted resistances, and consequently the standard error-of-fit has the dimensions of ohms. To calculate the equivalent variance in the temperature measurements the quadratic equation (2.44) must be solved for t for each value of R_i. The variance of the temperature deviations is then computed as

$$\sigma_t^2 = \frac{1}{N-3} \sum_{i=1}^{N} (t_i - t(R_i))^2 \tag{2.47}$$

where $t(R_i)$ is the inverse of the quadratic relationship. This is not the variance minimised by the least-squares fit; however, for equations which accurately describe real behaviour (as do calibration equations) the variance of the temperature errors is very nearly minimal and the bias in the estimates of the parameters is negligible. In principle the problem could be rewritten in terms of temperature; however, this leads to a non-linear least-squares problem which is extremely difficult to solve.

Finally the program reports

$$R(0°C) = a_0$$

$$A = a_1/a_0$$

$$B = a_2/a_0$$

Note that the standard deviation of the resistance errors is very close to 0.0029 Ω, which is the theoretical value for resistance measurements quantised to 0.01 Ω (equation (2.8)).

2.10.3 Correlated errors

Many times in this chapter we have assumed that errors are independent. While this is a satisfactory approximation for most cases, there are important instances where errors are *correlated* and more detailed analysis is required before the uncertainty can be estimated. By far the largest class of measurement in which correlation is important are those where a single thermometer is used to take a sequence of measurements. In these measurements the correlated fraction of the error is a systematic error. Before we look at the general result, let us first look at a simple measurement that exploits correlation in order to reduce the uncertainty.

SUMMARY FOR PLATINUM RESISTANCE THERMOMETER

Reading number	Measured resistance	Measured temperature	Predicted temperature	Residual error
1	100.0000	0.0000	0.0045	−0.0045
2	103.9000	10.0000	9.9971	0.0029
3	107.7900	20.0000	19.9942	0.0058
4	111.6700	30.0000	29.9958	0.0042
5	115.5400	40.0000	40.0020	−0.0020
6	119.4000	50.0000	50.0127	−0.0127
7	123.2400	60.0000	60.0020	−0.0120
8	127.0700	70.0000	69.9958	0.0042
9	130.8900	80.0000	79.9941	0.0059
10	134.7000	90.0000	89.9971	0.0029
11	138.5000	100.0000	100.0046	−0.0046

standard error-of-fit = 0.0025 ohms
standard deviation of residual errors = 0.0064°C
fitted parameters:
$a0 = 9.99982517e + 01 \pm 1.9e - 03$
$a1 = 3.90874126e - 01 \pm 8.8e - 05$
$a2 = -5.87412587e - 05 \pm 8.4e - 07$
$R(0°C) = 99.9983$
$A = 3.90881e - 03$
$B = -5.874e - 07$

Figure 2.13
Sample output of the least-squares fit program of Appendix A. The data are taken from the DIN 43760 standard for platinum resistance thermometers.

Example 2.13   Determining the uncertainty in a temperature difference

In Example 2.9 we found that the uncertainty in a temperature difference $T1 - T2$ was $(s_{T1}^2 + s_{T2}^2)^{1/2}$ where s_{T1} and s_{T2} are the uncertainties in $T1$ and $T2$ respectively. We did this calculation assuming that the errors in the measurement of $T1$ were independent of those in $T2$. However, when temperature differences are measured in practice, the same thermometer is often used for both measurements, so the errors are similar. Reconsider Example 2.9 assuming that the same thermometer is used and that the errors are correlated.

Consider the distribution of errors for measurements of $T1$. We expect that the errors (subscript e) will comprise a random component, T_{er}, which is apparent through fluctuations in the reading, and a systematic component, T_{es}, which represents the average error in the thermometer reading. If the two temperatures $T1$ and $T2$ are very close, we can expect that the systematic component will be the same for both measurements. Then the measured values of $T1$ and $T2$ will be

$$T1_m = T1 + T_{es} + T_{er1} \qquad (2.48)$$

and

$$T2_m = T2 + T_{es} + T_{er2}, \qquad (2.49)$$

Continued on page 76

Continued from page 75

where T_{es}, the unknown correlated (systematic) error, is characterised by the variance σ_{es}^2 and T_{er1} and T_{er2} are the two independent random errors characterised by the variance σ_{er}^2. The temperature difference is then

$$\Delta T = (T1 + T_{es} + T_{er1}) - (T2 + T_{es} + T_{er2})$$

$$= T1 - T2 + T_{er1} - T_{er2}. \tag{2.50}$$

That is, the systematic component does not affect the difference measurement. We can now use the result of Example 2.9 to determine the variance in the difference as

$$\sigma_{\Delta T}^2 = \sigma_{er1}^2 + \sigma_{er2}^2 = 2\sigma_{er}^2. \tag{2.51}$$

Since the contribution of the random-error component to the uncertainty is less than the total uncertainty in any measurement, it is practical to measure small temperature differences with a much lower uncertainty than for absolute temperatures. This is the principle which allows very high-resolution thermometers (i.e. those with a very low random error) to resolve temperature changes or differences of a few microkelvins, even though the temperature scale is only defined to about one millikelvin.

In difference measurements the high degree of correlation between the two measurements is used to advantage. Ratio-type radiation thermometers (see Section 8.7.2) also exploit correlation to reduce uncertainty. However, in many measurements correlation increases the uncertainty.

As might be expected, the mathematics for treating correlated errors is not as simple as that for independent errors. For any function of two variables of the form

$$z = f(u, v), \tag{2.52}$$

the uncertainties in u and v are propagated as

$$\sigma_z^2 = \left(\frac{\partial f}{\partial u}\right)^2 \sigma_u^2 + \left(\frac{\partial f}{\partial v}\right)^2 \sigma_v^2 + 2\left(\frac{\partial f}{\partial u}\right)\left(\frac{\partial f}{\partial v}\right)\sigma_{uv} \tag{2.53}$$

where σ_{uv} is known as the *covariance*. The covariance can be estimated from measurements as

$$\sigma_{uv} = \frac{1}{N}\sum(u_i - m_u)(v_i - m_v) \tag{2.54}$$

When a covariance is zero, the two errors are uncorrelated and said to be independent. With a covariance of zero, equation (2.53) reduces to the propagation-of-uncertainty formula (2.27) given in Section 2.10.1 for independent errors. Covariances are often expressed in terms of the correlation coefficient r, which is defined as

$$r = \frac{\sigma_{uv}}{\sigma_u \sigma_v}. \tag{2.55}$$

Depending on the degree of correlation, r varies between ± 1, with $r = 1$ for highly correlated variables and $r = 0$ for independent variables. Anticorrelation, when it occurs, results in negative values for r.

In practice equation (2.53) is of limited use in the analysis of results. In the temperature-difference example it is very easy to write $\sigma_{T1T2} = \sigma_{es}^2$ and derive the equation for the uncertainty (Exercise 2.13). It is, however, more difficult to determine a numeric value for σ_{es}^2; by its nature the correlated fraction of the error is often unknown and must be estimated. As with many aspects of uncertainty analysis, the greatest utility of equation (2.53) is in the analysis of measurement techniques.

Example 2.14   The uncertainty in average temperatures

In Exercise 2.8 you are asked to prove the result quoted above as equation (2.13), which is the so-called '1/N' rule for the variance in the mean:

$$\sigma_m^2 = \sigma^2/N \tag{2.56}$$

and where σ^2 is the variance of the N individual, and independent, measurements. Now let us look at an example where the errors are not independent. Consider the average of two measurements where, using the notation of Example 2.13,

$$T1_m = T1 + T_{es1} + T_{er1} \tag{2.57}$$

and

$$T2_m = T2 + T_{es2} + T_{er2} \tag{2.58}$$

where $T1$ and $T2$ are the true values.

Calculate the uncertainty in the average temperature T_{av}, given by $T_{av} = (T1_m + T2_m)/2$.

From equation (2.53),

$$\sigma_{av}^2 = \frac{1}{4}\left(\sigma_{es1}^2 + \sigma_{es2}^2 + \sigma_{er1}^2 + \sigma_{er2}^2\right)$$
$$+ \frac{1}{2}\left(\sigma_{es1,es2} + \sigma_{es1,er1} + \sigma_{es1,er2} + \sigma_{es2,er1} + \sigma_{es2,er2} + \sigma_{er1,er2}\right). \tag{2.59}$$

We have assumed that the uncertainties in the defined temperatures $T1$ and $T2$ are negligible. We will also assume that the random errors are independent of each other and of the systematic errors; and also that the correlation coefficient for the two systematic errors is 1.0. The uncertainty in the average is then

$$\sigma_{av}^2 = \frac{1}{4}\left(\sigma_{es1}^2 + \sigma_{es2}^2 + \sigma_{er1}^2 + \sigma_{er2}^2\right) + \frac{1}{2}(\sigma_{es1}\sigma_{es2}) \tag{2.60}$$

$$= \frac{1}{4}(\sigma_{es1} + \sigma_{es2})^2 + \frac{1}{4}\left(\sigma_{er1}^2 + \sigma_{er2}^2\right). \tag{2.61}$$

The result in the example above (equation (2.61)) is very interesting since it shows that the correlated systematic uncertainties (within the first set of parentheses) add linearly while the random uncertainties (second set of parentheses) add in quadrature. It was the occurrence of results like this that led to the practice of treating random and systematic uncertainties differently. However, in most measurements the various systematic components are independent. Further, where correlation exists, the correlation coefficient is not usually 1.0. The linear summation rule then leads to an unduly pessimistic estimate of uncertainty. For example, in the temperature-difference example (see Example 2.13) it would lead to an extremely pessimistic estimate of the uncertainty. There are also examples of distinctly random-error processes (see Example 2.16) where correlation occurs. Therefore the best practice is to treat random and systematic uncertainties identically, and where correlations are known to exist then equation (2.53) should be used to determine the total uncertainty.

Example 2.15   The uncertainty in repeated measurements

Continuing with Example 2.14, let us suppose that $T1 = T2$, and hence

$$\sigma_{es1} = \sigma_{es2} = \sigma_{es}$$

and

$$\sigma_{er1} = \sigma_{er2} = \sigma_{er}.$$

Calculate the variance in the average temperature and comment.

From equation (2.61) the variance in the average temperature is

$$\sigma_{av}^2 = \sigma_{es}^2 + \frac{1}{2}\sigma_{er}^2 \tag{2.62}$$

Thus averaging a set of measurements will reduce uncertainty, due to the random component of the error, but not the systematic component. The general result for N measurements is (Exercise 2.14)

$$\sigma_{av}^2 = \sigma_{es}^2 + \frac{1}{N}\sigma_{er}^2. \tag{2.63}$$

In rare instances the result of equation (2.63) can be used to justify the averaging of a very large number of results in order to make a more accurate measurement. To do so requires two conditions to be satisfied:

(1)   That to the level of accuracy required, the measurand is well described by a single true value, not a distribution.
(2)   That the errors in the measurement are to the level of accuracy required are entirely random. That is, the distribution for the errors must be known to have a zero mean.

It is generally rare that both of these conditions are satisfied, therefore as a general rule the standard error, as determined by the $1/N$ rule (equation (2.56)), is not a good estimator of uncertainty.

Exercise 2.13

(a)   Use equation (2.55) to show that the uncertainty in a temperature difference is

$$\sigma_{\Delta T}^2 = 2\sigma_T^2(1 - r)$$

where r is the correlation coefficient.

Continued on page 79

—— *Continued from page 78* ——

(b)   Hence, using $\sigma_{T1}^2 = \sigma_{T2}^2 = \sigma_T^2$, $\sigma_{T1T2} = \sigma_{es}^2$ prove equation (2.51):

$$\sigma_{\Delta T}^2 = 2\sigma_{er}^2.$$

Exercise 2.14

(a)   The multivariate version of equation (2.53) is

$$\text{for } z = f(u_1, u_2, \ldots, u_N)$$

$$\sigma_z^2 = \sum_{i=1}^{N} \sum_{j=1}^{N} \left(\frac{\partial f}{\partial u_i}\right) \left(\frac{\partial f}{\partial u_j}\right) \sigma_{u_i u_j},$$

where $\sigma_{u_i u_i} = \sigma_{u_i}^2$.

Use this result to show that the uncertainty in a measurement of average temperature is

$$\sigma_{av}^2 = \sigma_T^2 \left(r + \frac{1-r}{N}\right).$$

(b)   Hence prove equation (2.63).

$$\sigma_{av}^2 = \sigma_{es}^2 + \frac{\sigma_{er}^2}{N}.$$

Example 2.16   Correlation in random errors

For many measuring instruments the resolution is limited by random electrical *noise* originating in electrical components such as transistors and resistors, etc. Usually the contribution of the noise to the reading is limited by a simple low-pass filter which removes high-frequency noise components. Because the filter resists rapid changes it 'remembers' previous signals. The correlation coefficient of the random noise component of two successive measurements is

$$r = \exp(-\tau/\tau_F), \tag{2.64}$$

where τ and τ_F are the time between measurements and the time-constant of the filter respectively. It can be shown that when a large number of measurements, N, are taken at equal time intervals, τ, the variance in the mean of the measurements is

$$\sigma_m^2 = \frac{\sigma^2}{N} \coth\left(\frac{\tau}{2\tau_F}\right). \tag{2.65}$$

Discuss the implications of these two formulae.

This is a situation where the random error due to the noise in a single measurement is completely unpredictable and unbiased, yet when successive measurements are used to make averages, correlation affects the uncertainty.

—— *Continued on page 80* ——

Continued from page 79

> The coth function is always greater than 1.0 so the variance in the mean is always larger than what we would expect from the $1/N$ rule.
>
> This example highlights, again, the deficiencies in the $1/N$ rule for the variance in the mean, particularly for measurements made using data acquisition systems which, because of anti-aliasing filters, always exhibit correlation between successive samples.

In cases where there is correlation, the mathematics is often either trivial or very difficult. When reporting uncertainties in the difficult cases it may be sufficient to indicate that there is correlation between those correlated uncertainties and to simply add the variances as though they were independent.

Wherever possible the variance in the means for time averages, for example, should be determined experimentally by repeating the measurements, rather than by relying on the $1/N$ rule for determining the variance in the mean.

FURTHER READING

Guide to the Expression of Uncertainty in Measurement, ISO/IEC/OIML/BIPM (to be published).
This guide is endorsed by the International Organisation for Standardisation (ISO), the International Electrotechnical Commission (IEC), the International Organisation for Legal Metrology (OIML), and the International Bureau of Weights and Measures (BIPM).

An Introduction to Error Analysis, John R Taylor, Oxford University Press, Oxford (1982).
A text well suited to beginners, it is slightly dated in respect of its treatment of random and systematic errors but includes a lot of good examples and some additional topics.

Data Reduction and Error Analysis for the Physical Sciences, Philip R Bevington, McGraw-Hill, New York (1969).
Numerical Recipes, William R Press, Brian P Flannery, Saul A Teukolsky, and William T Vetterling, Cambridge University Press, Cambridge (1986).
These are two standard texts on numerical methods for analysing data and in particular on least squares fitting. Both are reasonably tutorial and contain example programs. Press et al. is also available with the programs on floppy disk in several computer languages.

Uncertainty, Calibration and Probability (2nd edition), C. F. Dietrich, Adam Hilger, Bristol (1990).
A very rigorous introduction to basic statistics and distribution theory.

3

Temperature Scale

3.1 INTRODUCTION

In principle, temperature is defined completely by thermodynamics — the science of heat. In practice, thermometers based on thermodynamic laws are neither convenient nor sufficiently accurate. Instead the international measurement community defines an empirical temperature scale that is sufficiently reproducible to satisfy the needs of science and trade. This scale is revised periodically to ensure that it covers as wide a range as is practical, and that it is as close as possible to the thermodynamic scale. The most recent revision was in 1990, so the current scale is known as the international temperature scale of 1990, or simply ITS-90.

ITS-90 differs significantly from the previous scale of 1968. While the scale is more complex the choice of temperature ranges is greater and more suited to real thermometry needs.

The main aim of this chapter is to provide a background to enable the more general thermometer user to understand the temperature scale. Several of the techniques and concepts will be of interest and directly applicable. It is particularly noteworthy that the fixed points by themselves can provide users with a convenient check on the stability of their thermometers without the need for a full calibration. Besides the fixed points which define the scale there are several other well-recognised fixed points. We detail the construction and use of an ice point because it is readily achieved with minimal resources. An ice point is an essential tool for all thermometer users to ensure traceability.

An interesting feature of ITS-90 is that it enables users requiring high-accuracy to establish the temperature scale for themselves over their range of interest. This process is facilitated by the ready commercial availability of the components necessary to establish extensive parts of the scale, as well as the relative ease of use of these components. As we expect progressively more users to be interested in adopting this route, we give an introduction to the scale and some of the procedures for its maintenance. This will help users to decide whether to use the scale directly. However, the chapter is not a handbook for the maintenance of the scale: if this is required, the official documents should be consulted (see References at the end of the chapter).

3.2 UNITS OF TEMPERATURE

The fundamental physical quantity known as *thermodynamic temperature* is usually represented by the symbol T, and its unit is the *kelvin*, symbol K. The unit of thermodynamic temperature is defined as the fraction $1/273.16$ of the thermodynamic temperature of the *triple point of water* (see Section 3.3). Note that this definition concerns the unit for temperature and not temperature itself.

A temperature is also commonly expressed in terms of its difference from 273.15 K, the ice point. To distinguish a thermodynamic temperature expressed this way, the temperature is known as a *Celsius temperature*, symbol t, defined by

$$t = T - 273.15 \text{ K}. \tag{3.1}$$

The unit of Celsius temperature is the *degree Celsius*, symbol °C (not deg C). By definition the unit is the same size as the kelvin and differences of temperature may be expressed in kelvins or degrees Celsius. When reporting temperatures, kelvins are generally used at low-temperatures, i.e. below 0°C, and degrees Celsius at higher temperature, but there is no hard-and-fast rule. Small temperature differences are more often expressed in millikelvins.

Example 3.1

Express the temperature of the ice point and the triple point of water in terms of the units K and °C.

The ice point is 0°C or 273.15 K.
The triple point of water is 0.01°C or 273.16 K.

It is often necessary to distinguish between *scale temperature* and *thermodynamic temperature* because they have the same name for their units. As seen in Section 1.2.4, the ITS-90 and thermodynamic temperature should be the same, but a defined scale will always differ by a small amount. The symbols T_{90} and t_{90} are used for the current scale, ITS-90, and previous scales are similarly denoted, e.g. T_{68} and t_{68} for IPTS-68.

3.2.1 Principles of ITS-90

The temperature range covered by ITS-90 is from 0.65 K up to the highest-temperature practicably measurable in terms of the Planck radiation law (see Section 3.6).

Figure 3.1 is a guide to the main features of ITS-90, outlining the fixed points, the types of thermometers used, and the ranges for which interpolation formulae are defined.

As seen in Figure 3.1, there are three basic stages in establishing the scale:

- *fixed points*, namely the melting point, freezing point and triple point of various substances, are constructed in accordance with BIPM Guidelines;
- the readings of thermometers of approved types are determined at each of the fixed points;
- temperatures over the various ranges are calculated from the determined values at the appropriate fixed points using the specified interpolation equations.

Fixed points are physical systems whose temperatures are fixed by some physical process and hence are universal and repeatable. The most successful systems for temperature have been phase transitions involving the major changes of state, e.g. liquid to solid or vapour to liquid. Under the proper conditions, i.e. a fixed point apparatus, the phase transition

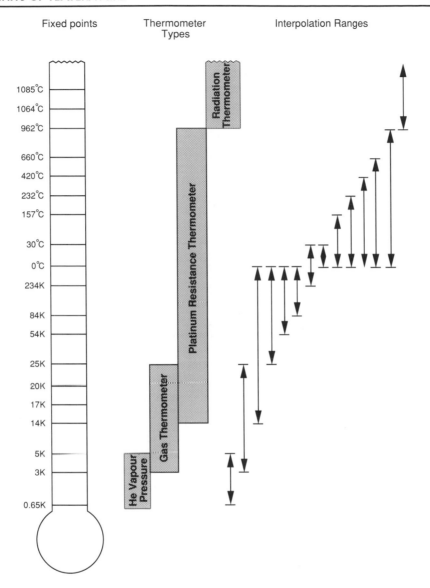

Figure 3.1
Simplified guide to the main features of ITS-90.

will occur at a single temperature which depends on the properties of the substance used and not on the apparatus. As the change involves *latent heat*, good temperature stability is possible. Latent heat is the amount of heat required to cause a change of state without a change in temperature. Thus a small amount of heat transfer between the substance and its surroundings will not cause a temperature change in the substance during the phase transition.

Freezing points of pure metals are controlled by the latent heat of fusion and provide highly repeatable temperatures. Triple points of many substances also make excellent fixed points since they represent balance between the three phases — solid, liquid and

Table 3.1. Defining fixed points of the ITS-90 scale.

Number	Temperature		Substance*	State†	$W_r(T_{90})$
	T_{90}/K	t_{90}/°C			
1	3 to 5	−270.15 to −268.15	He	V	
2	13.8033	−259.3467	e-H$_2$	T	0.001 190 07
3	~17	~ −256.15	e-H$_2$ (or He)	V (or G)	
4	~20.3	~ −252.85	e-H$_2$ (or He)	V (or G)	
5	24.5561	−248.5939	Ne	T	0.008 449 74
6	54.3584	−218.7916	O$_2$	T	0.091 718 04
7	83.8058	−189.3442	Ar	T	0.215 859 75
8	234.3156	−38.8344	Hg	T	0.844 142 11
9	273.16	0.01	H$_2$O	T	1.000 000 00
10	302.9146	29.7646	Ga	M	1.118 138 89
11	429.7485	156.5985	In	F	1.609 801 85
12	505.078	231.928	Sn	F	1.892 797 68
13	692.677	419.527	Zn	F	2.568 917 30
14	933.473	660.323	Al	F	3.376 008 60
15	1234.93	961.78	Ag	F	4.286 420 53
16	1337.33	1064.18	Au	F	
17	1357.77	1084.62	Cu	F	

*All substances except ^{3}He are of natural isotopic composition: e-H$_2$ is hydrogen at the equilibrium concentration of the ortho- and para-molecular forms.

†The symbols have the following meanings: V: vapour-pressure point; T: triple point (temperature at which the solid, liquid and vapour phases are in equilibrium); G: gas-thermometer point; M, F: melting point, freezing point (temperature, at a pressure of 101 325 Pa, at which the solid and liquid phases are in equilibrium).

vapour — at a single temperature and pressure. Table 3.1 gives the details of the systems and their defined values and Figure 3.2 illustrates an example of a commercially available fixed point. Boiling points are no longer used because of their dependence on the pressure. Section 3.4 examines the main types of fixed points.

Four classes of thermometers are used in establishing the scale:

- vapour-pressure thermometers;
- gas thermometers;
- platinum resistance thermometers; and
- radiation thermometers.

The class of platinum resistance thermometers is further split into three types:

- capsule thermometers, for 13.8 K to 157°C;
- long-stem thermometers for 84 K to 660°C;
- high-temperature thermometers for 0°C to 962°C.

A platinum resistance thermometer which is suitable to hold the scale will be denoted as an SPRT (standard platinum resistance thermometer). These are covered further in Section 3.5 and in Chapter 5. BIPM guidelines place specific restrictions on their construction.

The radiation, gas and vapour-pressure thermometers must be constructed according to established physical principles, but otherwise the BIPM guidelines place no other restrictions on them.

Figure 3.2
The gallium temperature standard, which includes the gallium cell and a fixed point apparatus, is a convenient means for realising and maintaining the liquid–solid equilibrium (melting point) of gallium. This precision instrument provides laboratories and manufacturers with a standard for the calibration of laboratory transfer and industrial thermometers, at a temperature (29.7646°C) which is a constant of nature in the biological temperature range.

Radiation and gas-thermometry interpolation formulae are based directly on thermodynamic equations but are referenced to defined temperatures on the scale. For vapour-pressure and resistance thermometry the equations are more empirical.

The resulting thermometers establish the primary temperature scale to which temperature measurements should be traceable. The fixed points should not be considered more fundamental than the scale thermometers, although they do provide convenient reference points for checking the performance of any thermometer.

The main use for the radiation, gas and vapour-pressure thermometers is to transfer the scale to more convenient reference devices, such as standard lamps or rhodium–iron resistance thermometers, with which calibrations can be made. Platinum resistance thermometers, on the other hand, can be used directly to calibrate a wide range of reference or working thermometers.

3.3 WATER TRIPLE POINT — DEFINING THE UNIT

The heart of the temperature scale is the water triple point, which is used to define the unit of thermodynamic temperature. The water triple point occurs at a single temperature and pressure when ice, water and water–vapour are in stable equilibrium with each other in a closed container. At other temperatures and pressures, only two phases can be in equilibrium, e.g. ice–water or water–vapour. In order to utilise this physical system as a precision temperature reference, special cells are used for the SPRTs (Figure 3.3). Cells can be made by a competent glassblower, paying particular attention to cleanliness.

The *ice point* at 0°C is closely allied to the water triple point. Historically this was a defining point for temperature scales until the more precise triple-point cells were developed. The ice point still has an important role in thermometry as it is an easily made reference point.

3.3.1 Water for triple-point cells

Water purity is an important consideration in the manufacture of a triple-point cell. Many of the observed variable properties of the triple point of water appear to arise from the impurities which get added as a result of attempts to clean the glassware or purify the water. Good procedures for freezing and using the cells should reduce the problem.

Impurities alter the triple point. The freezing-point depression constant of water is 1.86 K per mole of impurity in 1 kg of water. The impurity level in a triple-point cell can be readily controlled to achieve an accuracy of 0.1 mK.

The isotopic composition of the water can also alter the triple point. For example, the addition of 10 μmol fraction of heavy water (deuterium oxide) will raise the triple point by 40 μK. Natural water contains approximately 0.0015 mol fraction of heavy water. Variations in this concentration can easily cause 140 μK variation in a triple-point temperature. Purification of the water modifies the isotopic composition. Different purification processes can cause over 0.1 mK variation.

Most of the difference in temperature, 0.01 K, between the ice point and triple point is due to the difference in pressure between standard atmospheric pressure and the triple-point pressure, representing a 7.5 mK change in temperature, and the amount of dissolved air, which changes the freezing-point by 2.5 mK. Therefore the main impurity to be removed from the water is air, i.e. the preparation of triple-point cells requires a good degassing process such as shown in Figure 3.4.

3.3.2 Testing the triple-point cell

The triple point of water is a fixed point whose temperature is a matter of definition and therefore it cannot be measured. However, two tests can be applied to a cell to see if it will perform properly.

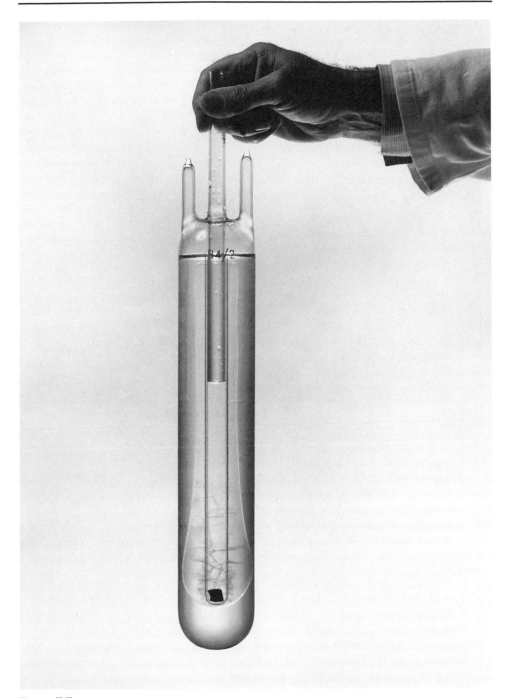

Figure 3.3
A triple-point cell showing the frozen ice mantle and thermometer-well containing some water and a small sponge. The cell is stored in ice during use.

Figure 3.4
Degassing of pure water for a triple-point cell.

(1) *Comparison of two cells* A cell is best checked for accuracy by comparison with a second cell or bank of cells. Certification of a cell by a National Standards Laboratory will follow this route, and where the user has only one or two cells, certification is essential. If the two cells are free from any air leaks or any other contamination then no temperature difference will be observed. If any difference is found then the cell of lower temperature contains more impurities. The defining temperature, 0.01°C, is a maximum value for the realisation of the triple point of natural water.

(2) *Test of an individual cell* Test for an air leak by tilting the cell, holding the seal-off tube down as in Figure 3.5. If there is no significant air contamination, an audible click is heard, and the trapped bubble in the seal-off tube will diminish in volume by a factor of three or more as the cell is tilted slowly to further compress the bubble. In most cases the bubble size will be reduced by a factor much greater than three. If a cell passes this test then any air present makes an insignificant difference, less than 50 μK, to the triple-point temperature. This test cannot reveal the presence of non-volatile impurities, however.

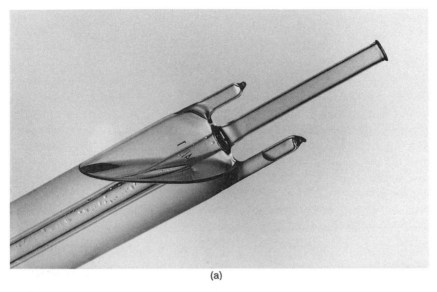

(a)

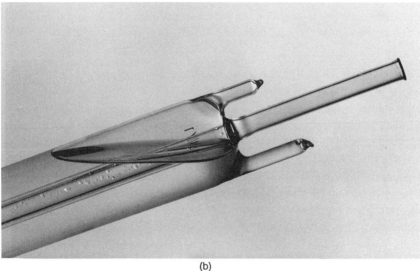

(b)

Figure 3.5
Air-bubble test for an air leak in a water triple-point cell: (a) air and water-vapour
are trapped in the seal-off tube; (b) tilting the cell further compresses the trapped air
bubble.

3.3.3 Freezing of the ice mantle

In principle the cell only needs to have ice, water and water-vapour present in order to
achieve the triple point. In practice we need to be able to immerse a thermometer in

this system and to avoid any impurity problems. Immersion is best achieved with the ice mantle surrounding the well of the thermometer and with a thin water interface between the ice and the well. This ensures that the triple point is not affected by mechanical pressure from the well, and that the water–ice interface is close to the thermometer. If the well were surrounded by one phase, i.e. ice or water, then the temperature in the well would rely on thermal conductivity and not on the balance between the phases. Impurity variations can be minimised by ensuring the freeze process starts from the well and proceeds outwards slowly. This is because the growth of ice crystals is a purification process so that the ice will be purer than the surrounding water. Thus when the well is warmed to produce a thin water film, the water will be purer than that in the rest of the cell and so the effects of any contamination originally present or leached from the walls is minimised.

   Details of the freezing process depends on the type of refrigerant readily available to the user. An outline follows:

- Precool the cell in ice or a refrigerator for a few hours.

- Hold the cell vertical. Once ice has formed any tilting needs to be done with extreme care or the stress on the well may cause it to break.

- Fill the thermometer-well uniformly with refrigerant: for example, 'dry ice' (solid CO_2) in alcohol or cold gas through a tube from a liquid-nitrogen dewar.

- Gently rotate the cell as the ice mantle freezes to obtain a uniform mantle.

- The outside of the cell at the water–vapour interface should be warmed to prevent ice from freezing to the outer walls. Rapid ice expansion may crack the cell.

- Stop the freeze once the ice mantle is large enough, that is, remove the refrigerant. Refractive index effects of the curved cell make the ice mantle appear larger than it is, but the ice should fill three-quarters of the space if the triple point is to have a reasonable lifetime.

- Half fill the thermometer-well with water. This will ensure good thermal contact when a thermometer is inserted. A small rubber sponge at the bottom of the well will insure against breakage (see Figure 3.3).

- Store the cell in an appropriate storage vessel, e.g. a copper sleeve inside a vessel packed with crushed or shaved ice. Commercial storage units are available. Well-designed systems can keep the cell frozen for months. The cell should be stored for 24 hours before it is used, to allow strain in the ice to relax.

- The cell's useful life depends on the storage method and ends when the ice mantle no longer completely encloses the thermometer well. Add ice to the storage vessel if appropriate and check the size of the ice mantle. When stored in ice with low heat leaks the mantle will grow slowly. Check the cell periodically and free the ice mantle so that it rotates independently.

3.3.4 Use of the triple-point cell

A cell is usually frozen and kept for months or for the period it will be in use. For high-accuracy work this allows very convenient checks to be made before and after use of the SPRT. The main expense is the means of storage of the frozen cell.

The main points to follow are:

- Leave the cell to settle for 24 hours after first freezing to allow stress in the ice to relax (see Figure 3.3).

- A small rubber bung can be used on the bottom of the well for protection.

- Fill the well with water for good thermal contact between thermometer and cell.

- Free the ice mantle from the well by inserting a metal rod into the well. The water film around the well should allow the ice to float freely when the cell is gently rotated. This ensures that the water–ice interface is close to the thermometer.

- Ensure that a minimum of heat is introduced into the cell from outside, such as from leads, the body of the thermometer and radiant energy sources. The cell is stored in ice during use. Thermometers can be pre-cooled before insertion.

- For very high-accuracy, a temperature correction is needed to account for the head of water above the centre of the sensor. The triple point is given by

$$t = 0.0100 - 7 \times 10^{-4} \times h °C \qquad (3.2)$$

where h is the height in metres from the centre of the sensor to the surface of the water in the cell.

- Provide adequate time (15 to 20 minutes) for thermal equilibrium to be reached.

Triple-point measurements should be reproducible to better than 0.15 mK.

3.3.5 The ice point

The ice point is the equilibrium between melting ice and air-saturated water. It still has a major role in thermometry since it is a fixed point that can be readily achieved by almost any laboratory with a minimal outlay of resources. It is a *must* for people who take their temperature measurements at all seriously. Whether the accuracy required is ±100°C or ±0.01°C, the ice point is an invaluable aid in ensuring that a thermometer is functioning correctly. Its limitations should, however, be recognised. Although an ice-point apparatus can be made to achieve an accuracy of better than ±0.01°C, it has to be very carefully constructed to do so. If the accuracy requirements are usually ±0.01°C or better, then the triple-point cell should be used. If used as a 'poor man's triple point' to achieve ±1 mK then very close adherence is needed to the procedure below and an independent verification must be made that the accuracy has been achieved.

Ice is such a fundamental substance for good thermometry that a laboratory should invest in the means to obtain shaved ice in adequate quantities rapidly.

The ice-point procedure given below is suitable for a reference standard and the user should become sufficiently familiar with it to place a high level of confidence in it. Most thermometers have a 0°C point and, where possible, only those that do so should be purchased. A thermometer check at 0°C is a very convenient way to see if the thermometer is performing correctly and that the calibration certificate is still valid. Hence the ice point is essential to demonstrate traceability. If there is a discrepancy at 0°C some remedial action can be taken. For liquid-in-glass thermometers, this means adjusting the certificate correction terms, but for other thermometers it may mean a full check to determine if the difference is due to the sensor or signal processor. Thus, confidence in the instrument can be built up and a maintenance and recalibration schedule can be established.

An ice-point apparatus can be easily assembled (Figure 3.6) and consists of a container, a siphon tube, ice, and distilled water. A *dewar flask* (vacuum-insulated flask), approximately 30–40 cm deep and 8–10 cm in diameter, can serve as a container for the ice. A vessel of this type is preferable, since the melting of the ice is retarded by its insulating properties. A siphon is placed in the flask to permit removal of excess water as the ice melts. Clear or transparent commercially purchased ice, or ice made from distilled water, can be used. The ice is shaved into small chips measuring less than 1 mm. The flask is first one-third filled with distilled water before the shaved ice is added. This mixture is compressed to form a tightly packed slush and any excess water is siphoned off. Before the bath is used, adequate time (15–20 min) should be given for the mixture to reach a constant temperature throughout.

Ideally, as much ice should be packed into the flask as possible, so that the small spaces between the chips contain mainly distilled water with a little air to ensure that the water is air-saturated. It will be necessary to add ice periodically and to remove the excess water while the bath is being used in order to maintain this ideal consistency. If

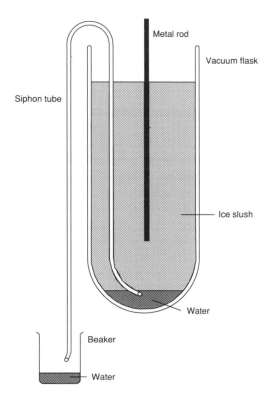

Figure 3.6
ice-point apparatus for calibrating thermometers or
for checking their stability.

care is taken to prevent contamination of the ice and water, the ice point can be readily realised to better than 0.01°C.

Some notes on the above procedure

- *Distilled water*   Good, clean tap water may be sufficient; at 10°C its resistivity should be higher than 0.5×10^6 Ω m. Some tap water will be completely unsatisfactory in this respect and it pays to check. De-ionised water is usually satisfactory.

    Contamination by salts or organic solvents should be avoided by good housekeeping procedures.

- *Dewar flask*   The dimensions given above cater for a wide range of thermometers; obviously shorter dewars can be used for smaller thermometers. For metal thermometer probes, an adequate immersion depth is essential. Other well-insulated vessels can serve the same purpose if a dewar is not available — for example, expanded polystyrene serves quite well.

- *Excess water*   Some of the depth of the flask is required to take up the melted ice. If the depth is not sufficient, siphoning will need to be frequent. The water level should not rise to the level of the thermometer.

- *Ice*   See comments on distilled water. Many commercial ice-making machines purify the water by the freezing process and thus serve as a satisfactory source of ice. Check the resistivity of the melted ice.

- *Shaved*   Shaved ice is preferred for tight packing around a metal probe. For glass thermometers crushed ice up to 5 mm is adequate. The ice can be crushed with a mallet or hammer after putting it inside a protective bag. A more satisfactory method to produce fine ice is to use a mechanical ice shaver or grinder.

- *Tightly packed*   It is essential to pack the ice tightly to achieve good thermal contact between the ice and the thermometer. Lack of this thermal contact can be the main source of error in using an ice point. In order to protect a delicate thermometer a clean rod of aluminium or stainless steel can be used to form a hole in the tightly packed ice. After the thermometer has been inserted, the ice should be pressed tightly around it. For metal probes shaved ice is essential to achieve good contact.

- *Slush*   When ice is frozen its temperature is usually lower than 0°C. For the ice point all of the ice needs to be at 0°C; this is one of the reasons for shaving it into small pieces. Crushed ice at a temperature lower than 0°C will appear white due to frosting on its surface, i.e. due to the condensation of water-vapour from the air onto the cold ice. Formation of the slush is to ensure that all the ice is at 0°C, i.e. just on the point of melting; hence it will appear clear (see Figure 3.7).

Although the above ice point is suitable for a thermocouple reference junction, it may not be suitable for general application (see Chapter 7). If there is a large heat input due to many thermocouples or a heavy thermocouple, then good thermal contact with the reference junction is essential. In this instance a well-stirred ice–water mixture or a commercial ice-point apparatus may be more suitable. The ice–water mixture is susceptible to temperature stratification, i.e. ice at 0°C floating on top of the water and water at 4°C (the temperature at which water is most dense) at the bottom of the container. For these reasons the ice–water mixture cannot be considered a reference temperature

Figure 3.7
Shaved ice, with frosty ice on the left and ice after slushing on the right.

and its quoted traceability is through a measurement of the ice–water temperature with a calibrated thermometer. This helps confirm that the water was kept well stirred and there was no excessive heat loading. If electrical insulation from the water is required, an oil-filled thermowell may be inserted into the ice.

3.4 FIXED POINTS — DEFINING THE SCALE

Fixed points based on a variety of materials form the backbone of the temperature scale. While not as inexpensive or as easy to use as the water fixed points, they can be used conveniently where high reliability and accuracy are required. Metal fixed points rely on the latent heat of the metal liquid-to-solid transition to give sufficient heat to enable a stable temperature to be reached. Gas fixed points with lower latent heats require higher-quality temperature control to achieve their accuracy. Both types of fixed point are good examples of well-engineered systems which enable the user to achieve a high thermometric accuracy and hence repay careful study.

3.4.1 Metal freezing and melting points

Many pure metals have freezing points which can be repeated to within 1 mK. The metal needs to be very pure, typically 99.9999% free from solution impurities, and it must be contained in specialised apparatus with low thermal gradients so that the user can observe the change in phase from liquid to solid. Melting points of the metals serve the same purpose as they are also controlled by the same latent-heat-of-fusion process, but they are more susceptible to impurities and therefore require extremely pure metals.

Most of the metal fixed points for ITS-90 are available commercially in convenient cells with furnaces and control gear. A major cost in a fixed point is the cost of the very pure metal.

The detailed procedures for achieving a satisfactory freeze condition depend on the metals used. We outline a procedure for the zinc point in order to highlight the salient features of a metal freeze calibration.

Figure 3.8 shows the main components for a metal fixed point cell for use above 200°C.

The system comprises the following:

- A high-purity graphite crucible and thermometer well. Graphite is not soluble in the metals used and provides good thermal conduction with enough strength to withstand the freeze and any thermal stresses.

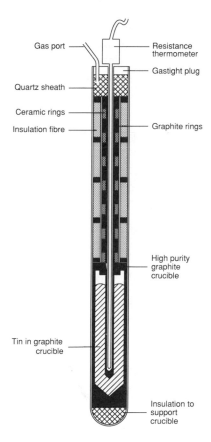

Figure 3.8
Basic construction for a metal freezing-point cell.

- A high-purity metal ingot to give an immersion depth around 20 cm with a volume of around 100 to 150 cm^3.

- Insulation to control the heat transfer along the cell and hence allow the furnace heaters or temperature bath to maintain a uniform temperature around the crucible.

- Graphite rings to serve as thermal shunts from the thermometer-well extension to the heater, so that the temperature gradient along the thermometer stem is controlled by the heaters.

- An inert atmosphere to prevent oxidation of the graphite and the metal. A dry gas is needed if the metal reacts with water, e.g. Al, Ag and Cu. A tight seal is not needed, provided that the gas flow is low enough not to upset the thermal balance. The gas pressure must also be kept at one standard atmosphere (see Table 3.2) because the freeze temperature is pressure sensitive. Cells are sometimes available as completely sealed units, which makes them more convenient to operate and provides better integrity of the cell. The disadvantage is that a leak may not be detected, and so on heating the pressure in the cell will alter the freeze temperature. Transfer of heat by radiation may cause heat loss through the sheath if quartz or glass are used. Sandblasting of the sheath can help prevent this (see Section 5.4.3).

From the above it can be seen that the fixed point cell has to be carefully designed to ensure that the thermometer is in good thermal contact with the freezing-metal system. We now examine the main steps in obtaining a satisfactory freeze with such a cell. Figure 3.9 outlines the main course of events for a system where the freezing point is above room temperature.

The procedure is as follows:

- The cell is placed in a uniform temperature enclosure.

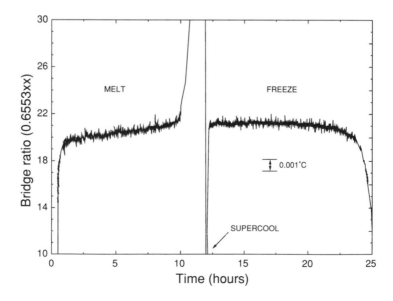

Figure 3.9
The melting and freezing curves obtained with a zinc point.

- A known reference thermometer (SPRT) is placed in the thermometer-well as a monitor. Any SPRT should be treated to reduce radiation effects as outlined in Section 5.4.3.

- The temperature of the enclosure is raised to about 10 K above the melting point to ensure a complete melt.

- From the graph of temperature vs. time a melt plateau is observed. Ideally this should be flat, but in practice there will be a small slope (see Figure 3.9) because any impure metal will start melting at a lower temperature and then dissolve the more pure metal. Also any stress or crystal defect structure may alter the melt point.

- The enclosure temperature is lowered until it is 1–2 K below the freeze point.

- A freeze is initiated. Ideally a continuous liquid–solid interface should enclose the well of the thermometer. Two liquid–solid interfaces are actually required: a fixed metal crust around the well and an expanding metal crust around the walls of the crucible, thus enclosing the liquid. The crust on the well can be made by carefully removing the monitoring thermometer from the cell, allowing it to cool a little, and then returning it to the well. The crust on the crucible walls should grow by itself due to the colder surroundings, provided that the metal does not exhibit a large amount of supercooling, i.e. greater than 4 K. For a metal such as tin which has a large supercool, thermal shock is used to commence the freeze. This can be achieved by removing the complete cell from the furnace and allowing it to cool for a minute before returning it. Inducing a satisfactory freeze still appears to be more of an art than a science, with every fixed point requiring slightly different procedures. Therefore you should follow specific instructions for your apparatus and the BIPM guidelines.

- The freeze plateau should last for at least three hours, allowing several pre-heated thermometers to be calibrated. Always check that the plateau still exists after each measurement by returning the monitoring thermometer to the well.

- In contrast to the melting curve, the plateau for freezing should be very flat since the freeze or crystallisation is a purification process with only the pure metal freezing (see Figure 3.9). The temperature of the freeze drops when the impurity concentration in the remaining liquid is too large for the purification to proceed. The temperature of the freeze plateau should be the same as that of the melt plateau. Any differences can be taken as an indication of impurities and thus provides a monitor on the integrity of the cell.

- The freeze temperature will be raised a little because of the pressure head due to the molten metal. Table 3.2 gives suggested correction factors. Corrections may also be needed if the gas is not at atmospheric pressure.

Unlike the other metal fixed points, gallium is realised as a melting point. The melt plateau is used because problems with the supercool make it difficult to obtain a successful freeze plateau. Gallium can be obtained as an extremely high-purity metal with less than 1 part in 10^7 impurity.

3.4.2 Triple points of gases

One of the big advances in reference standards for low-temperature thermometry is the increasing availability and use of sealed triple-point cells of gases, especially H_2, Ne,

Table 3.2. Effect of pressure on the temperature of some fixed points*.

Substance	Temperature T_{90}/K	Change of temperature with pressure, p $(\mathrm{d}T/\mathrm{d}p)/(10^{-8}\ \mathrm{K\ Pa^{-1}})^{\dagger}$	Variation with depth, l $(\mathrm{d}T/\mathrm{d}l)/(10^{-3}\ \mathrm{K\ m^{-1}})^{\ddagger}$
e-Hydrogen (T)	13.8033	34	0.25
Neon (T)	24.5561	16	1.9
Oxygen (T)	54.3584	12	1.5
Argon (T)	83.8058	25	3.3
Mercury (T)	234.3156	5.4	7.1
Water (T)	273.16	−7.5	−0.73
Gallium (M)	302.9146	−2.0	−1.2
Indium (F)	429.7485	4.9	3.3
Tin (F)	505.078	3.3	2.2
Cadmium§ (F)	549.219	6.2	4.8
Zinc (F)	692.677	4.3	2.7
Aluminium (F)	933.473	7.0	1.6
Silver (F)	1234.93	6.0	5.4
Gold (F)	1337.33	6.1	10
Copper (F)	1357.77	3.3	2.6

†Equivalent to millikelvins per standard atmosphere

‡Equivalent to millikelvins per metre of liquid

*The reference pressure for melting and freezing-points is the standard atmosphere (p_0 = 101 325 Pa). For triple points (T) the pressure effect is a consequence only of the hydrostatic head of liquid in the cell.

§Not a defining fixed point (see Table 3.1)

O_2 and Ar. Such cells can be obtained through some National Standards Laboratories who also provide procedures for their use. We outline a general procedure to illustrate how it differs from procedures for the other fixed points. The main difference lies in the relatively small amount of gas used. Considerably less material is available and it is not possible to rely on its latent heat alone to hold the temperature. A very well-controlled cryostat is needed along with special procedures to approach the triple point.

The cells must be filled with sufficient gas to achieve the triple-point pressure and this means that the room-temperature pressure must be in the range 0.5–10 MPa. Cells should not be overheated, partly because of the pressure and partly to preserve the seal. Cells should be held vertically and can contain more than one capsule thermometer. Figure 3.10 illustrates only one of the many types of cells available. At least one argon-point cell is also available for long-stem SPRTs.

Gas cells must be fully enclosed in a heat shield of high thermal conductivity that is independently temperature-regulated so that it is about the same temperature as the cell, and has a uniformity better than 0.1 K. Residual temperature drift just below and just above the melting plateau should be increasing at a rate less than 5 mK/h. Thermometer leads should be of 0.1 mm diameter copper wire, thermally anchored to the cell, with a minimum length of 30 cm to the cryostat to give sufficient thermal isolation.

The cell is cooled so that the gas condenses all around the thermometer block; this prevents drops of liquid or solid falling down during melting. The solidification should be slow, to allow uniform crystallisation with little stress and to limit the temperature gradient in the solid. The shield is adjusted to a few tenths of a kelvin below the triple-point temperature and held there for stress annealing to occur.

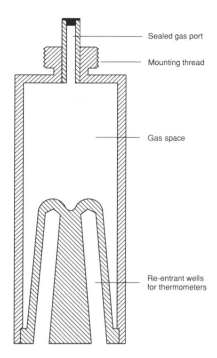

Figure 3.10
Sealed gas cell of a type suitable for establishing a gas triple point; the thermometer block can hold 3 capsule thermometers.

The triple point is realised by controlled melting of the solid. Figure 3.11 shows the effect of the heating on the melt plateau. After each heating step, time is allowed for equilibrium to be established. A check is made to adjust the shield for the low drift rate required. Heating steps are applied for melted fractions (F) 5%, 10%, 15%, 20%, 30%, 40%, 50% and 70%. For each step an overheating is observed, initially around 0.1 mK and increasing to several millikelvins for the later steps. After the heating step the cell returns to an equilibrium value T_n, which increases as more solid melts. Extrapolation procedures are used to give a value of T_n for a 100% melt, which is considered the true triple-point temperature. While there is some uncertainty in the procedure it is unlikely to be greater than 0.2 mK.

3.5 PLATINUM RESISTANCE THERMOMETRY

In Section 3.2.1 we introduced the three main types of standard platinum resistance thermometers (SPRT) that had sufficient precision and stability to hold the scale. SPRTs can be obtained commercially and should be purchased with a calibration and with the specification that they meet the requirements of ITS-90.

For platinum resistance thermometry the quantity of interest is not the absolute resistance, $R(T_{90})$, but a resistance ratio, $W(T_{90})$, where

$$W(T_{90}) = R(T_{90})/R(273.16 \text{ K}), \tag{3.3}$$

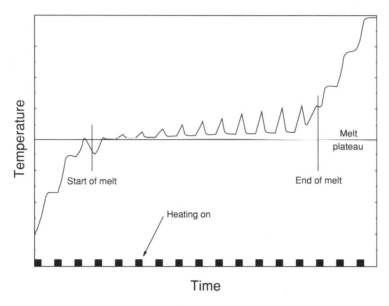

Figure 3.11
A schematic representation of a melt observed in a sealed gas triple-point cell.

i.e. the resistance ratio with respect to the water triple point. This is different from previous scales where the ice point, 0.01 K lower, was used as the reference temperature for the ratio function.

An SPRT needs to be made out of sufficiently pure platinum such that

$$W(29.7646°C) \geq 1.110\ 08 \qquad (3.4)$$

or

$$W(-38.8344°C) \leq 0.844\ 235. \qquad (3.5)$$

In addition, if the SPRT is to be used up to the silver point,

$$W(961.78°C) \geq 4.2844. \qquad (3.6)$$

This ensures that the calibration equations for all SPRTs are similar, and reduces the uncertainty in the realisation of the temperature scale.

Instead of specifying an empirical formula to describe the resistance–temperature relationship, as in previous scales, ITS-90 uses a defined function, $W_r(T_{90})$, from which deviations are calculated. IPTS-68 used a similar function for the lower temperature range. ITS-90 specifies the function in two parts: one for the range 13.8033 K to 273.16 K and the other for 0°C to 961.78°C.

In the range 13.8033 K to 273.16 K the reference function $W_r(T_{90})$ is defined by the equation

$$\ln[W_r(T_{90})] = A_0 + \sum_{i=1}^{12} A_i([\ln(T_{90}/273.16\ \text{K}) + 1.5]/1.5)^i \qquad (3.7)$$

and the corresponding inverse function

$$T_{90}/273.16 \text{ K} = B_0 + \sum_{i=1}^{15} B_i([W_r(T_{90})^{1/6} - 0.65]/0.35)^i. \qquad (3.8)$$

In the range from $0°C$ to $961.78°C$ the reference equation is

$$W_r(T_{90}) = C_0 + \sum_{i=1}^{9} C_i([T_{90}/K - 754.15]/481)^i. \qquad (3.9)$$

with the corresponding inverse function

$$T_{90}/K - 273.15 = D_0 + \sum_{i=1}^{9} D_i([W_r(T_{90}) - 2.64]/1.64)^i, \qquad (3.10)$$

where the coefficients A_0, A_i, B_0, B_i, C_0, C_i, D_0 and D_i are set out in Table 3.3. The reference and inverse equations can be considered equivalent to within ± 0.1 mK.

The two reference functions can be considered to represent idealised platinum thermometers; indeed, they were derived from the real data for two thermometers. They describe approximately the behaviour of all SPRTs. Two functions are used because no single platinum thermometer can cover the whole range, from 13.8 K to 962°C. This approach is possible because platinum has a very repeatable behaviour if sufficiently pure and free from undue mechanical stress. Deviations from the ideal can be measured at

Table 3.3. The constants A_0, A_i; B_0, B_i; C_0, C_i; D_0 and D_i in the reference functions of equations (3.7), (3.8), (3.9) and (3.10) respectively.

A_0	−2.135 347 29	B_0	0.183 324 722	B_{13}	−0.091 173 542
A_1	3.183 247 20	B_1	0.240 975 303	B_{14}	0.001 317 696
A_2	−1.801 435 97	B_2	0.209 108 771	B_{15}	0.026 025 526
A_3	0.717 272 04	B_3	0.190 439 972		
A_4	0.503 440 27	B_4	0.142 648 498		
A_5	−0.618 993 95	B_5	0.077 993 465		
A_6	−0.053 323 22	B_6	0.012 475 611		
A_7	0.280 213 62	B_7	−0.032 267 127		
A_8	0.107 152 24	B_8	−0.075 291 522		
A_9	−0.293 028 65	B_9	−0.056 470 670		
A_{10}	0.044 598 72	B_{10}	0.076 201 285		
A_{11}	0.118 686 32	B_{11}	0.123 893 204		
A_{12}	−0.052 481 34	B_{12}	−0.029 201 193		
C_0	2.781 572 54	D_0	439.932 854		
C_1	1.646 509 16	D_1	472.418 020		
C_2	−0.137 143 90	D_2	37.684 494		
C_3	−0.006 497 67	D_3	7.472 018		
C_4	−0.002 344 44	D_4	2.920 828		
C_5	0.005 118 68	D_5	0.005 184		
C_6	0.001 879 82	D_6	−0.963 864		
C_7	−0.002 044 72	D_7	−0.188 732		
C_8	−0.000 461 22	D_8	0.191 203		
C_9	0.000 457 24	D_9	0.049 025		

the fixed points and used to calculate the coefficients in an approved deviation function. There are three generic deviation functions to fit the whole range, which has 11 subranges that may overlap. Table 3.4 details the subranges and the appropriate deviation functions to use. The deviation function gives the difference

$$W(T_{90}) - W_r(T_{90}). \tag{3.11}$$

In use the value of the resistance ratio, $W(T_{90})$, is measured. Using the calibration constants for the deviation function from the certificate, a value of the reference function, $W_r(T_{90})$, is found that can be substituted into the appropriate reference function equation to give T_{90} directly.

While at first sight all these equations may seem overly complex, the subranges make the scale more practical for a user wishing to implement the range of interest. For example, a user requiring temperature measurements from $0°C$ to $100°C$ would need only two fixed points to cover the $0°C$ to $156°C$ range. The maximum temperature the thermometer would be exposed to is around $156°C$. IPTS-68 would have required three fixed points and exposure of the thermometer to temperatures around $420°C$.

The use of overlapping ranges leads to a difficulty in that the resulting temperature depends on the calibration route taken, even for the same thermometer, let alone if

Table 3.4. Deviation functions and calibration points for platinum resistance thermometers in the various ranges in which they define T_{90}.

Ranges with an upper limit of 273.16 K

Lower temperature limit T/K	Deviation functions	Calibration points (see Table 3.1)
13.8033	$a[W(T_{90}) - 1] + b[W(T_{90}) - 1]^2$ $+ \sum_{i=1}^{5} c_i[\ln W(T_{90})]^{i+n}$, $n = 2$	2–9
24.5561	As for above with $c_4 = c_5 = n = 0$	2, 5–9
54.3584	As for above with $c_2 = c_3 = c_4 = c_5 = 0$, $n = 1$	6–9
83.8058	$a[W(T_{90}) - 1] + b[W(T_{90}) - 1] \ln W(T_{90})$	7–9

Ranges with a lower limit of $0°C$

Upper temperature limit t/°C	Deviation functions	Calibration points (see Table 3.1)
961.78*	$a[W(T_{90}) - 1] + b[W(T_{90}) - 1]^2$ $+c[W(T_{90}) - 1]^3$ $+d[W(T_{90}) - W(660.323°C)]^2$	9, 12–15
660.323	As for above with $d = 0$	9, 12–14
419.527	As for above with $c = d = 0$	9, 12, 13
231.928	As for above with $c = d = 0$	9, 11, 12
156.5985	As for above with $b = c = d = 0$	9, 11
29.7646	As for above with $b = c = d = 0$	9, 10

Range from 234.3156 K ($-38.8344°C$) to 29.7646°C

	Deviation functions	Calibration points
	As for above with $c = d = 0$	8–10

*Calibration points 9, 12–14 are used with $d = 0$ for $t_{90} \leq 660.323°C$; the values of a, b and c thus obtained are retained for $t_{90} > 660.323°C$, with d being determined from calibration point 15.

different thermometers are used. Below 660°C this may lead to differences of 1 mK, but more typically 0.5 mK, which for many purposes is negligible.

Because the SPRT is itself a practical thermometer, it should be used if uncertainties better than ±10 mK are sought. We outline here the essential features in the use of a SPRT; further details should be found in the manufacturer's instruction book or the BIPM documents.

Three essential components need to be purchased:

- an accurate a.c. resistance bridge with a temperature-controlled reference resistor;
- a water triple-point cell; and
- a standard platinum resistance thermometer (SPRT).

Accurate measurement of the temperature depends critically on accurate resistance measurement, which is covered more fully in Chapter 5. An automatic a.c. bridge is generally used with 4 leads, 7 display digits and a variable sensing current, which is nominally 1 mA, although up to 5 mA may be required for very low-temperatures and up to 10 mA for high-temperature SPRTs. While 7-digit accuracy is preferred, adequate readings can be made with 6 digits if the reference resistor is specifically chosen to match the SPRT resistance over the range used. D.C. resistance measurements can readily give the precision but considerable care is needed if the accuracy is to be achieved. Several suitable bridges are available commercially and the bridge will be the largest part of the investment.

While the SPRT will be purchased with a calibration, it is necessary to check its performance before and after each use to ensure the accuracy. The check is best made with a water triple-point cell, although a very carefully made ice point may do (see Section 3.3). Equipment for the water triple point is relatively inexpensive and time-saving in use. The ice point is even less expensive to establish but is more time-consuming if many measurements are made.

A capsule-type SPRT should be used for low temperatures down to 13.8 K (see Figure 3.12). The capsule comprises the sensor only, unlike a long-stem SPRT which also includes lead wires. The transducer is the platinum resistance, nominally 25 Ω, which is thermally connected to the system under investigation by means of helium gas whose pressure is 30 kPa at room temperature, and a platinum sheath. The capsule needs to be totally immersed, using suitable grease, in a well in the copper block whose temperature is being measured. The four short capsule lead-wires, which pass through a platinum/glass seal, are connected to longer leads of fine insulated copper wires which are thermally anchored to prevent heat transfer to the capsule. The glass seal and the gas pressure limit the ultimate useful high temperature. For high accuracy the capsule should not be used above 30°C, but its use can be extended to 231°C if reduced accuracy and lifetime are acceptable.

A higher-temperature SPRT needs to have a longer stem so that the seal is close to room temperature (see Figure 3.13). The expansion and contraction of the lead wires in the stem now become an important design factor. Sufficient allowance is required for this to occur over the temperature range and immersion depth used.

Glass or quartz sheaths are used to lower the thermal conduction along the stem, but radiation can be piped along the transparent sheath, upsetting the thermal balance. For example, room lighting can readily raise the apparent temperature of a water triple-point cell by 0.2 mK. To avoid radiation piping the sheath can be sand-blasted just above the sensor region or coated with graphite paint (see Section 5.4.3).

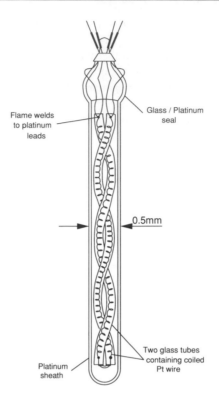

Flame welds
to platinum
leads

Glass / Platinum
seal

0.5mm

Two glass tubes
containing coiled
Pt wire

Platinum
sheath

Figure 3.12
A typical 25 Ω capsule platinum resistance thermometer. The platinum sheath
is 5 mm in diameter and 50 mm long.

Chemical changes are an important consideration in the design of the higher-temperature SPRTs. A partial pressure of 2 kPa of oxygen is used as it is low enough to prevent platinum oxidation but high enough to prevent impurity oxides from breaking down to metals which can contaminate the platinum. Contamination from, and breakdown of, the supports or sheath also limit the acceptable upper temperature. For example, for maximum stability the mica support is best not taken over 450°C. Electrical leakage also becomes a problem at very high temperatures. ITS-90 is the first scale to use high-temperature SPRTs, and as a consequence these thermometers are still undergoing some development and the user will need to follow the manufacturer's recommendations closely in order to obtain the best performance. Of particular concern is the porosity of the quartz to some metal vapours which can contaminate the platinum wire. A platinum sheath over the sensor end of the quartz sheath is highly desirable.

Mechanical vibration can cause strain and work-hardening of the platinum wire and hence an increase in the resistance at the water triple point. Large knocks have been known to cause errors of the order of 10 mK. Annealing above 450°C for several hours and gentle cooling to room temperature will usually restore the original resistance. The resistance should be repeatable to 7 digits on a resistance ratio bridge, i.e. to higher

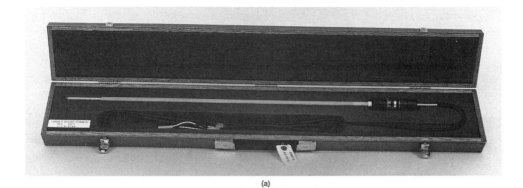

(a)

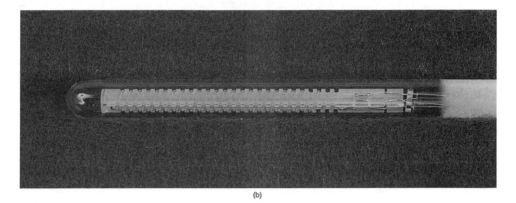

(b)

Figure 3.13
Standard platinum resistance thermometer (SPRT) made to ITS-90 requirements: (a) complete assembly in a carrying case; (b) tip showing details of construction of the resistance element.

precision than the accuracy of the reference resistor. A capsule SPRT cannot be annealed and therefore should not be used for long periods where there is any vibration, e.g. in a stirred bath.

Strain due to thermal shock can also upset the resistance thermometer. SPRTs should be inserted slowly into higher temperatures. A rate of 50°C per minute is a good guide. The use of preheating furnaces is useful here. The thermometer should also be removed with care. Table 3.5 gives recommended cooling rates for SPRTs.

Table 3.5. A typical cooling schedule for SPRTs. The SPRT may be cooled gradually or allowed to anneal at the lowest temperature of each of the three highest ranges.

Range	Cooling rate	Period
From 960°C down to 850°C	25°C per hour	4 hours
From 850°C down to 630°C	100°C per hour	2 hours
From 630°C down to 540°C	400°C per hour	30 minutes
From 450°C to room temperature	50°C per minute	10 minutes

The immersion depth for SPRTs is large, in part because of the precision and in part because of the length of the sensing element. SPRTs which have a high self-heating constant (see Section 5.4.4) appear to require greater immersion. Adequate immersion depths are typically 15 cm to 20 cm at room temperature and up to 27 cm at 200°C and above. If in doubt, perform a temperature profile versus immersion depth to give an indication of the required immersion depth.

3.6 RADIATION TEMPERATURE SCALE

Above the freezing point of silver, 961.78°C, ITS-90 uses the Planck *blackbody radiation law* (see Chapter 8) to define the ratio between the spectral radiance at the temperature T_{90}, $L_\lambda(T_{90})$, and the spectral radiance at a fixed point temperature, $L_\lambda(T_{90}(X))$, where $T_{90}(X)$ is the freezing point of silver, gold or copper. Thus

$$\frac{L_\lambda(T_{90})}{L_\lambda(T_{90}(X))} = \frac{\exp(c_2/(\lambda T_{90}(X))) - 1}{\exp(c_2/(\lambda T_{90})) - 1} \tag{3.12}$$

where λ is the wavelength in vacuum and $c_2 = 0.014\ 388$ m.K.

There is no specified way to achieve the scale apart from good measurement practice. To illustrate the steps required we follow the measurement model of Figure 1.10. More details on radiation thermometry are covered in Chapter 8.

The physical system being measured is a metal freeze as covered in Section 3.4.1. In particular the solid–liquid interface should enclose as much of the sensor as possible. The sensor is a blackbody usually formed as a cavity in high-purity graphite (see Figure 3.14); it is typically 50 to 80 mm long with a 1 to 6 mm aperture.

There must be good thermal conduction from the metal crust and graphite to the cavity surface. The cavity surface converts the thermal energy to radiant energy in an approximation to the Planck radiation law. The cavity and the aperture shape are made to ensure that the radiant signal leaving the blackbody is accurately represented by the Planck radiation law for a given wavelength range.

The radiation will generally propagate through a gas rather than a vacuum and a correction may be needed to account for this. No windows should be in the radiation path.

A radiometer is needed to measure accurately the radiance of the blackbody at the wavelength selected. The radiometer is, of course, a measuring instrument in its own right and should not be considered a thermometer any more than a resistance bridge is.

Radiation thermometry has an accuracy of 0.1 K at the silver point where it meets the platinum resistance scale, which is itself accurate to around 0.01 K at this point. However, the precision of the radiation measurements needs to be at the 0.01 K level. At this level of accuracy the blackbody cavity must behave as an ideal blackbody to within 1 part in 10^4. This in turn puts a restriction on the maximum aperture size. On the other hand, too small an aperture will cause optical errors in the use of the radiometer. Intensity ratios (equation (3.12)) of the order of 1 in 10^4 are required to be measured, hence linearity corrections to the radiometer response are essential.

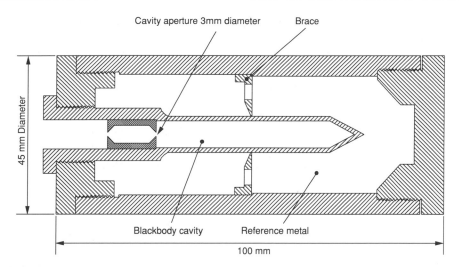

Figure 3.14
Metal fixed point with a blackbody cavity suitable for a radiation standard.

Although the scale is defined in terms of a spectral radiance ratio at a single wavelength, practical radiometers must operate over a finite bandwidth. Typical bandwidths are 10–100 nm. One of the more difficult tasks is to determine the effective wavelength of the radiometer, which needs to be known to within 0.02 nm, assuming that the measurements are made in the 600–900 nm range typical of modern radiometers. This requires the spectral response of the whole radiometer to be determined. Hence the effective wavelength involves an average of the radiometer spectral response with the spectral shape of the Planck radiation. As the Planck function changes with temperature so too does the effective wavelength.

Two types of reference radiometers can be distinguished. The first is the comparator type which compares the blackbody directly with another radiant source and hence only needs a limited stability with time in order to carry out the comparison. The other type is a transfer standard of good long-term stability. Neither of these devices is suitable for use by the general user.

3.7 CRYOGENIC THERMOMETRY

Very low-temperature measurement techniques are not considered in detail in this text because of their more specialised nature and requirements. Further information can be found from the references given at the end of the chapter. Here we outline only how the low-temperature scale can be transferred to suitable reference thermometers as this gives a good example of how to solve thermometric problems.

As seen in Figure 3.1, there are two types of thermometers for low-temperature work: vapour-pressure thermometers and gas thermometers.

- *Vapour-pressure thermometers*   Helium vapour-pressure thermometers do not need any fixed points. ITS-90 defines the numerical relationship between the temperature and the vapour pressure. While a thermodynamic equation can be given for the vapour pressure, its solution is not always as exact as is required. The relationship between temperature and pressure depends on experimental data. Two different liquids can be used:

— ^{4}He with equations for 1.25 K to 2.2 K and 2.2 K to 5 K;

— ^{3}He with an equation for 0.65 K to 3.2 K.

The vapour-pressure thermometer is designed to allow two phases of helium, the pure liquid and vapour phases, to come to thermal equilibrium. The absolute pressure at the interface between liquid and vapour is then measured. Heat losses and thermal gradients need to be well taken care of at these low-temperatures where the properties of materials are very different from those at room temperature. With care an accuracy of around ±0.5 mK is possible.

- *Gas thermometers* Gas thermometry using helium, either ^{3}He or ^{4}He, covers the range 3 K to 24.6 K (the neon triple point), and requires calibration at three temperatures. Modern gas thermometers have been used to determine the thermodynamic temperature relative to a single fixed point (usually the neon triple point). They require considerable care if accurate results are to be achieved over a wide range of low-temperatures. However, the measurement difficulties are reduced by using three fixed points well placed over a narrow range. Accuracies of around ±0.1 mK can be achieved. Over the range 4.2 K to 24.6 K for ^{4}He, a simple quadratic is used:

$$T_{90} = a + bp + cp^2 \qquad (3.13)$$

where p is the measured pressure and a, b and c are coefficients determined at the fixed points of Table 3.1. One of the points should be a temperature between 4.2 K and 5 K as measured by a vapour-pressure thermometer. With ^{3}He as the thermometer gas or with ^{4}He over the range 3.0 K to 24.6 K, an additional term is specified by ITS-90 to cope with the temperature variation of the second virial coefficient. This term requires some knowledge of the volume of the gas thermometer. The equations are firmly thermodynamically based with corrections to account for the known departures from ideal gas behaviour, e.g. finite atomic size and bonding.

Both the gas and vapour thermometers use specialised cryogenic equipment which has been designed to ensure optimal thermal coupling between the physical system and the thermometer being calibrated. As an example, consider the measurement process for the vapour-pressure thermometer. The physical system whose temperature is required is the region between the liquid and vapour in equilibrium. The vapour serves as the sensor because it is coupled to both the physical system and the pressure signal. That is, the vapour converts a temperature to a pressure. By attaching an accurate pressure gauge via a tube it is possible to interpret the pressure readings as temperatures using the ITS-90 formulae and the BIPM supplementary information.

These thermometers are used to transfer the scale to more suitable temperature sensors. The most stable and reliable sensor is considered to be the rhodium–iron resistance thermometer, which is constructed similarly to the platinum capsule thermometer (Figure 3.12). The thermometer is made from rhodium wire doped with 0.5% of iron to give a resistance between 20 Ω and 50 Ω at 0°C. The rhodium–iron thermometer is preferred over the range 0.5 K to 30 K. It is still useful up to room temperatures, but the platinum thermometer gives superior performance. A suitable calibration equation is of the form

$$R = \sum_{i=0}^{n} b_i [\ln(T + g)]^i, \tag{3.14}$$

where the b_i are to be determined and g is a constant between 0 K and 10 K. With $n = 6$ a residual of 0.3 mK is possible. Thin-film versions of these thermometers are now available.

While the rhodium–iron thermometers give very good stability they may be too large for some applications. Also, higher sensitivities may be required. Germanium thermometers can be used for these cases (see Figure 3.15). A single crystal of germanium is used with 'n' or 'p' doping and four leads are attached in a can filled with ^{4}He or ^{3}He gas. Individual germanium thermometers can exhibit good stability but instabilities often arise from thermal cycling or from sources which are unknown. A variety of types are

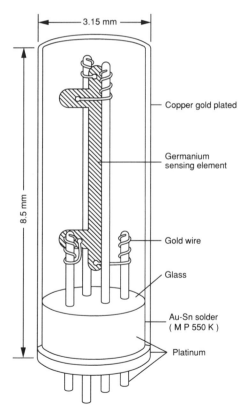

Figure 3.15
Example of one type of construction for a germanium thermometer. The germanium is in the form of a bridge, with current contacts on the ends and potential contacts on side arms.

available and are best used for narrow temperature ranges below 30 K. They are very non-linear and a suitable calibration equation is of the form

$$\ln T = \sum_{i=0}^{n} A_i [(\ln R - M)/N]^4, \tag{3.15}$$

where N and M are suitable constants and the A_i are to be determined by curve fitting and thus require $3n$ calibration points; n is about 12 for a wide temperature range and 5 for a narrow range. As with all resistance thermometers, self-heating errors occur. Because the resistance can be very high there are also leakage resistance problems, and the a.c. and d.c. resistances are different.

A variety of other electrically- based thermometers are available for low temperatures: resistors, capacitors, and thermocouples. The literature should be consulted for details. In particular, where high magnetic fields are used at low temperatures, suitable temperature sensors are essential because all electrical sensors interact with the magnetic field but the degree of interaction varies widely, e.g. capacitors are very little affected but are not good thermometers.

FURTHER READING

Supplementary Information for the International Temperature Scale of 1990, Working Group 1, BIPM, Sèvres (1990).
Contains the essential information for realising the scale and also an official corrected version of ITS-90.

Techniques for Approximating the International Temperature Scale of 1990, Working Group 2, BIPM, Sèvres (1990).
Expands the information for use of temperature sensors and overlaps the content of this text. Gives more details on low-temperature thermometry.

Experimental Techniques in Low Temperature Physics, (3rd edition with corrections), G. K. White, Clarendon Press, Oxford (1987).
Gives the essential experimental information for low-temperature thermometry.

Temperature (2nd edition), P. J. Quinn, Academic Press, London (1990).
Extensive coverage of temperature scales and the physical basis for higher precision thermometry.

Modern Gas-based Temperature and Pressure Measurements, F. Pavese and G. Molinar, Plenum Press, New York (1992).
A good coverage of the gas triple-point cells as well as other gas thermometry.

4

Calibration

4.1 INTRODUCTION

Most of us have at some time tried to interpolate an analogue scale to one-tenth of a scale division, believing that this is the accuracy of the instrument. For thermometers at least this is hopelessly optimistic; errors of two to five scale divisions are usually within the manufacturer's specifications. But worse than that, the experience of most calibration laboratories is that about 15% of all instruments, including thermometers, are outside the manufacturers' specifications. This 1-in-6 failure rate is almost independent of cost, manufacturer and model.

We are often oblivious to the errors and faults in our measuring instruments, only becoming aware of them when we compare two instruments and find that the readings differ. Suddenly our faith in their reliability is shaken. Usually our first reaction is to ask, 'Which is the more reliable?', the more reliable of the instruments being the more accurate, more stable and more trustworthy. These attributes are the essence of traceability, the attributes that a calibration laboratory looks for and quantifies in a calibration. Calibrations provide objective answers to the questions: 'What is the accuracy of the instrument?','Does it conform to the manufacturer's specification or documentary standard?' and 'Can I trust it?'.

In this chapter we consider calibration in detail including the meaning of calibration, how to design, carry out, record, report and maintain calibrations. The chapter is relevant not only to suppliers and users of calibrations but also to thermometrists who wish to improve the reliability of their measurements. We will provide thorough examples for specific types of thermometer in later chapters.

4.2 THE MEANING OF CALIBRATION

Chapter 1 defined calibration as:

> The set of operations which establish, under specified conditions, the relationship between values indicated by a measuring instrument or measuring system, or values represented by a material measure, and the corresponding known values of the measurand.

To satisfy this definition for a thermometer the calibration laboratory must supply three items:

(1) The means to relate the thermometer readings to the ITS-90 definition of temperature.

(2)   The uncertainty in the ITS-90 relationship. Specifically, the expected uncertainty in temperatures measured by the thermometer.

These two items characterise the thermometer readings in the same way that measurements are characterised by a result and uncertainty (Chapter 2). The third item distinguishes calibrations from other measurements:

(3)   Assurance that, subject to appropriate care of the thermometer, both the ITS-90 relationship and the uncertainty will be valid for some reasonable period beyond the calibration date.

The ITS-90 relationship

The ITS-90 relationship is the means to relate the thermometer readings to the true temperature. For direct-reading thermometers the relationship is usually a set of corrections, and for resistance thermometers it is an equation relating resistance to temperature. In a very real sense this fulfils the old definition of a calibration: determining the calibre of the instrument.

Since thermometers are sensitive to factors other than temperature the relationship must be established under well-defined conditions. Further, these conditions must be readily accessible to the user so that the same relationship can be established in the

Figure 4.1
When is a calibration certificate not a calibration certificate? The type of document shown here provides no measure of the instrument's relationship to the ITS-90 – it does not mention temperature at all. It is really a manufacturer's warranty rather than a calibration certificate.

user's laboratory. The specified conditions should also eliminate as many of the sources of error as is practical.

For example, resistance thermometers may be calibrated at zero current to eliminate the self-heating error. However, the zero-current resistance–temperature relationship is only accessible to a client with a bridge with the facility to vary the sensing current. Without such a bridge all measurements will be subject to an unknown self-heating error. Ideally the calibration conditions should be as near as practical to the client's normal use. Such conditions might be: a sensing current of 1 mA and 150 mm immersion in a stirred bath. Specification of these conditions will ensure that the self-heating error will be the same in use as in calibration.

The uncertainty

The uncertainty is arguably the most important part of the ITS-90 relationship. Since the main purpose of the calibration is to characterise the accuracy of the thermometer, the ITS-90 relationship is incomplete without the uncertainty, which is defined as

> **Uncertainty:**
> an estimate characterising the range of values within which the true temperature lies.

In most measurements there is more than one source of uncertainty. There may be the uncertainty in the readings of the thermometer, uncertainty due to poor immersion, bath non-uniformity, and any other known experimental error. Unlike the other experimental uncertainties, the uncertainty in the thermometer readings affects every measurement. Further, it cannot be determined by any ordinary subsidiary experiment, it can only be determined by comparison to a better thermometer or by direct comparison to the ITS-90, namely by calibration. Likewise, the uncertainty in the readings of the better thermometer can be determined only by calibration.

Thus the provision of the uncertainty in the thermometer's readings is essential to every calibration. If the uncertainty is not supplied at any point in the calibration chain, then all downstream users will be deprived of information essential for the estimation of the uncertainty in their measurements.

Occasionally calibration certificates report uncertainties which are of no practical use to the client. The most common example is the certificate which reports either the uncertainty in the reference thermometer used in the calibration or the least uncertainty according to the laboratory's accreditation. Neither provide the user with any information about the likely error in the calibrated thermometer.

Another example of a poor statement of uncertainty is one based on an incorrect standard deviation. The mistake in this practice is a little more subtle. Here we must make the distinction between the standard deviation of the error distribution and the standard deviation of the estimate of the mean (see discussion in Section 2.6). The standard deviation of the mean, which characterises the uncertainty in the correction, decreases as the number of calibration points increases and can be made arbitrarily small. It is the standard deviation of the error which characterises the residual errors in the corrected readings, and hence characterises 'the range within which the true temperature lies'.

There remains the question of what is meant by the accuracy of a thermometer. With most thermometers the accuracy is dependent on use, maintenance, and how the readings are interpreted. Once again, for the uncertainty to be meaningful the calibration laboratory must assume the best conditions accessible to the user. Therefore the calibration laboratory must eliminate all errors that the user can readily eliminate or assess, and must include in the estimate of the uncertainty, the uncertainty caused by all the errors that the user will be unable to control.

Assurance of reliability

Most clients buy a calibration for the assurance of reliability. Few buy the calibration for the improvements in accuracy alone. With this in mind it is somewhat surprising that some certificates have statements like, 'These results are valid only at the time of test.' A paraphrase might read, 'This instrument may not have sufficient stability to maintain traceability.' Clearly such a certificate is of no use to a client who wishes to assert reliability or to interpret accurately the readings of the thermometer. By issuing a calibration certificate a calibration laboratory is supplying assurance that both the ITS-90 relationship and the uncertainty will be valid for a reasonable period of time.

An assurance of reliability begs the obvious question, 'How, on the basis of a calibration performed over a period of a few days, can we assess the likely stability of the instrument over the next month, year, or even 5 years?' The answer is a two-stage process that places considerable demands on both the calibration supplier and the client.

The calibration supplier must:

(1) Rely on the history of many similar thermometers which proved stable over long periods when subject to normal usage and reasonable care.

(2) Show that the thermometer under calibration is no different from those with the established history.

(3) Assert that the thermometer under calibration will have a similar stability to those with the established history so long as it is subject to the same usage and care.

The client or user (who may also be the supplier for in-house calibrations) must:

(1) Demonstrate, through regular ice-point checks, or other simple verification checks, that the instrument continues to behave as it did at the time of calibration.

(2) Demonstrate that the thermometer has not been exposed, during use or storage, to conditions that may adversely affect its performance.

The two components of this process are essentially histories: firstly, the collected knowledge on the typical behaviour of similar thermometers, which we call the *generic history* of the thermometer; and secondly, the calibrations, and service records relating to a single thermometer which we call the *specific history*.

4.3 ASSEMBLING A CALIBRATION

4.3.1 Establishing reliability — the generic history

Generic history is the collected and documented knowledge of typical behaviour for a particular type of thermometer. The most general aspects of generic history include:

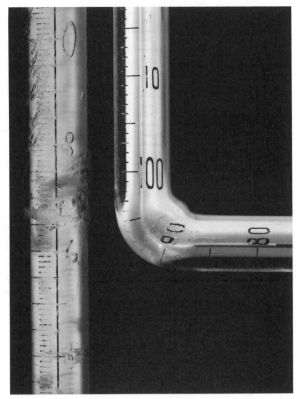

Figure 4.2
Two thermometers which have clearly been exposed to conditions that might adversely affect their reliability. One has a waist ground into the stem, the other has been bent to allow horizontal reading of the scale. Neither should be calibrated.

- the typical relationship between the response of the sensor and the true temperature;
- the typical accuracy of the thermometer;
- the typical stability of the thermometer;
- the typical construction of the thermometer;
- the typical errors and faults in the thermometer;
- the typical usage and non-usage of the thermometer.

The presence or absence of these factors is a signature of good behaviour for a thermometer. Departure from this signature is deviant behaviour and therefore an important indicator of potential errors or faults. Calibrations are designed to verify the signature for each thermometer.

- *Typical relationship*   All thermometers are based on sensors, namely devices which have a physical property that changes with temperature. For example, the volume of mercury changes with temperature, as does the resistance of platinum wire. For a thermometer to be a suitable candidate for maintaining a temperature scale, the relationship between the physical property and the temperature must be well-established.

That is, it must be known and understood by a reasonable fraction of the temperature measurement community. This relationship in part determines how the thermometer is used to best advantage, and hence the calibration must be designed with the relationship in mind. For example, a platinum resistance thermometer calibration should be designed to find the best values for the constants in the mathematical relationship of resistance versus temperature, and to confirm that the thermometer does in fact follow the expected relationship. Platinum thermometers with excessive departures from the expected relationship have often been found to be contaminated and are unsuitable for maintaining a temperature scale.

- *Typical accuracy* The accuracy of a thermometer normally depends on a large number of factors: the method of construction, the temperature range, the environment it is exposed to, and how it is used. A part of the generic history of a reliable thermometer is that its accuracy will fall within a well-known range provided that it is constructed along certain well-known guidelines, its exposure restricted to a certain range and environment, and it is used with due care. An accuracy which falls outside this range is usually an indicator of a faulty thermometer. For example, most calibrated mercury-in-glass thermometers can be expected to be accurate to between one-fifth and one-half of a scale division. Large uncertainties are normally indicative of non-uniform bores or insufficient annealing of the glass.

- *Typical stability* Stability is the most important part of the generic history. It is impractical to hold a thermometer for two years just to prove that it has a certain stability over this period. Instead we must rely on the history of many similar thermometers which proved stable over periods of two years. The evidence and criteria relating to stability must be well documented in the measurement literature. For example, over the last decade or more there have been significant advances in the manufacturing techniques for thermistors. Thus their generic history includes papers describing glass-encapsulated thermistors which are stable to a few millikelvins over periods in excess of a year.

- *Typical construction* One of the key factors in the development of thermometers is that there should be a means of constructing them in such a way that they are least affected by environmental conditions other than temperature. Eliminating the air-pressure effect in early gas thermometers, and radiation errors in air-temperature measurements, are two examples. Thermometers which do not adhere to proven design and construction practices almost certainly compromise both their short-term and long-term accuracy.

- *Typical errors and faults* Each type of thermometer, because of its construction and materials, is prone to particular manufacturing defects or physical damage. Thermometers with indications of these defects are likely to exhibit high uncertainty and long-term instability. For example, a low insulation resistance between the detector element and the sheath of a resistance thermometer indicates that it has excess moisture in the probe assembly and will give unreliable temperature measurements.

- *Typical usage* Unlike many measuring instruments thermometers must be able to withstand large changes in their environment, and in particular large changes in temperature. We do not expect other precision measuring instruments to withstand massive cycling in temperature. More than most instruments, thermometers suffer simply because they are used. In order to withstand very high or low-temperature exposure, compromises must be made in their construction. Again the generic history of types of thermometers includes the various constructions and purposes for which they are intended. Thermometers manufactured for one purpose may be quite unsuitable for another. Both the thermometer and its calibration must be appropriate for its intended purpose. The need to match construction and calibration with use arises often with thermocouples which will maintain calibration only under very specific conditions.

One thing that is sure to make a calibration supplier uneasy is a home-made instrument. Home-made instruments completely lacking in generic history have no basis for assuring reliability; a certificate cannot be issued. Similarly, a thermometer assembled using inappropriate materials and techniques, or to be used in an environment where it has a high likelihood of being damaged, is not suitable for maintaining the temperature scale. Fortunately many of these cases sort themselves out: instruments that are poorly made usually fail the short-term tests. Instruments that perform well enough to pass the short-term tests are usually manufactured according to accepted design principles and therefore possess some generic history. These problems are not unique to home-made instruments. Every time a new model instrument is released on the market, calibration suppliers must carry out extra tests to prove the reliability of the new model. Likewise, as with any new instrument, the client must treat the instrument with a little more scepticism until a reasonable specific history is established.

Clearly to make the best use of the generic history the calibration laboratory must have some knowledge of how the thermometer is to be used. Questions to ask the client include:

- What are the temperature range and accuracy required?
- How is the thermometer to be used, e.g. as a reference or a working thermometer?
- Must it conform to any documentary standards?
- Will it be exposed to any difficult environments, e.g. corrosive chemicals, vibration, pressure, moisture, rapid cycling, or electromagnetic or ionising radiations.
- Are there any departures from normal usage, e.g. in respect of immersion, response times, and other sources of error normally excluded from measurements?

One of the main functions of this book is to provide those who carry out calibrations with the necessary generic history of the thermometers currently used to maintain temperature scales. As a general rule, the more precise the calibration, the greater the knowledge and expertise required of the calibration laboratory.

4.3.2 Establishing the SI relationship

At a superficial level the establishment of a calibration equation for a thermometer can be accomplished by intercomparison with a reference thermometer. However, this only satisfies the old definition of calibration — determining the calibre of the instrument.

Modern traceability requirements also demand assurance that the measured relationship will be a good indicator of the thermometer's performance for some time. Thus the calibration must show that the measured relationship is a stable feature of the thermometer. This is the generic history argument: use a calibration equation that is well-established, and show that this thermometer follows the equation within accepted limits. Proving the validity of an equation is a long-drawn-out process because the establishment of an equation requires application to many thermometers, from different manufacturers, and by a number of independent workers.

In Chapter 2 we introduced the method of least-squares fitting for determining the co-efficients in a calibration equation. Least-squares analysis is a means of determining the best values for the constants in a calibration equation from a large number of calibration data. In Chapter 2 we recommended that a minimum of 3 or 4 calibration points per unknown constant are used for the fit. Least-squares fitting serves two purposes. First, it demonstrates the suitability of the calibration equation as a description of the thermometer, and hence shows that interpolation is valid; and secondly, and most importantly, it demonstrates the suitability of the thermometer for maintaining a scale. A thermometer which has large residual errors in the fit to a recognised equation, or whose fitted values lie well away from accepted values, is untrustworthy.

4.3.3 Estimating the uncertainty

Estimating the uncertainty in a calibration can be difficult. The uncertainty must be an accurate assessment of the likely error of the thermometer when it is used with reasonable care, and over a reasonable period. There are five main factors that contribute to the calibration uncertainty:

- uncertainty in the reference-thermometer readings;
- variations in the stability and uniformity of the calibration medium;
- departures from the determined ITS-90 relationship;
- uncertainty due to hysteresis;
- uncertainty due to drift.

The total uncertainty is determined by combining these uncertainties together with any other uncertainties specific to the type of thermometer.

These factors are discussed in more detail below:

Uncertainty in the reference-thermometer readings   The reference thermometer is the link between the thermometer and the ITS-90; any errors in the scale of the reference thermometer will be transferred to the newly calibrated thermometer. The uncertainty in the reference thermometer, which should be on the certificate for the reference thermometer, must therefore be included in the total uncertainty of the calibrated thermometer. If the reference certificate does not report the correct uncertainty or reports it at different confidence limits, then additional work may be required to determine the uncertainty or to scale it to the correct confidence limits.

Variations in the stability and uniformity of the calibration medium   Throughout the calibration we assume that the reference thermometer and the thermometer under test should be at the same temperature. However, no matter how well-controlled the cal-ibration bath, furnace or cryostat, there will always be residual spatial and temporal fluctuations in the temperature which lead to differences in the temperatures of the two thermometers. The distribution of these differences has two components. Firstly, a fluctuating component that will cause random variations in the readings of both thermometers; and secondly a steady component due to temperature gradients within the medium that leads to a systematic variation in the thermometers' readings.

Random fluctuations in bath temperature cause random differences in the two thermometers' readings and already contribute to the uncertainty in the calibration by affecting the error-of-fit. Therefore it is unnecessary to add to the total a specific estimate of the uncertainty due to the fluctuations.

However, the systematic part of the error, due to the non-uniformity of the bath, is not directly apparent in the calibration results and some assessment must be made of the additional uncertainty. In extreme cases a correction may have to be included as well. In a well-designed calibration the correction should be negligible.

While we do not need to know the variance for the random error caused by the fluctuations, we do need to ensure that it is not a major contributor to the total uncertainty. For this reason records should be kept which describe the commissioning of all calibration baths and furnaces.

Departures from the determined ITS-90 relationship All thermometers with a well-established generic history have one or more accepted calibration equations. These equations and the typical values for the constants in the equations describe the expected relationship to the temperature scale. For example, for platinum resistance thermometers above $0°C$, the relationship is a quadratic equation relating resistance and temperature.

However, all such relationships are idealised. Small departures from the accepted relationship occur for many reasons. In most cases the equation is an approximation to very complex real behaviour which is also affected by material variations in the construction of the thermometer. In any case these non-idealities in the thermometer's behaviour lead to small and generally unpredictable departures from the behaviour reported on the calibration certificate. The uncertainty which characterises these errors is usually the main contributor to the error-of-fit that is calculated from the measurements.

Uncertainty due to hysteresis Hysteresis is a property of a thermometer whereby the readings depend on previous exposure to different temperatures (Example 2.6). It is usually impractical to eliminate the effects of hysteresis from a thermometer's calibration. Not only would the calibration time become excessive, because of the long preconditioning required for each measurement, but the procedures for the usage of the thermometer would become so restrictive as to make the thermometer useless.

To minimise the effects of hysteresis, reference thermometers may be used so that the measured temperature is approached from room temperature. This effectively halves the contributing uncertainty (see Section 2.7) and fixes the calibration sequence. However, in adopting this procedure the hysteresis effects are hidden but still affect measurements. In order to assess the resulting uncertainty at least one measurement must be made which assesses the width of the hysteresis loop. In some cases the difference between two ice-point measurements, made before and immediately after the intercomparison, may provide sufficient information to allow an assessment of the hysteresis.

Uncertainty due to drift Drift in thermometers usually arises because of dimensional changes or compositional changes with time. The changes may occur with time only, as with bulb contraction in liquid-in-glass thermometers, or it may be dependent on use. Drift assessments fall into three categories: where no assessment is necessary; where there is specific information on the drift rate of a thermometer; and where there is only generic information on the drift rate. Mercury-in-glass thermometers fall into the first category since there is an accepted procedure for measuring and correcting for long-term drift. Such procedures eliminate the need to include drift in the total uncertainty.

Where correction procedures are not available and the thermometer is expected to drift, the estimate of total uncertainty should include allowance for drift. The question is: how much and how should it be assessed?

A reasonable approach is to estimate the likely drift in one year's usage and to use this figure to derive an uncertainty at the appropriate confidence level. In cases where a reference thermometer will be used to carry out the annual calibration of other thermometers immediately after calibration, an uncertainty based on a 30-day or 90-day period will be more appropriate. Where there is specific information on the drift-rate of the thermometer from previous calibrations, the assessment is straightforward. For brand-new thermometers it is necessary to predict drift on the basis of the behaviour of similar thermometers.

Guidelines on drift assessments for each type of thermometer can be found in the respective chapters.

The total uncertainty   Once all these uncertainties, and others specific to the particular type of thermometer, have been considered, the total uncertainty can be determined according to equation (2.21) as

$$U_{cal}^2 = U_{ref}^2 + U_{bath}^2 + U_{fit}^2 + U_{hys}^2 + U_{drift}^2 + \cdots . \qquad (4.1)$$

The uncertainties should all be determined and reported for the same confidence limits. If the client has not requested any particular confidence level then they may be reported at any appropriate confidence level, with 95% or 99% being preferred. The confidence level should also be reported.

Note that all of the uncertainties, except the uncertainty in the fit, are Type B uncertainties. As with all Type B uncertainty assessments, considerable knowledge and understanding of the errors is necessary in order to make accurate assessments. The procedures and assumptions required for the assessment of all Type B uncertainties should be well documented.

4.3.4 Choosing the reference thermometer

Equation (4.1) has two important features which influence the choice of reference thermometer for a calibration. Firstly, the uncertainty assigned to a thermometer, U_{cal}, is always greater than that of the reference thermometer used to calibrate it, U_{ref}. Or equivalently, the uncertainty in a calibration (or any other measurement) increases as the number of links in the calibration chain is increased.

Secondly, all but the first two terms of equation (4.1) are properties of the thermometer under test. That is, if we can make the uncertainties due to the reference thermometer and the bath so small that they are negligible, then the calibration uncertainty is a property of the thermometer only and independent of the calibration laboratory. If this can be satisfied, then the calibration uncertainty is genuinely a measure of the accuracy of the thermometer.

Let us say, for argument's sake, that contribution of the reference and bath uncertainties should be less than 10% of the overall calibration uncertainty. Then it follows that

$$U_{\text{ref}}, \; U_{\text{bath}} < \frac{1}{3} \left(U_{\text{fit}}^2 + U_{\text{hys}}^2 + U_{\text{drift}}^2 + \cdots \right)^{1/2}. \tag{4.2}$$

That is, the uncertainties due to the reference thermometer and the bath must be at least a factor of three less than the expected uncertainty of the thermometer. The factor of 3 is a useful rule-of-thumb for determining the best uncertainty in a calibration. It is sometimes called the *3X rule*.

It should be recognised that the 3X rule is based on a 'fair trading' argument which is in turn based on the client's reasonable expectations that the calibration of a thermometer does not depend on how it was calibrated. The 3X rule is not a definition of what is technically feasible. A smaller factor than 3 is acceptable for in-house calibrations or where a very high level of transfer is required and there is no alternative supplier.

4.3.5 The calibration certificate

As with any formal report of a measurement, the calibration certificate must include sufficient information to: uniquely identify the equipment tested; uniquely identify the nature of the tests; and present an unambiguous statement of the results. ISO Guide 25 provides a useful check list for all types of tests. The items relevant to thermometer calibrations are

(1) The name, address and location of the calibrating laboratory.

(2) The name, address and location of the client.

(3) The means to uniquely identify the certificate, usually a number which is traceable to the measurement records.

(4) The means to uniquely identify the equipment tested, and the manufacturer, model and serial numbers of all items submitted for test.

(5) The date(s) when the calibration(s) were carried out.

(6) Where relevant, a description of the calibration method. This applies particularly where there are departures from standard methods.

(7) The results and observations, and/or results and conclusions derived from the results.

(8) A statement of traceability to the appropriate scale (ITS-90 or IPTS-68).

(9) A statement of the estimated uncertainties in the readings of the thermometer.

(10) The author of the report and/or personnel responsible for the tests.

(11) The conditions under which the report may be reproduced.

(12) An endorsement by an independent accrediting body.

This may look like an excessive amount of information. However, all of it is necessary to ensure traceability, and to avoid all possible confusion. In many cases it will all fit on a single A4 sheet of paper. Several examples of completed certificates are given later.

In studying the above checklist, note firstly that it is not necessary to identify the test equipment and reference thermometer used for the calibration on the certificate. This information is not important to the client. For a calibration to be traceable it is sufficient that the test equipment be identified in the test record held at the calibration laboratory. If necessary, the information can be traced through the report number (item (3) above).

Secondly, the statement of traceability to ITS-90 is required only to resolve ambiguities as to which scale is used. The certificate is itself a statement that all measurements reported on the certificate are traceable to the appropriate National Standards. Endorsement of the certificate by an independent accrediting body (item (12)) is an assurance that all measurements are indeed traceable and that all appropriate records have been kept.

Exercise 4.1

(a) Inspect the example certificates given in this book and decide if all of the requirements of ISO 25 (Section 4.3.5) are met.
     (Hint: Give special attention to Figure 4.1.)

(b) Locate some certificates in your laboratory and see if they satisfy the ISO 25 guide.

4.3.6 A calibration procedure

Different laboratories and Standards organisations have different calibration procedures, depending on their individual quality assurance procedures. The 8 steps given below are intended to be an outline of a suitable calibration procedure. Before using it, we suggest you incorporate into this procedure those requirements specified by your own organisation.

Step 1 — Initiate record-keeping

The calibration begins formally with the receipt of the instrument and the order for the work to commence. A test record is started with the client's name, address, order number and a complete description of the thermometer submitted for testing, including the make, model and serial numbers. Any specific requirements of the client should also be noted, such as the range and accuracy required, particular temperatures of importance, relevant documentary standards, the intended use of the thermometer and any potentially damaging factors in the working environment.

The file should be continually updated to include copies or summaries of test records, calibration results and a copy of the certificate, if one is issued. In some cases the information may be most appropriately stored in laboratory notebooks or computer files. In any case the information must be secure and readily retrievable.

Step 2 — General visual inspection

Immediately on receipt of the thermometer a simple visual check should be performed, to record the state of the instrument on receipt. The thermometer should also be examined for any damage that may have occurred during shipping.

Step 3 — Conditioning and adjustment (if required)

For many thermometers, some form of conditioning or precalibration adjustment is required. For example, rare-metal thermocouples and standard platinum resistance thermometers (SPRTs) require annealing. Similarly, many electronic thermometers benefit from adjustments for offset, range and linearity.

A calibration does not normally cover the servicing or repair of the thermometer. If the client expects this, then it should be sought from the manufacturer of the thermometer. It would be unusual for the calibration supplier to have the expertise and equipment required. Also, service by a person not approved by the manufacturer may invalidate any warranty. As a matter of procedure instruments which do have user-serviceable adjustments (ice points, range and linearity) should *not* be adjusted except in consultation with the client. Changing the adjustments will prevent the client from using the calibration retrospectively and may interrupt the ice-point record which is the client's proof of stability. The client should always carry out an ice-point check immediately before and after calibration, firstly to check that the thermometer has survived shipment, and secondly, to ensure that the ice-point record is continuous despite any required adjustments.

Step 4 — Generic checks

At this stage it is necessary to carry out all of the subsidiary checks required to establish consistency of the thermometer with generic history. The checks fall into two main categories:

- Detailed inspection for faults and damage. These may be indicative of faulty manufacture or mistreatment and may be evidence to explain why a thermometer does not perform as well as it should or, in extreme cases, why it should not be calibrated at all.

- Measurements which will provide information that cannot be obtained during the intercomparison proper. Usually the measurements are used in the assessment of uncertainties, for example estimates of drift and hysteresis.

Many of these measurements may be carried out at any time; some, such as flare assessments of radiation thermometers, are best carried out at the end of a calibration; others, such as checks of the insulation resistance of resistance thermometers, are more efficiently carried out before the intercomparison.

One of the most important of the generic tests is that for stability. Where possible, ice points should be determined before and after all intercomparisons, since this may be the only specific information available on the stability of a thermometer. It is sometimes necessary (for example with mercury-in-glass thermometers) to allow time for relaxation to occur after the intercomparison.

Step 5 — Intercomparison

This is the part of the calibration which gathers the data for determining the ITS-90 relationship for the thermometer. The intercomparison must be designed to determine the relationship accurately, ensure that the thermometer obeys the expected relationship, and cover the range required by the client. Enough measurements should be taken to ensure reasonable confidence in the determined relationship and uncertainty. The intercomparison should also be carried out in such a way as to avoid as many systematic errors as is practical while ensuring that the conditions are still readily accessible to the user.

Step 6 — Analysis

Once all the data have been gathered, the results are processed to determine the best ITS-90 relationship for the thermometer.

Step 7 — Uncertainties

The uncertainty for the calibration is established by considering the five general sources of error described in Section 4.3.3 as well as uncertainties specific to the thermometer being calibrated.

Step 8 — Complete records

It is first necessary to decide, on the basis of the results, if the thermometer should be certificated. If at any of the steps 2–7 there is evidence that the thermometer deviates strongly from expected behaviour, then a certificate should not be issued. If, for example, the precalibration ice-point check shows that the thermometer is outside specifications, the remaining steps 5–7 can be missed. In cases where a certificate is not to be issued a covering letter should be supplied to the client explaining why. This information may be invaluable in uncovering poor handling and making warranty claims. All information relating to a thermometer, whether a certificate is issued or not, should be kept for a reasonable period. The ISO Guide 25 has good recommendations as to what information should be stored and how.

It is worth noting that the most time-consuming calibrations are those for thermometers that fail to meet the requirements. When faults are uncovered there is usually a lot more double-checking and repetition of measurements. The better specified the calibration procedure, the less the results will be questioned. One very useful time-saving device is a list comprising the criteria for failure. For a liquid-in-glass thermometer the list may include quality of marking, uniformity of bore, as well as performance-related criteria such as maximum error and maximum rate of change of error.

Once the thermometer has been pronounced satisfactory, a calibration certificate is prepared, a copy placed in the file, the client invoiced and the file closed.

4.3.7 Maintaining a calibration — recalibration and specific history

By issuing a certificate the calibration laboratory is asserting that the thermometer is capable of long-term stability. However, the calibration laboratory cannot control the way in which the thermometer is used. Proof of the long-term validity of the certificate rests almost entirely with the user. The user of the thermometer must be able to demonstrate that the thermometer is continuing to behave in the same way as other thermometers of that type, and the same as it did on the day of calibration.

To prove the validity of the certificate the client must:

(1) demonstrate through regular ice-point checks, or other simple verification checks, that the instrument continues to behave as it did at the time of calibration;
(2) demonstrate that the thermometer has not been exposed to use or conditions that may adversely affect its performance.

The first requirement is the single most important factor in the proof of validity of a certificate. For most thermometers, about 95% of all possible faults appear as a change in the ice-point readings. For thermometers, such as thermocouples and radiation thermometers, where an ice point is absent or provides little information about the integrity of the thermometer, regular verification checks against other thermometers or fixed points are required.

To meet the second requirement the user(s) of the thermometer must be able to demonstrate that the thermometer has always been used with due care. Here the ISO 25 Guide has very strong recommendations based on an equipment log which includes:

- Full description of the instrument.
- Procedures for use, including a copy of the manufacturer's instructions.
- The complete calibration and verification history of the instrument.
- Dates when the instrument is due for recalibration and service.
- A complete history of service and repair.
- Restrictions on the use of the equipment to approved sites and approved personnel.

All this serves to demonstrate that the instrument is in good repair, behaves as expected, and has never been exposed to use or conditions that may adversely affect its performance. This information constitutes the specific history of the thermometer.

The question of when to recalibrate is one of the more confused areas of calibration. The answer is really very simple — the calibration certificate is valid so long as the user is able to demonstrate its validity. If the ice-point record suggests that the thermometer has drifted too far for comfort then it is time for the thermometer to be recalibrated.

As a rough guide, thermometers should be calibrated as new, after one year of use, and then as necessary, up to a maximum period of 5 years. If at any time the thermometer ice-point reading drifts by much more than half of the reported uncertainty, it is also due for recalibration. If the rate of drift is excessive, then the thermometer may need adjustment or service. Thermocouples and radiation thermometers that do not have the ice point or triple point within their range may need to be recalibrated more frequently.

4.4 REASONABLE CARE — ELIMINATING THE OBVIOUS SOURCES OF ERROR

An essential feature of all calibrations is the elimination of the most obvious and most easily handled errors. In thermometry a number of errors occur in almost every measurement. Technically competent calibration suppliers and clients understand these errors and know how to either eliminate them or assess their contribution to the uncertainty.

The one common feature of all of these errors is that it is a simple matter to vary the measurement conditions to establish whether there is a problem and, if so, to give an indication of the magnitude of the problem. Also remember that the thermometer reads its own temperature; it is the user's responsibility to ensure and demonstrate that the thermometer is at the same temperature as the system of interest.

4.4.1 Immersion errors

The immersion problem appears in many guises and is known by many names: thermal anchoring, self-cooling, stem losses, heat-sinking, to name a few. The general problem

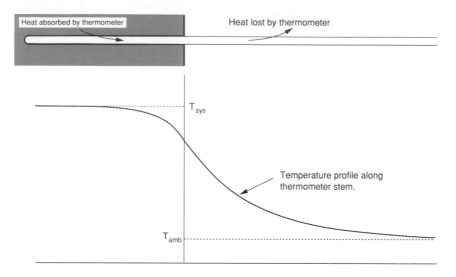

Figure 4.3
The flow of heat down the stem of a thermometer causes the thermometer to indicate temperatures slightly lower than that of the medium of interest.

occurs because there is a continuous flow of heat along the stem of a thermometer between the medium of interest and the outside world. Since heat can flow only where there is a temperature difference, the flow of heat is evidence that the tip of the thermometer is at a slightly different temperature than the medium of interest. This is shown graphically in Figure 4.3.

A simple model of this effect relates the error in the thermometer reading to the length of immersion by

$$\Delta T_m = (T_{amb} - T_{sys})k \exp \left(\frac{-L}{D_{eff}} \right), \tag{4.3}$$

where T_{sys} and T_{amb} are the system and ambient temperatures respectively, L is the length of immersion, D_{eff} is the effective diameter of the thermometer, and k is a constant approximately equal to, but less than, 1. This equation, which is plotted in Figure 4.4 for $k = 1$, is very useful for determining the minimum immersion which will ensure that the error is negligible.

Example 4.1   Determining minimum immersion

Determine the minimum immersion for a 4 mm diameter sheathed thermometer with the detecting element occupying the last 40 mm of the sheath. The measurement should have immersion errors of less than 0.01°C for temperatures up to 100°C.

First we determine the relative accuracy in the measurement as

$$\left| \frac{\Delta T_m}{T_{sys} - T_{amb}} \right| = \frac{0.01}{100 - 20} \approx 0.01\%.$$

Continued on page 127

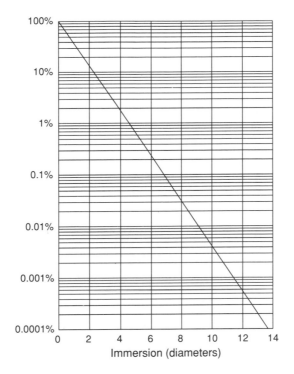

Figure 4.4
The relative temperature error $|\Delta T_m/(T_{sys} - T_{amb})|$ plotted against thermometer immersion length in diameters.

Continued from page 126

Then, referring to Figure 4.4, we find that the minimum immersion is a little more than 9 diameters. To be conservative we will immerse the thermometer to 10 diameters beyond the detector element, i.e. 80 mm total immersion. Ten diameters is a useful rule-of-thumb for accurate thermometry ($\sim \pm 0.01\%$). Five diameters is more typical of industrial thermometry ($\sim \pm 1\%$).

Exercise 4.2

Find the minimum immersion for a 6 mm diameter probe in a 10 mm diameter thermowell at 800°C such that the immersion error is less than 1°C.

The main problem with equation (4.3) is that the two constants, k and D_{eff}, are unknown and dependent on the thermometer's surroundings as well as on the thermometer. This variable behaviour is demonstrated in Figure 4.5. In situations where the medium is well stirred, such as in an oil bath, the equation works well if you use the actual diameter of the probe. However, in situations where the medium is not stirred or there is poor thermal contact, the effective diameter can be much larger than the actual diameter of the probe. Other problems include uncertainty in the location of the detector element and difficulty in defining the diameter, for example with multiple sheaths or thermowells. In all

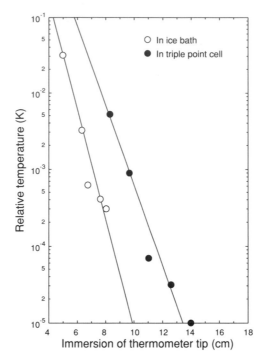

Figure 4.5
The immersion characteristics of a standard platinum resistance thermometer in an ice bath and
triple-point-of-water cell.

these cases it pays to be pessimistic and add the detector length to the length determined
from Figure 4.4 and use the outside diameter of any sheath or thermowell assembly.

The most difficult immersion problems occur when making measurements of air
and surface temperatures. For air-temperature measurements the effective diameters of
probes may be as large as ten times the actual diameter of the probe; a probe requiring
10 diameters immersion in the calibration bath may require more than 100 diameters
immersion in air.

The fundamental problem with surface temperature measurements is that, since a
surface is an infinitely thin boundary, there is no 'system' into which you can immerse
the thermometer. With surface-temperature measurements, the answer to the measurement
problem often lies in analysing the reason for making the temperature measurement in
the first place. For example, if we need to know how much energy the surface is radiating
we should use a radiation thermometer; if we want to know the likelihood of the surface
posing a human burn risk then we should use a standard finger as specified by a safety
standard; and if we require a non-intrusive measurement of the temperature of the object
behind the surface, then a measurement using one of the techniques in Figure 4.6 may
be the answer. Assessment of the uncertainties in surface measurements is also difficult
because of the number of sources of error present.

In all cases where immersion errors are suspected it is a very simple matter to vary
the immersion length by one or two diameters to see if the reading changes. As a crude

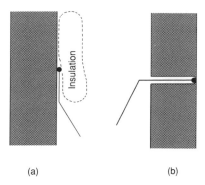

(a)   (b)

Figure 4.6
Two solutions to the problem of surface temperature measurement: (a) attaching a length of the probe to the surface can approximate immersion. In some cases insulation may be helpful in reducing heat losses by radiation or convection, although it can cause the surface to become hotter; (b) approaching the surface from the side which has the least temperature gradient will give the least error.

approximation about 60% of the total error is eliminated each time the immersion is increased by one effective diameter. In some cases it may be practical to estimate the true temperature from a sequence of measurements at different immersions (see Exercise 4.3).

Exercise 4.3

(a) Suppose that 3 measurements are made at immersion depths of L_1, L_2 and L_3 where $L_2 - L_1 = L_3 - L_2 = \Delta L$, and the resulting temperature readings are T_1, T_2 and T_3 respectively. By manipulating equation (4.3) show that

$$T_{sys} = T_1 + \frac{(T_2 - T_1)^2}{2(T_2 - T_1) - (T_3 - T_1)}$$

and

$$D_{eff} = \Delta L / \ln \left(\frac{T_{sys} - T_1}{T_{sys} - T_2} \right)$$

(b) If 3 measurements are made at immersion depths of 3, 4, 5 cm, giving temperatures of 115, 119 and 121°C, what is the system temperature and effective diameter of the thermometer?

4.4.2 Heat capacity errors

Whenever a thermometer is immersed in a system, the system must supply or absorb heat so that the thermometer comes to the same temperature as the system. Depending on the complexity of the measurement there are four aspects of the measurement that may be affected by the heat capacity of the thermometer. The first is due to the heat capacity itself; the others — settling, lag and dynamic errors — are all due to the thermometer's time-constant. Heat capacity corrections are dealt with here, and the other factors are considered in the sections below.

In a system where there is no mechanism for replacing the heat, either through a control mechanism or through utilising a physical phenomenon such as an ice point, the removal or addition of heat will permanently affect the temperature of the system. Measuring the temperature of a beaker of hot water with a large mercury-in-glass thermometer is a simple example. Again a simple experiment will expose the problem: withdraw the thermometer, allow it to cool to room temperature, and re-immerse it. If there is a problem with the heat capacity of the thermometer the two temperature readings will be different. This technique may be used to estimate the error and to estimate the temperature before the thermometer was immersed.

Example 4.2   *Estimating the heat capacity error*

A thermometer of unknown heat capacity is inserted into a large vacuum flask of hot fluid and indicates a temperature of 84.3°C. After withdrawing the thermometer, allowing it to cool to ambient temperature and reinserting it, the reading is 83.8°C. Estimate the initial temperature of the flask of fluid.

We assume that the change in temperature on the first immersion is the same as that on the second immersion, and that the temperature would otherwise be constant. The change on second immersion was:

$$\Delta T_m = 83.8 - 84.3 = -0.5°C.$$

The initial fluid temperature is the first recorded temperature plus the correction for the error, hence

$$T_{sys} = 84.3 + 0.5 = 84.8°C.$$

An estimate of the size of a heat-capacity error can also be based on estimates of the heat capacity of the thermometer and that of the beaker of hot water (see Exercise 4.4).

The heat capacity of most solids and liquids varies between those of water, $4.2 \text{ J}° \text{ C}^{-1}\text{cm}^{-3}$, and oil, $1 \text{ J}° \text{ C}^{-1}\text{cm}^{-3}$. A value of $2 \text{ J}° \text{ C}^{-1}\text{cm}^{-3}$ is a reasonable estimate where no other data are available.

Exercise 4.4

(a)   Show that the heat capacity error in an uncontrolled system is

$$\Delta T_m = \frac{-C_t}{C_t + C_{sys}}(T_{sys} - T_{init})$$

where C_t and C_{sys} are the heat capacities of the thermometer and system respectively, T_{sys} is the system temperature, and T_{init} is the initial thermometer temperature.

Continued on page 131

___ *Continued from page 130* ___

(b) Find the heat capacity error that occurs when a large mercury-in-glass thermometer ($C_t \sim 20$ J/$^\circ$C) is used to measure the temperature of a hot cup of coffee. Assume 1 cup of coffee is equivalent to 250 ml water, hence $C_{sys} = 1000$ J/$^\circ$C, and that $T_{sys} = 90^\circ$C.

4.4.3 Settling response errors

In systems where there is some temperature-control mechanism, or the system is very large, the heat-capacity error is absent or negligible. However, it will take time for the system to replace the heat lost in heating the thermometer, and for the thermometer to settle to the temperature of the system. If insufficient time is allowed for either of these processes to occur then there will be an error in the thermometer reading. The response of the thermometer is shown graphically in Figure 4.7.

A very simple model estimates the error as

$$\Delta T_m = (T_{init} - T_{sys}) \exp(-\tau/\tau_0) \tag{4.4}$$

where T_{init} and T_{sys} are the initial temperatures of the thermometer and the system respectively, τ is the time between immersion and reading, and τ_0 is the *1/e time-constant* of the thermometer or of the control system, whichever is the slowest. This equation allows us to estimate the minimum wait before we can read the thermometer with negligible error. To simplify calculations the equation is plotted in Figure 4.8.

Example 4.3   Estimating minimum measurement time

Given a system with a response such as shown in Figure 4.7, estimate the minimum measurement time required to achieve an accuracy of 0.5°C at temperatures near 150°C. Assume that the initial temperature of the thermometer is 25°C.

The relative error is required to be less than $0.5/(150-25) = 0.4\%$. Referring to Figure 4.8, it is found that at least $5.5\tau_0$ seconds must elapse before the error is less than 0.4%. The time-constant of the thermometer is 20 seconds, hence the minimum measurement time is 110 seconds.

As with the immersion problem, the most difficult time-constant problems occur in air temperature measurements. The time-constant of a thermometer in air may be 10 or 20 times that in a calibration bath. For some particularly heavy thermometer assemblies the time-constant may be 10 minutes or more, and require an hour to settle for a single measurement.

An additional problem with time-constants is that the assumptions leading to equation (4.4) and Figure 4.8 are optimistic. There are some situations and probe designs where there is more than one time-constant involved; a thermometer immersed in a thermowell measuring the temperature of a controlled process may have three time-constants characterising the overall thermometer response. In these cases there is simply no alternative to experimentation in order to expose potential errors in the indicated temperature.

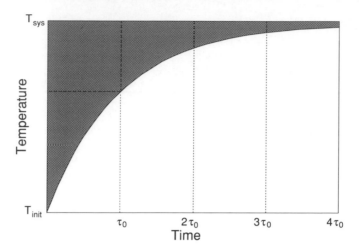

Figure 4.7
The settling response of a thermometer assuming that a single time-constant, τ_0, is dominant. After each interval of τ_0 seconds the error is reduced by about 63%.

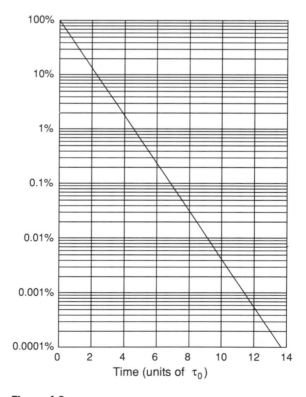

Figure 4.8
The relative temperature error $|\Delta T/(T_{sys} - T_{init})|$ plotted against measurement time in multiples of the time-constant, τ_0.

4.4.4 Lag errors with steadily changing temperatures

In systems where the temperature is changing at a constant rate the settling response of the thermometer causes a more serious error. The situation is shown graphically in Figure 4.9. There are two components to the error. The first component, the shaded portion of Figure 4.9, is the same as the time-constant error as discussed above and will gradually decrease to a negligible value. The main error is the 'lag error', which is proportional to the time-constant and the rate of change of the bath temperature:

$$\text{lag error} = -\tau_0 \times \text{rate of change in temperature.} \qquad (4.5)$$

The effect of the error is to cause the thermometer reading to lag τ_0 seconds behind the bath temperature.

Example 4.4 Estimating lag error

Lag errors can be a serious problem in process control where both the time-constants and the rates of change are large. Calculate the lag error when a thermometer with a time-constant of 20 seconds monitors a process temperature changing at 3°C per minute.

From equation (4.5) the lag error is

$$\Delta T_m = \frac{-20 \times 3}{60} = -1°C.$$

In systems where the thermometers are mounted in thermowells the errors may be much larger.

Example 4.5 Minimising lag errors in calibration

A very important example of lag error occurs in the rising-temperature method of calibration (see Section 4.5). Consider, for example, the situation where we wish to calibrate a set of working thermometers with time-constants of 5 seconds against reference thermometers which have time-constants of 7 seconds. What is the maximum rate of rise in the calibration bath temperature if we require the lag errors to be less than 0.01°C?

The lag error for the reference thermometers is

$$\Delta T_r = -7 \times \text{rate of rise,}$$

and the error for the working thermometers is

$$\Delta T_w = -5 \times \text{rate of rise.}$$

Hence the error in the transfer is

$$\Delta T_{cal} = (7 - 5) \times \text{rate of rise.}$$

Since we require this error to be less than 0.01°C, the maximum rate of temperature rise is

$$\text{max rate of rise} = 0.01/(7 - 5) = 0.005°C/s = 0.3°C/min.$$

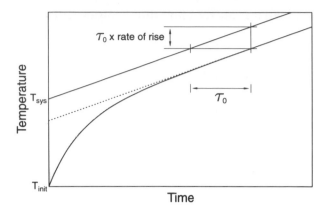

Figure 4.9
The temperature error due to the thermometer's time-constant
in a system with a steadily increasing temperature.

It is instructive to investigate how much a 5 mK/s rate of rise restricts the design and operation of a calibration bath. For a 25 l water bath, with no other heat losses, and heated by a 100 W heater, the rate of rise is approximately 1 mK/s. On paper quite high-accuracy calibrations should be feasible using the rising-temperature technique. In practice there are, however, additional complications in ensuring that the bath temperature is uniform. The maximum rate of rise of 5 mK/s determined in the example above is very much a maximum. A more practical and conservative design figure would be a third or a fifth of that value.

4.4.5 Dynamic measurement errors

In more difficult cases where the temperature is changing unpredictably, it becomes practically impossible to estimate the errors. However, it is possible to gain a qualitative picture of the thermometer's behaviour under these circumstances. By considering the effect of the time-constant on a periodic temperature variation it can be shown that the thermometer's response is less than the actual variation by the factor

$$G(f) = \frac{1}{\left(1 + 4\pi^2 \tau_0^2 f^2\right)^{1/2}} \tag{4.6}$$

where f is the frequency of the periodic variation. Those familiar with electronics will recognise equation (4.6) as the response of a first-order filter. At frequencies less than about $1/(5\tau_0)$ the thermometer will follow the changes in temperature fairly well. Variations at higher frequencies are effectively filtered out by the thermometer response.

This has practical consequences when choosing thermometers for applications where a fast response or detection of short-term events is required. A simple rule-of-thumb is to choose thermometers with time-constants five times faster than the event to be measured. This is the same as the rule-of-thumb inferred from Example 4.3.

One of the advantages of thermometers with long time-constants is that they can be used to measure average temperatures. The meteorological air-temperature measurement described in Chapter 1 is an example.

4.4.6 Radiation errors and shielding

Heat can be transferred by any of three mechanisms:

- *conduction* — for example, heat is conducted along a metal bar;
- *convection* — for example, heat is transferred by the movement of air or other fluids;
- *radiation* — for example, heat is radiated by lamps, radiant heaters, and the sun.

Radiation is one of the most insidious sources of error in thermometry. We often fail to recognise the physical connection between the radiant source and the thermometer and overlook it as a source of error. Radiation errors are a particular problem in air and surface thermometry where there is nothing to obscure or shield the source, and where the thermal contact with the object of interest is already weak. Examples of troublesome radiant sources include lamps, boilers, furnaces, flames, electrical heaters and the Sun. A particularly common problem to watch for is the use of incandescent lamps when reading thermometers. If you must use a lamp, then use a low-power fluorescent lamp which will radiate very little in the infra-red portion of the spectrum.

With more difficult measurements, such as air and surface temperatures, anything at a different temperature which has a line-of-sight to the thermometer is a source of error. This includes cold objects such as freezers which act as radiation sinks and absorb radiation emitted by the thermometer. To put things in perspective, remember that at room temperature everything radiates (and absorbs from its neighbours) about 500 watts per square metre of surface area, so the radiative contact between objects is far greater than we would expect intuitively. In a room near a large boiler a mercury-in-glass thermometer may exhibit an error of several degrees.

There are two basic strategies when you are faced with a measurement that may be affected by radiation. Firstly, remove the source; and secondly, shield the source. Removing the source is obviously the most effective strategy if this is possible. However, the thermometry is very often required in association with the source, particularly in temperature-control applications. In these cases it may be possible to change the source in a way which will give an indication of the magnitude of the error.

If you are unable to remove the radiation source then shielding is the only resort. A typical radiation shield is a highly reflective, usually polished, metal tube which is placed over the thermometer. The shield reflects most of the radiation away from the thermometer and itself. An example is shown in Figure 4.10. The shield will usually reduce the error by a factor of about 3 to 5. The change in the thermometer reading when the shield is deployed will give a good indication of the magnitude of the error and whether more effort is required. Successive shields will help but will not be as effective as the first. Suitable trial shields are clean, shiny metal cans and aluminium foil.

The disadvantage of using a radiation shield in air-temperature measurements is that the movement of air around the thermometer is greatly restricted, further weakening the thermal contact between the air and the thermometer. The problem is compounded if the shield is warmed by the radiation and conducts the heat to the stagnant air inside the shield. Therefore, to be effective the shield must allow free movement of air as much as possible. In some cases a fan may be needed to improve thermal contact by drawing air over the sensor, and to keep the shields cool. Note that the fan should not be used to push the air as the air will be heated by the fan motor and friction from the blades.

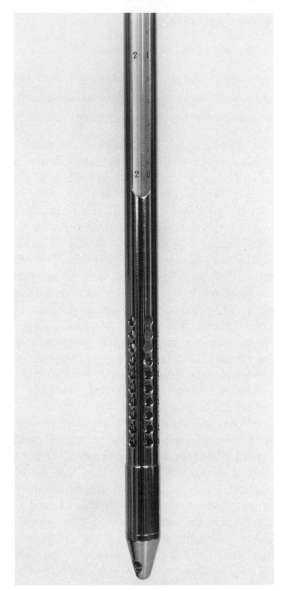

Figure 4.10
An example of a radiation shield for a mercury-in-glass
thermometer.

4.5 THE RISING-TEMPERATURE INTERCOMPARISON METHOD

Ideally, mercury-in-glass thermometer calibrations should be performed in calibration baths with controllers that enable the set-point to be changed by a fraction of a scale division. This allows an accurate assessment of the quality of the thermometer bore and markings near the calibration point. The method described here achieves the same end without the use of a sophisticated controller. The method is based on a well-stirred bath with a simple heater powered from a variable power supply such as a variable a.c. transformer.

The power provided for the heater needs to be a few watts more than is required to keep the bath stable. In this way the bath temperature rises very slowly and steadily. By placing the thermometers in the bath and reading them in a correctly timed sequence it is possible to ensure that the average reading for all the thermometers is the same. Figure 4.11 shows a graphical representation of the technique.

The process shown in Figure 4.11 is repeated three times for each calibration point in order to build up statistical information about the distribution of the errors near the calibration point.

The technique has a number of advantages over fixed-temperature calibrations:

- By design, the readings are taken at random over several scale divisions, ensuring that the bore and scale markings are well sampled.
- The mercury column rises steadily as the temperature is increased, ensuring that the mercury meniscus is properly shaped.
- The technique has a relatively low cost and is quicker than the fixed-temperature technique described in Section 4.7.
- The technique also provides ready access to temperatures down to −80°C through the use of stirred-alcohol and dry-ice baths.

The technique has a number of disadvantages and potentially suffers from a number of small systematic errors:

- A variable rate of rise of the bath temperature will lead to the average temperatures not being the same for all sets of readings. The problem is not serious so long

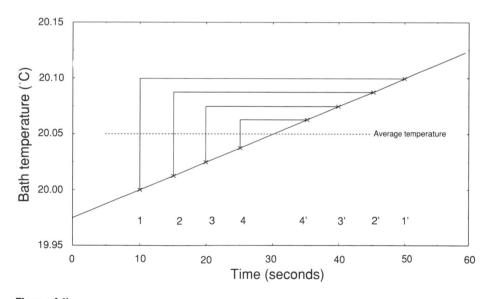

Figure 4.11
The rising-temperature calibration method. In this calibration run the four thermometers are read at 5-second intervals with a slightly longer delay before they are read at 5-second intervals in the opposite sequence. The average bath temperature should be the same for all the thermometers.

as the variations do not correlate with the movement of the operator reading the thermometers, and the resulting errors should be random over the whole set of data.

- If the readings are taken at variable or irregular intervals, the effect on the results may be similar to that caused by a variable temperature rise.

- When thermometers with different time-constants are used they will lag behind the bath temperature by different amounts, leading to a systematic error in the calibration. Example 4.5 considers this problem and yields a useful rule-of-thumb for the calibration of mercury thermometers, namely: the rate of rise should be less than 1 mK/s.

- The strict timing requirements are quite demanding for the operator who must read the thermometer, record the results, and move the viewing telescope to the next thermometer in time to take the next reading. Careful planning is required to avoid reading and transcription errors, particularly if the reference thermometers and thermometers under test have different scale markings.

- The steadily rising mercury column will almost certainly suffer from stiction; the mercury moving up in fits and starts. It is important that the thermometer is tapped lightly immediately before the reading to encourage the mercury to move to its equilibrium level and so minimise the stiction error.

- It is likely that the uniformity and stability in a bath that is not in equilibrium are less than in a fixed-temperature bath.

- As we shall see in the following example, this method does not provide as much information about the distribution of the errors as does the fixed-temperature calibration.

Example 4.6   Calibrating mercury-in-glass thermometers using the rising-temperature method

An order is received from ACME Thermometer Co, for the calibration of two mercury-in-glass thermometers, namely ASTM 121C kinematic viscosity thermometers. The procedure given in Section 4.3.6 is followed, using the rising-temperature method described in Section 4.5 above. Describe the procedure and present a calibration certificate in the name of your (imaginary) firm: Calvin, DeGries & Co.

Step 1 — Initiate record-keeping

A file is opened with a code number for the job. This file contains the client's address; the contact person; the contact telephone and/or fax number; a copy of the order; a complete description of the thermometers including the manufacturer, type number and serial numbers; and the calibration points required. The file will be continually updated to include summaries of the test records, calibration results and a copy of the certificate if one is issued.

In this case the thermometers are short-range thermometers with an auxiliary ice-point scale and a main scale covering the range 98.5°C to 101.5°C. The thermometers are marked to 0.05°C with calibration required at 0°C, 100°C and 101°C.

Step 2 — General visual inspection

The thermometers are unpacked and inspected immediately on receipt. They are found to be in good condition. The packaging is satisfactory.

___ *Continued on page 139* ___

Continued from page 138

Step 3 — Conditioning and adjustment

In this case the thermometers do not require conditioning or adjustment other than a three-day wait at room temperature, which is normal for this type of thermometer. They are stored horizontally in a secure cabinet to protect them from exposure to mechanical and thermal shock.

Step 4 — Generic checks

Detailed inspection shows that the markings on both thermometers are clear, well formed and unambiguous. There are no impediments, constrictions or obstructions in the bore. The mercury column is intact with no signs of mercury in any of the chambers above the meniscus.

An ice-point measurement is made so that it can later be compared with a post-calibration ice point to check on the thermometer's stability. The ice-point reading is found, as required by the ASTM, to be within two scale divisions of 0°C. The ice points on the two reference thermometers that will be used in the intercomparison are also checked now.

Step 5 — Intercomparison

The intercomparison follows the guide in Section 6.5.2 for the short-range calibration of liquid-in-glass thermometers.

The ASTM standard requires intercomparison at 100°C and 101°C. The rising-temperature technique is used at both points. The procedure for the 100°C point is as follows:

The calibration bath is first prewarmed to a couple of degrees below 100°C and the thermometers are then located in the bath as indicated in Figure 4.12, with the mercury just visible above the surface of the oil. Once the telescope has been positioned and the operator is ready to record the results, the bath heater is adjusted to bring the bath slowly through the 100°C mark. The heat capacity of oil is about one-quarter that of water so the bath requires about 1 watt per litre of oil in excess of the bath losses to rise at 1 mK/s (see Example 4.5).

Readings commence once the reference thermometers indicate that the temperature is within about 3 scale divisions of 100°C. A typical record is shown below, where the values of the reference thermometer are tabulated as 'Ref' and the thermometers being calibrated are tabulated as 'working' thermometers. The record is arranged in the same sequence as the sequence of thermometers in the calibration bath to help prevent transcription errors, and the readings are taken in the order indicated by the arrows.

Reading		Ref 1	Working 1	Working 2	Ref 2
Set 1	→	99.740	99.740	99.720	99.770
	←	99.750	99.750	99.720	99.780
Set 2	→	100.040	100.050	100.030	100.078
	←	100.042	100.045	100.030	100.078
Set 3	→	100.180	100.190	100.175	100.220
	←	100.182	100.190	100.180	100.224

Once both intercomparisons are complete the thermometers are removed from the bath, cleaned to remove the oil, and returned to the cabinet to allow recovery from the high-temperature exposure. An ice-point measurement is carried out at this time to help assess the recovery of the thermometer.

Continued on page 140

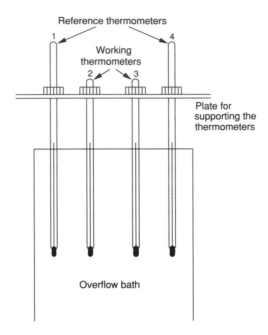

Reference thermometers

Working
thermometers

Plate for
supporting the
thermometers

Overflow bath

Figure 4.12
The thermometers are placed in the calibration bath
in the sequence in which they will be read, to prevent
recording errors.

Continued from page 139

Step 6 — Analysis

The first step in the analysis is to calculate the average reading for each set of results, retaining guard figures so as to avoid round-off errors during calculation.

Reading	Ref 1	Working 1	Working 2	Ref 2
Set 1	99.745	99.745	99.720	99.775
Set 2	100.041	100.0475	100.030	100.078
Set 3	100.181	100.190	100.1775	100.222

Once this has been done the reference-thermometer readings are corrected using their calibration certificates and averaged to determine the calibration temperatures. For example, the results at 100°C are:

Correction for Ref 1 at 100°C = +0.045
Correction for Ref 2 at 100°C = +0.010

Corrected Readings	Ref 1	Ref 2	Average	Difference
Set 1	99.790	99.785	99.7875	+ 0.005
Set 2	100.086	100.088	100.087	− 0.002
Set 3	100.226	100.232	100.229	− 0.006

Continued on page 141

Continued from page 140

The reference thermometers disagree by at most +0.005°C and 0.006°C, which is quite good agreement for thermometers marked with 0.02°C divisions. The corrections for the two working thermometers are now determined along with the average and standard deviations.

Corrections to working thermometers
(correction = true temperature − mean reading)

	Temperature	Working 1	Working 2
Set 1	99.7875	0.0425	0.0675
Set 2	100.087	0.0395	0.057
Set 3	100.229	0.0490	0.0615
Average	100.0378	0.0437	0.0559
standard deviation		0.0048	0.0102

These three tables are calculated for each calibration point and a summary prepared for each thermometer. The summary for working thermometer 2 is as follows:

Summary for working thermometer 2

Temperature	Reading	Correction	Standard deviation
0 (ice point)	− 0.020	+ 0.02	−
100.04	−	+ 0.056	0.0102
101.02	−	+ 0.047	0.0071
cumulative standard deviation			0.0086

The cumulative standard deviation is calculated from the variance of the residual errors of both calibration points. The total number of degrees of freedom used to calculate the variance is equal to 4, calculated as the number of measurements of error, 6 (three per point) minus the number of corrections calculated, 2 (1 per point).

Step 7 — Uncertainties

In order to determine the total uncertainty the various contributing factors identified in Section 4.3.3 are evaluated.

- *Uncertainty in the reference-thermometer readings*   This is read directly off the calibration certificates for the two reference thermometers, and is already reported at the 95% confidence limit:

$$U_{ref} = 0.008(95\% \text{ CL})$$

- *Variations in the stability and uniformity of the calibration medium*   It is known from commissioning tests that the bath non-uniformity is no greater than 0.005°C per 200 mm. Since the thermometers have been placed in the bath within 100 mm of each other, the non-uniformity can be treated as the semi-range on a rectangular distribution and the uncertainty is estimated as

$$U_{bath} = 0.0025(95\% \text{ CL})$$

Continued on page 142

Continued from page 141

- _Departures from the determined ITS-90 relationship_   The error-of-fit for the measurement is 0.0086°C. This is a Type A uncertainty with 4 degrees of freedom. The 95% confidence limits are found by multiplying this value by 2.78, which is the k value from Student's t-distribution (Table 2.2) corresponding to $P = 95\%$ and $\nu = 4$. Hence

$$U_{\text{fit}} = 0.024(95\% \text{ CL})$$

- _Uncertainty due to hysteresis_   The likely uncertainty due to hysteresis is indicated by the difference between the pre-calibration and post-calibration ice points, in this case 0.005°C. Treating this as a rectangular distribution the 95% confidence limit is estimated as half of the total range:

$$U_{\text{hys}} = 0.0025(95\% \text{ CL}).$$

- _Uncertainty due to drift_   For mercury-in-glass thermometers any drifts with time show as changes in the ice-point reading. This so-called secular change is used to correct for drift so long as it is not excessive, and this usage is known by the client. Thus

$$U_{\text{drift}} = 0$$

- _Total uncertainty_   The total uncertainty is the quadrature sum of the values of the individual uncertainties

$$U_{\text{tot}} = [8^2 + 2.5^2 + 24^2 + 2.5^2]^{1/2} \text{ mK}$$
$$= 25 \text{ mK } (95\% \text{ CL}).$$

In this case the total uncertainty is very nearly equal to that for the greatest contributor. On the certificate the uncertainty will be quoted as 0.025°C, which is equivalent to half a scale division and typical of this type of thermometer.

Step 8 — Complete records

After comparison of the results with ASTM specifications the decision is made that the thermometer warrants a certificate and it is prepared with the results rounded to the appropriate decimal place. The completed certificate for working thermometer 2 is shown in Figure 4.13.

4.6 CALIBRATING DIRECT-READING ELECTRONIC THERMOMETERS

Direct-reading thermometers form a group which have errors and faults in common despite the wide variety of sensors that are employed. These errors, which centre on the electronic processing of the sensor response, have considerable influence on the design of the calibration. These considerations are additional to those for the sensor itself.

Direct-reading electronic thermometers are all based on electronic sensors: typical examples are platinum resistance thermometers, thermistors, thermocouples and radiation

CALVIN, DeGRIES AND CO

1 Traceability Place, P O Box 31-310, Lower Hutt, New Zealand
Telephone (64) 4 566-6919   Fax (64) 4 569-0003

CALIBRATION CERTIFICATE

Report No.:	T92-2001.
Client:	ACME Thermometer Co, 100 Celsius Avenue, P O Box 27-315, Wellington, New Zealand
Description of Thermometer:	ASTM 121C kinematic viscosity thermometer divided to 0.05°C. Serial number 2925, manufactured by Zeal.
Date of Calibration:	22 to July 1992.
Method:	The thermometer was compared with standard thermometers held by this laboratory. The temperature scale used is ITS-90.
Conditions:	The thermometer was calibrated in total immersion.

Results:

Thermometer reading (°C)	Correction (°C)
0 (ice point)	+0.02
100.04	+0.06
101.02	+0.05

Note: Corrections are added to the reading to obtain the true temperature.

Accuracy: The uncertainty in the corrected thermometer readings is estimated to be ±0.025°C at the 95% confidence level.

Checked: _____   Signed: _____
         W Thomson   R Hooke

This report may only be reproduced in full.

Figure 4.13
A typical calibration certificate for a mercury-in-glass thermometer, based on the information discussed in Example 4.6.

detectors. For all of these sensors the response is a non-linear function of temperature, that is the relationship between temperature and resistance, for example, would not be a straight line on a graph. In order to convert the sensor response to an electrical signal which is linearly related to temperature, some form of linearisation is required. There are three basic categories of linearisation:

- *Segmented linearisation*   These techniques approximate the response of the sensor by a series of straight lines. This was one of the earliest techniques and is not particularly common nowadays. The residual error from this technique has a rather jagged shape that does not lend itself to accurate interpolation between calibration points (see Figure 4.14).

- *Analogue linearisation*   Nowadays, linearisation is accomplished using a variety of non-linear electronic circuits such as function generators, negative resistance circuits, analogue multipliers and logarithmic amplifiers to make smooth approximations to the sensor response. For example, a common approximation is based on the equation

$$L(T) = k\frac{R(T) - a}{R(T) - b} \tag{4.7}$$

where $R(T)$ is the sensor response, $L(T)$ is the linearised response and a, b and k are constants corresponding to offset, linearity and range adjustments.

Modern electronic technology makes this a low-cost option and moderately accurate. The most important feature from the calibration point of view is that, although

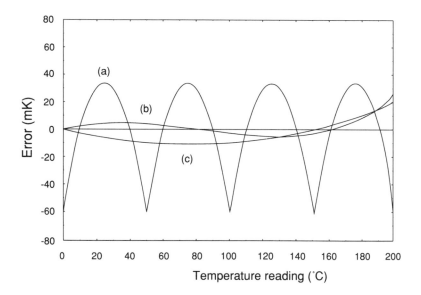

Figure 4.14
Linearisation strategies for a platinum resistance thermometer: (a) segmented linearisation leaves a non-smooth error curve so that interpolation between calibration points may not be valid; (b) analogue linearisation matches the general shape of the curve at the expense of more complex but smaller and smoother errors; (c) the residual errors from microprocessor linearisation change very slowly with the reading.

the linearisation is not perfect, the residual error curve is very smooth. This ensures that accurate interpolation between calibration points is practical.

- *Microprocessor linearisation* Microprocessor linearisation is common amongst the better quality electronic thermometers. The microprocessor is able to perform all the mathematical functions which relate resistance, radiance or voltage to temperature. The residual errors due to the electronic processing are usually negligible. Since the microprocessor is usually programmed to linearise a thermometer which conforms to the documentary standard, the largest errors tend to be due to departures of the sensors from the standard.

When we calibrate an electronic thermometer we expect to sample the residual errors and determine corrections which will enable interpolation. One means is to use our knowledge of the likely errors (a part of the generic history), propose a simple mathematical form for the relationship between error and reading, and use the calibration to determine the relationship in detail.

There are two linear errors that occur in all electronic thermometers. The first, the offset error, is constant for all temperatures, and is similar to the ice-point shift in mercury-in-glass thermometers. The second linear error is proportional to the temperature and is essentially a range error. For these two errors we expect the error to take the form:

$$\text{linear error} = -A - B \times \text{reading} \tag{4.8}$$

where A and B are constants to be determined in the calibration.

Non-linear errors also occur in two forms. The first is known as even-order non-linearity and causes a U-shape in the thermometer's error curve. The second is the odd-order error which introduces an S-shape in the error curve. So long as the errors are not severe they are well approximated by

$$\text{non-linear error} = -C \times (\text{reading})^2 - D \times (\text{reading})^3 \tag{4.9}$$

where C and D are coefficients for the even-order and odd-order errors respectively, and are to be determined by calibration. This equation is particularly appropriate for platinum resistance thermometer instruments since most linearisation techniques are designed to remove the second-order (term in t^2) behaviour at the expense of a smaller third-order (term in t^3) non-linearity. Equation (4.9) would describe the residual t^2 errors and the additional t^3 errors.

By combining the two equations and changing from error to correction we obtain the calibration equation:

$$\Delta T_{\text{cor}} = A + B \times \text{reading} + C \times (\text{reading})^2 + D \times (\text{reading})^3 \tag{4.10}$$

where ΔT_{cor} is the correction to the reading.

The four constants can be determined by a least-squares fit to the intercomparison data.

This equation with the fitted values for the coefficients is known as the *deviation function* for the thermometer. The deviation function style of calibration may also be used to determine the departure of thermocouples and resistance thermometers from standard tables of voltage or resistance versus temperature. This effectively uses the tables as a standard direct-reading thermometer.

4.7 THE FIXED-TEMPERATURE CALIBRATION METHOD

For calibrations to have the highest accuracy, all settling and response errors must be completely eliminated. This can be achieved only by the fixed-temperature calibration method. The equipment required is a bath (furnace, cryostat) with high stability and uniformity. The temperature of the bath should be controlled by a controller with a high short-term stability so that the uncertainty due to bath-temperature fluctuations is negligible. For mercury thermometer calibrations the controller should also have a sufficiently fine sct-point adjustment to enable the small increments in temperature which are required to assess the bore and scale markings.

In a fixed-temperature calibration all the calibration temperatures are planned in advance and the controller set-point is set to each of the nominal calibration temperatures in turn. The bath and thermometers are allowed to settle for several minutes (or longer as required) once the bath has reached the set-point. The calibration readings are then taken and the bath moved to the next temperature.

The advantages of the fixed-temperature method are:

- settling and response errors are eliminated;
- the bath has higher uniformity than with a rising-temperature calibration;
- there is more accurate control over the calibration temperatures;
- the operator has greater flexibility over when the readings are taken, and this results in fewer recording and transcription errors;
- this method provides more information than the rising-temperature method for the same number of measurements.

Disadvantages of the method include:

- the cost of bath and controller are higher;
- the calibration time is much longer because of the additional settling time.

Example 4.7   Calibrating an electronic thermometer

A platinum resistance thermometer with a resolution of $0.01°C$ is received from ACME Thermometer Co. and requires calibration between $-20°C$ and $180°C$. The thermometer is calibrated against a standard thermometer (SPRT) with a computer used to analyse the results. By following the steps of the procedure given in Section 4.3.6, describe the procedure and present a certificate in the name of your (imaginary) firm, Calvin, deGries & Co.

Step 1 — Initiate record-keeping

As with Example 4.6, all the relevant details on the client and the thermometer are recorded.

Step 2 — General visual inspection

As for Example 4.6.

Continued on page 147

Continued from page 146

Step 3 — Conditioning and adjustment

Although the operator manuals for such thermometers include instructions for resetting the ice point, range and linearity immediately prior to calibration, this is not carried out since the client has asked that the instrument not be adjusted unless the ice point is in error by more than 0.05°C.

Step 4 — Generic checks

Four checks are carried out on the thermometers: a detailed visual inspection; an insulation resistance check to confirm that the probe is free of moisture; an ice-point check before and after the intercomparison to confirm the short-term stability of the thermometer; and a measurement of the hysteresis.

- *Detailed visual inspection* The instrument is inspected for bends and dents in the probe, damage to the leads, plugs, sockets, cable strain relief, etc. The electronic unit is found to be well maintained and nothing is loose or broken. The general condition of the instrument is consistent with its age and usage and indicates that the instrument is well enough maintained to be trusted to hold its calibration.

- *Insulation check* The probe assembly is disconnected from the instrument and the insulation resistance between the steel sheath and one of the four lead wires measured. The resistance is found to be in excess of 1 GΩ, which is typical of probes assembled using alumina insulation.

- *Ice-point check* Carrying out ice-point checks on stainless steel sheathed probes can be quite difficult and errors of several hundredths of a degree are possible. The high thermal conductivity and thermal mass of the stainless steel probe make it difficult to keep the ice well packed and in contact with the sheath. This is aggravated by heat being dissipated in the sensing element. Particularly with instruments that read to 0.01°C or better, it is extremely important to use very finely shaved ice. The probe is thus allowed to settle for at least 10 minutes, and the ice is pushed firmly down around the probe immediately before reading.

- *Hysteresis check* The hysteresis is assessed by comparing readings before and after exposure to high temperatures. In this case the range extends below 0°C so that the change in ice point before and after the intercomparison is indicative of the width of the hysteresis loop.

Step 5 — Intercomparison

Since the intercomparison is to provide the data for a least-squares fit to a calibration equation (equation (4.10)) with four unknown constants, a total of 17 points, are measured giving more than 4 data points per constant. These points are distributed over the calibration range −20°C to 180°C. Columns 1, 2 and 3 in the following table summarise the intercomparison. The reference SPRT is interfaced to a computer through a high-accuracy a.c. bridge. At each calibration point the thermometer under test is read by the operator and the result typed into the computer. The computer then interrogates the bridge and calculates the temperature. Readings reported to 0.005°C are the average of two equally frequent readings. The two ice-point readings are added to the table after the intercomparison has been completed.

Reading no.	Temperature (°C)	Reading (°C)	Correction (°C)	Residual errors (°C)
1	−19.9504	−19.96	0.00	+0.0096
2	−7.4467	−7.46	+0.01	+0.0033

Continued on page 148

Continued from page 147

Reading no.	Temperature (°C)	Reading (°C)	Correction (°C)	Residual errors (°C)
3	5.0430	5.045	0.00	−0.0020
4	17.5320	17.535	0.00	−0.0030
5	30.0153	30.015	0.00	+0.0003
6	42.4994	42.50	0.00	−0.0006
7	43.9758	54.975	−0.01	+0.0108
8	67.5422	67.55	−0.01	+0.0022
9	80.0084	80.005	−0.01	+0.0034
10	92.4734	92.49	−0.01	−0.0066
11	104.9527	104.97	−0.02	+0.0027
12	117.4225	117.44	−0.02	+0.0025
13	129.8958	129.915	−0.02	+0.0008
14	142.3688	142.39	−0.01	−0.0062
15	154.8518	154.86	−0.01	+0.0018
16	167.3067	167.305	0.00	+0.0017
17	179.7642	179.755	+0.01	−0.0008
18	0.0000	0.00	0.00	0.0000
19	0.0000	0.005	0.00	−0.0050

Ice-point shift = 0.0050°C

Standard deviation of residuals = 0.0048°C

Step 6 — Analysis

The readings and temperatures recorded in the first three columns of the table are now analysed. The computer carries out a least-squares fit on the results, including the two ice points. As described in Section 2.10.2, the fit determines the values of the constants in equation (4.10) that best describe the measured data. Based on this equation the computer calculates the correction that should be applied to the readings at all the calibration points, rounding them to the nearest 0.01°C, the resolution of the thermometer. The corrections are listed in column 4 of the table and the residual errors in the corrected readings are listed in column 5.

Figure 4.15 graphs the results of the intercomparison and gives a visual display of the accuracy of the thermometer. There are two notable features of the calibration curve. Firstly, the non-linearity is quite evident, with some even (U-shaped) and some odd (S-shaped) non-linearity. Secondly, there is a 0.01°C step in the data at 0°C. This step feature is quite common in electronic thermometers that display both +0.00 and − 0.00, effectively adding an extra reading to the scale.

The table and graph are also examined in order to answer the following questions:

- Are there any large residual errors in the residual column that would indicate an incorrect reading or gross misbehaviour of the sensor?

- Are the residual errors of random sign? Randomness is a good indicator that the thermometer behaves as expected. A regular pattern of + and − signs is indicative of a resistance thermometer that has been damaged through poisoning or excessive moisture.

- What is the overall shape of the error curve? If the error is too great at any point the thermometer may need adjustment. This is also often evident from a large ice-point correction.

Continued on page 149

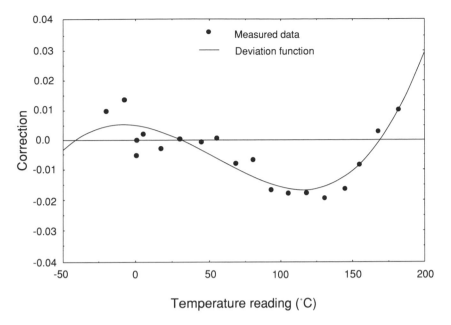

Figure 4.15
The calibration data and fitted deviation function for the thermometer of Example 4.7.

Continued from page 148

- How large is the standard deviation of the residual errors? The value should be typically between
  0.3 digits and 2 or 3 digits. At 0.3 digits the residual errors are entirely due to quantisation. At
  3 digits the errors are probably too large. Large and random residuals may be indicative of a poorly
  stirred bath or a faulty thermometer.

Step 7 — Uncertainties

The uncertainties are now analysed as discussed in Section 4.3.3.

- *Uncertainty in the reference-thermometer reading*   The uncertainty of the reference SPRT is estimated to be 2 mK;

$$U_{ref} = 0.002°C \ (95\% \ CL)$$

- *Variations in the stability and uniformity of the calibration bath*   From commissioning tests for the
  bath it is known that the gradients are less than 2 mK over the 300 mm wide controlled volume.
  Since the reference thermometer and the thermometer under test were within 100 mm, the maximum
  error is 0.6 mK, and the uncertainty is characterised as

$$U_{bath} = 0.0006°C \ (95\% \ CL)$$

- *Departures from the determined ITS-90 relationship*   This uncertainty is assessed as the standard
  error-of-fit, multiplied by the appropriate k value from Student's t-distribution corresponding to 95%

Continued on page 151

CALVIN, DeGRIES AND CO

1 Traceability Place, P O Box 31-310, Lower Hutt, New Zealand
Telephone (64) 4 566-6919   Fax (64) 4 569-0003

CALIBRATION CERTIFICATE

Report No.:   T92-2002.

Client: ACME Thermometer Co, 100 Celsius Avenue, P O Box 27-315,
                                   Wellington, New Zealand

Description of Thermometer: An electronic platinum resistance thermometer model RT200,
                                   manufactured by PEL, serial number 001, probe serial number
                                   SDL11.

Date of Calibration: 13 to 16 January 1992.

Method: The thermometer was compared with standard thermometers held
                                   by this laboratory. The temperature scale used is ITS-90.

Conditions: The probe was immersed to a minimum depth of 200 mm.

Results:

Thermometer reading (°C)	Correction (°C)
−19.6	0.00
−7.46	+0.01
5.04	0.00
17.54	0.00
30.01	0.00
42.50	0.00
54.97	−0.01
67.55	−0.01
80.00	−0.01
92.49	−0.01
104.97	−0.02
117.44	−0.02
129.91	−0.02
142.39	−0.01
154.86	−0.01
167.30	0.00
179.75	+0.01
0 (ice point)	0.00

Note: Corrections are added to the reading to obtain the true temperature.

Accuracy: The uncertainty in the corrected thermometer readings is estimated to be ±0.015°C
                at the 95% confidence level.

Checked: _____   **Signed:**   _____
                W Thomson   R Hooke

This report may only be reproduced in full.

Figure 4.16
A calibration certificate for an electronic reference thermometer based on the information in
Example 4.7.

— *Continued from page 149* —

confidence limits. There were a total of 19 data points and 4 unknown parameters, thus $\nu = 15$. It is found from Table 2.2 that $k = 2.3$, hence

$$U_{\text{fit}} = 0.0102°C \ (95\% \ \text{CL})$$

- *Uncertainty due to hysteresis and drift*   For platinum resistance thermometers operated at temperatures less than 200°C drift and hysteresis are caused by the same mechanism. The change in ice point before and after the calibration is therefore a good indicator of both (see Chapter 5). For this thermometer the measured change in the ice point was 5 mK, hence

$$U_{\text{drift}} = U_{\text{hyst}} = 0.005°C \ (95\% \ \text{CL})$$

- *Self-heating*   For platinum resistance thermometers which are not calibrated at zero current there is an additional uncertainty due to the likely variation in the self-heating between the calibration bath and the media in which the thermometer may be used (Section 5.6). For 100 Ω sheathed elements operated at 1 mA sensing current the variations are usually less than 2 mK, hence

$$U_{\text{self-heating}} = 0.002°C \ (95\% \ \text{CL})$$

- *Total uncertainty*   Summing all of these terms the total uncertainty is

$$U_{\text{tot}} = [2^2 + 0.6^2 + 10.2^2 + 5^2 + 5^2 + 2^2]^{1/2} = 13 \text{ mK}.$$

For the presentation in the certificate this is rounded to 0.015°C.

Step 8 — Complete records

The entire performance of the thermometer is reviewed before the decision is made to issue a certificate. The thermometer is found satisfactory and a certificate prepared, as shown in Figure 4.16.

FURTHER READING

Thermometer Calibration: A Model for State Calibration Laboratories, J. A. Wise and R. J. Soulen, NBS Monograph 174, US Department of Commerce (1986).
A useful guide for the establishment of calibration laboratories with information on laboratory equipment, layout and procedures.

ISO Guide 25   *Requirements for Technical Competence of calibration and Testing Laboratories.*
ISO 10012-1   *1991 Quality Assurance Requirements for Measuring Equipment Part 1: Management of Measuring Equipment.*

5

Platinum Resistance Thermometry

5.1 INTRODUCTION

Platinum resistance thermometers (PRTs) are remarkable instruments. In various forms they measure temperature over the range 14 K to 960°C, to an accuracy approaching 1 mK. They can be cycled repeatedly over several hundred degrees and still provide a very severe test of the best resistance bridges. Few material artefacts can be treated in this manner and remain as stable.

A wide range of PRTs is available, from the standard PRT (SPRT) defined by ITS-90 to more robust industrial PRTs which may be accurate to only a few tenths of a degree. The lower overall cost and higher accuracy of platinum thermometry make it the preferred means to measure temperature for many applications. In precision applications their accuracy is second to none.

Resistance thermometers are unlike other temperature sensors in that they require external stimulation in the form of a measuring current or voltage. This gives rise to errors associated with resistance-measuring instruments that must be considered in addition to those due to the sensor itself. Therefore this chapter covers resistance measurement as well as the construction, use and calibration of resistance thermometers. We include advice on the use and calibration of SPRTs, but this chapter is primarily about industrial PRTs. Those readers involved in scale maintenance should refer to Chapter 3. The two BIPM publications (see references at the end of the chapter) contain more detailed advice on scale maintenance and excellent bibliographies. The proposed IEC table giving the resistance as a function of t_{90} for industrial PRTs is incorporated as Appendix C.

The nature of electrical resistance in metals is discussed first, as this will aid the understanding of the properties and limitations of resistance thermometers.

5.1.1 Resistance in metals

All metals are good electrical conductors. This is because some of the electrons in metals are not bound to atoms and so are able to move freely. In other materials such as insulators, electrons are unable to move so freely, if at all. Let us consider a simple model which will help explain the electrical properties of metals.

Imagine the inside of a section of platinum wire as a huge lattice of platinum atoms (actually charged positive ions) all in neat rows in three dimensions. Amongst these atoms are electrons moving about at random until you apply a voltage, when they move to the positive terminal of the voltage source. The moving electrons constitute an electric current.

In a perfect lattice the electrons would be completely unimpeded in their movement, so that a perfect metal crystal would have no scattering and therefore zero electrical resistance. This is never observed in practice because there are two basic mechanisms that scatter the electrons and restrict their movement. One mechanism is due to temperature; the other is due to impurities and lattice defects.

5.1.2 The effects of temperature on resistance

The temperature of any material is a measure of the energy of motion of the atoms and electrons. In a crystal lattice the movement of atoms is very restricted: they cannot easily change their position in the lattice. They can, however, vibrate about their mean positions. It is this temperature-related vibration which is the major cause of electron scattering. As the temperature of the lattice increases, the vibrations increase and the scattering of the electrons increases.

When a voltage is applied to the ends of the wire the electrons move towards the positive terminal, all the time colliding with the vibrating atoms. The greater the voltage across the wire, the faster the electrons move, the greater the nett flow of electrons, and the greater the current. For all metals the relationship between the voltage, V, and current, I, follows:

$$I = V/R \qquad (5.1)$$

where R is a constant, known as the resistance, which depends on the amount of electron scattering. This relationship, known as *Ohm's law*, is followed so accurately by metals that it is possible to define and measure resistances to a few parts per billion (a few parts in 10^9).

As the temperature of the lattice increases, the vibrations, and hence the resistance to the flow of electrons, increase in proportion to the absolute temperature. The resistance–temperature relationship is usually written in terms of the Celsius temperature:

$$R(t) = R(0°C)(1 + \alpha t) \qquad (5.2)$$

where α is the temperature coefficient of resistance, approximately equal to $1/273.15 = 3.66 \times 10^{-3}\text{K}^{-1}$. Figure 5.1 shows that as a first approximation the model is very good.

5.1.3 The effects of impurities on resistance

Detailed understanding of resistivity beyond this simple model is extremely complicated. It is, for example, very difficult to predict the curvature in the resistance–temperature curves of Figure 5.1. One of the few simple improvements we can make to the model is to include the effects of impurities. If a different atom is placed in the lattice it will cause additional scattering of the electrons, in a manner which is almost independent of temperature. That is, impurities in the lattice tend to increase the resistance by a constant amount

$$R'(t) = R(t) + \Delta R \qquad (5.3)$$

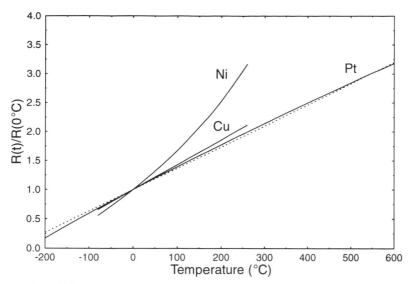

Figure 5.1
The resistance of nickel, copper and platinum as a function of temperature. The
dotted line is the resistance according to simple theory (equation (5.2)).

where ΔR is the resistance due to the impurities. This equation, known as *Mathiessen's rule*, can be rewritten

$$R'(t) = R'(0°C)(1 + \alpha't) \tag{5.4}$$

where

$$\alpha' = \alpha \frac{R(0°C)}{R(0°C) + \Delta R}. \tag{5.5}$$

That is, impurities increase the ice-point resistance and decrease the temperature coefficient of resistance. In turn this means that the higher the temperature coefficient of the metal, the purer it is.

The most important impurities are defects, i.e. points or planes in the lattice where atoms are missing or doubled-up. These are always present even in the purest of metals. Defects can be created very easily by working the metal, for example by bending, drawing or hammering: this breaks and distorts the lattice, forcing some atoms to become misplaced. This process also causes the metal to become harder and resist further deformation, and is therefore known as *work-hardening*.

A large proportion of defects can be removed relatively easily by annealing. This is accomplished by heating the lattice and exciting the atoms sufficiently so that they can fall back into place. However, if the lattice is heated further, atoms can jump out of place and create defects. This affects thermometers used at high temperatures. The concentration of defects quickly reaches a state of equilibrium where the rate of creation is equal to the rate of removal by annealing. Because the equilibrium concentration of defects increases with temperature, thermometers used at high temperatures must be cooled very slowly to ensure that these thermal defects are not quenched into the lattice and allowed to affect the resistance at lower temperatures.

5.1.4 Platinum resistance thermometers

All metals behave very much as the simple model suggests, but few metals are suitable as resistance thermometers. A good thermometer must be able to withstand high temperatures, be chemically inert, and be relatively easy to obtain in a pure form. Platinum is one of the few suitable metals.

In the early days of platinum thermometry, Callendar found that the resistance of platinum was fairly accurately described by a simple quadratic equation with constants A and B,

$$R(t)' = R(0°C)(1 + At + Bt^2).\tag{5.6}$$

Historically this was rewritten in an alternative form:

$$R(t) = R(0°C)\left(1 + \alpha t + \alpha\delta\left(\frac{t}{100}\right)\left(1 - \frac{t}{100}\right)\right)\tag{5.7}$$

which simplified the calculations required to determine the calibration constants α and δ. This form also explicitly defined the *alpha value* of the thermometer:

$$\alpha = \frac{R(100°C) - R(0°C)}{100R(0°C)}\tag{5.8}$$

which was readily determined from measurements at the ice point (0°C) and the steam point — then defined to be 100°C.

This measure of the α value is still used today as a measure of the purity of platinum and to define the various grades of platinum thermometer. Because the steam point is no longer defined by ITS-90, the alpha value is likely to be replaced by an alternative measure of purity, namely ρ (rho):

$$\rho = \frac{R(29.7646°C)}{R(0°C)}\tag{5.9}$$

where 29.7646°C is the melting point of gallium (see Figure 3.2).

It was later found by van Dusen that an additional correction term was required to describe the resistance–temperature relationship below 0°C:

$$R(t) = R(0°C)(1 + At + Bt^2 + C(t - 100)t^3)\tag{5.10}$$

where C is zero above 0°C.

This equation, known as the *Callendar–van Dusen equation*, was the basis for the temperature scales of 1927, 1948 and 1968, and continues to be used to define the resistance–temperature relationship for industrial resistance thermometers. Typical values for the coefficients for a standard PRT are

$$A = 3.985 \times 10^{-3}/°C$$
$$B = -5.85 \times 10^{-7}/°C^2$$
$$C = 4.27 \times 10^{-12}/°C^4$$

$$\alpha = 3.927 \times 10^{-3}/°C$$

$$\rho = 1.11814.$$

For an industrial PRT more typical values are

$$A = 3.908 \times 10^{-3}/°C$$

$$B = -5.80 \times 10^{-7}/°C^2$$

$$C = 4.27 \times 10^{-12}/°C^4$$

$$\alpha = 3.85 \times 13^{-3}/°C$$

$$\rho = 1.1158.$$

The constants are similar for different grades of platinum, with the α-value varying between the two values shown above. The α-value may also be expressed in several different ways, for example:

$$0.385 \ \Omega/°C \text{ for a } 100 \ \Omega \text{ PRT}$$

$$3.85 \times 10^{-3}/°C$$

$$0.385\%/°C$$

$$3850 \text{ ppm}/°C$$

all of which are equivalent. We later use an approximation $\alpha = 4 \times 10^{-3}/°C$ to estimate the magnitude of some of the errors in resistance thermometry.

5.2 CONSTRUCTION OF PLATINUM THERMOMETERS

The main aim when assembling a resistance thermometer is to ensure that the metal is allowed to respond to the temperature, while being unaffected by all other environmental factors. Other factors might be; corrosive chemicals, vibration, pressure, and humidity. For platinum thermometry the most serious concern is instability caused by mechanical shock and thermal expansion.

In its simplest form a resistance thermometer is a coil of wire loosely mounted on an insulating support. However, the thermometer is susceptible to mechanical shock when mounted in this way. Small knocks and vibration cause the unsupported parts of the wire to flex. This works the wire, introduces defects and increases the resistance.

The logical solution to this flexing problem is to support the wire fully by mounting it on a solid bobbin so that it is unable to flex. Now there is a different problem. When the thermometer is heated, the wire and the bobbin expand at different rates, causing the wire to be stretched or compressed. If the strain is small, the resulting deformation of the wire will be elastic and temporary — as with a rubber band. If the strain is too large then the deformation will be plastic — as with putty — and any dimensional changes will be permanent. This process also work-hardens the wire and further increases the resistance.

Platinum resistance thermometers therefore have many different forms, which make a compromise between mechanical robustness on one hand, and precision and susceptibility to mechanical shock on the other.

5.2.1 Standard platinum resistance thermometers

The construction and limitations of the three forms of SPRT are described in detail in Section 3.5. The long-stem SPRT, which is the most common, is basically a coil of very pure platinum wire loosely supported on a mica or quartz cross, and sheathed in a glass or quartz tube. Cleanliness of the various components is absolutely critical for these thermometers, especially at high temperatures, where contaminants migrate very quickly.

Standard thermometers are extremely delicate instruments: shock, vibration or any acceleration that causes the wire to flex will strain the wire and change its resistance. Large knocks have been known to cause errors of the order of 10 mK, while long exposure to vibration may cause errors as large as 100 mK. However, with care a SPRT can be used regularly for periods well in excess of a year with cumulative drifts of less than 1 mK.

5.2.2 Partially supported platinum thermometers

The extreme fragility of SPRTs generally limits their use to maintenance of the ITS-90 scale, calibration and the very highest accuracy applications. The first step in making a more robust PRT is to support the wire as much as possible while still allowing the wire to move. Two successful industrial resistance-element designs are shown in Figure 5.2. The first uses a bobbin formed from high-purity alumina to support a tightly wound helix of the platinum wire. The second supports the tightly wound helix inside the bore of a high-purity alumina insulator. The wire may be restrained further with alumina powder,

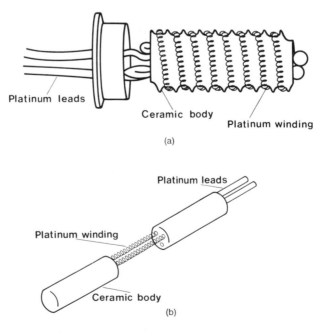

Figure 5.2
Two practical designs for partially supported PRTs.

which fills the spaces in the bores. In some designs cement fills about one-third of the insulator bores to further restrain the wire movement. In elements designed for aerospace applications, the case or sheath may be oil-filled to dampen vibrations.

According to the intended application and accuracy, there are three basic grades of wire used in partially supported PRTs. The user should be aware, however, that there are two internationally accepted documentary standards for platinum resistance elements and at least five national standards, all of which differ slightly. This is complicated by the re-publication and in some cases changing of the standards to conform to the ITS-90 definition of the scale. The wire can be manufactured as the highest purity grade, or manufactured from a moderately high-purity grade and doped with the required concentration of impurities. The three basic grades of wire are distinguished by their nominal alpha values.

- $\alpha = 3.926 \times 10^{-3}/^{\circ}C$ By using the same grade of wire as required by the ITS-90 these thermometers comply with the ITS-90 definition of an SPRT. However, the additional support of the wire degrades the performance of the thermometers to about ± 5 mK and reduces the maximum upper temperature exposure to about 500°C, depending on the sheath material.

- $\alpha = 3.916 \times 10^{-3}/^{\circ}C$ This grade is a compromise between the SPRT grade and the more common industrial grade. It is primarily an American standard for laboratory instruments. The main advantage over lower-grade industrial thermometers is the higher reproducibility between thermometers.

- $\alpha = 3.85 \times 10^{-3}/^{\circ}C$ This is the grade of wire used most commonly for industrial PRTs. The exact temperature dependence of the wire depends considerably on what metals are used to dope the wire. In most cases a rare metal from the same chemical family as platinum is used, so that the shape of the resistance–temperature curve and other physical properties are not too dissimilar to those for pure platinum. Nevertheless the curve is sufficiently different from the ITS-90 reference function to limit the fitting of calibration equations to about ± 10 mK. Usually a simple quadratic (the Callendar equation) is almost as good a fit as the ITS-90 function. Some of the best of these elements withstand intermittent use to 850°C with accuracies of a few tenths of a degree.

Overall the partially supported PRTs achieve typical accuracies between 2 and 20 mK, with a variable sensitivity to vibration and shock depending on the degree to which the wire is supported. There is often a small amount of hysteresis and drift caused by the different thermal expansion of the wire and ceramic substrate. The ice-point resistance values for partially supported thermometers are normally in the range 10–500 Ω, with the 100 Ω units being the most common. The dimensions of the elements are also varied with diameters between 0.9 and 4 mm and lengths between 6 and 50 mm.

5.2.3 Fully supported platinum thermometers

The most robust of the PRTs are fully supported elements mounted either in glass or alumina ceramic. If the wire is completely encapsulated the susceptibility to mechanical vibration and mechanical shock is made minimal. The penalty for increased robustness is a much lower long-term stability and large hysteresis due to thermal expansion and contraction.

Almost all fully supported PRTs are manufactured with the $\alpha = 3.85 \times 10^{-3}/°C$ grade wire. As with the partially supported PRTs, the temperature range depends very strongly on the sheath material, which is a major source of contaminants. Ceramic substrates have a temperature coefficient of expansion which is closer to that of platinum than glass is, so ceramic-based PRTs exhibit less hysteresis than glass elements. However, the cement used to bond the wire in ceramic elements is often porous, as is the ceramic itself, and so ceramic elements can be more susceptible to contamination. Ceramic elements are sometimes encapsulated in glass to overcome the porosity.

Glass elements, although low-cost and impervious to fluids, have a number of serious drawbacks. At high temperatures glasses undergo a rapid change in their coefficient of expansion which is associated with the softening of the glass. This causes hysteresis and work-hardening due the greatly increased strain on the platinum, as well as an increase in the mobility of the metal components in the glass (sodium, lead, boron, etc.). This in turn causes the glass to become electrically conductive, especially to a.c. current (see Section 5.4.11), and allows the metal atoms to contaminate the platinum.

Overall the fully supported PRTs achieve typical accuracies of 20–200 mK with minimal sensitivity to mechanical shock and vibration. The range of fully supported elements available is very similar to that for partially supported elements with ice-point resistances in the range 10–1000 Ω, with the 100 Ω units again being the most common. The elements also have diameters between 0.9 and 4 mm and lengths between 6 and 50 mm.

5.2.4 Platinum film thermometers

One of the disadvantages of the PRT elements manufactured from wire is that the construction does not lend itself to automation. Another disadvantage in some applications is the moderately long time-constant of 2–6 seconds. For some applications such as the control elements in household irons, where speed and low cost are important, the thick film element is an attractive alternative. Film elements are made most commonly by sputtering platinum onto an alumina substrate in a meandering pattern. The resistance is then trimmed to the nominal value, and the element coated with a glaze which provides protection. Film elements are about one-third of the cost of other elements and have time-constants as low as 0.2 seconds.

Flexible platinum film elements can also be made and are also very useful for applications requiring a fast response and for surface temperature measurements. In these elements the platinum is printed onto a high-temperature plastic substrate which allows the entire element to flex. Usually they are available with an adhesive back so that they can be attached to the surface. The temperature range is limited by the highest temperature the plastic will withstand, typically 150–200°C. As with all film PRTs, the lead wires have a tendency to break free, and must be restrained when the element is installed.

Overall the accuracy is similar to that of the fully supported elements, but over a slightly reduced temperature range. Because the wire is bonded to the substrate they are more susceptible to thermal expansion effects. The increased strain in film elements may also cause large departures from the resistance–temperature tables at high temperatures.

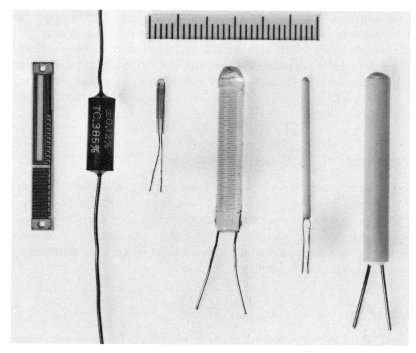

Figure 5.3
Examples of industrial platinum resistance thermometer elements. From the left: a thick film element, a PRT designed for printed circuit mounting, two glass PRTs and two ceramic PRTs.

Film thermometers usually have higher resistances than other types, ranging from 100 to 2000 Ω. The dimensions are highly variable, from 2 mm square to some in excess of 100 mm long, and there is also a range compatible with cylindrical fully supported elements.

5.2.5 Sheathing

The choice of sheath for platinum resistance elements is a key factor in determining the range and accuracy of the thermometer. There are two basic classes of sheathing materials: metallic such as stainless steel or inconel; and non-metallic, including glass, alumina, and quartz.

Metallic sheaths are the least fragile and easiest to manufacture. However, they are limited to use below 450°C and preferably below 250°C. At higher temperatures the metal atoms become very mobile and can contaminate the platinum wire. For use at temperatures above 250°C, stainless steel and inconel sheaths should be heat-treated in air or oxygen before assembly, to build an impervious layer of oxide on the inside of the sheath. This treatment prolongs the useful life of the assembly considerably. Glass elements and glass-encapsulated ceramic elements, which are less susceptible to contamination by the sheath, are more suited to operation above 250°C.

At temperatures above 450°C all platinum elements become increasingly susceptible to contamination and any metallic component of an assembly should be viewed as a source of impurities. As the temperature is increased above 450°C the sheath material must be correspondingly cleaner. At the highest temperatures only quartz and high-purity alumina are suitable sheaths. Both are normally baked at 1100°C to drive off impurities before the thermometer is assembled. Quartz sheaths do have additional problems with devitrification: some impurities cause the quartz to form small crystals rather than stay as a glass. This crystalline form is porous, as is alumina ceramic, and will allow contamination of the platinum.

5.2.6 Lead wires

As with the sheath, the lead wires which conduct the current to and from the thermometer should also be seen as a potential source of impurities. For the highest-temperature applications platinum is the only suitable lead wire, but for most applications the cost is not warranted. At low-temperatures (up to 250°C) glass-insulated copper or silver wire is used; the glass prevents oxidation of the wire. For high-temperature assemblies, nickel alloy or platinum-coated nickel wires are often used.

5.2.7 Insulation

The insulation for the lead wires is also a crucial component in the thermometer assembly. For the highest accuracy and stability, quartz spacers and supports, as used in high-temperature long-stem PRTs, are best. Partially supported PRTs are often assembled using four-bore alumina insulators, sometimes with alumina powder to restrict their movement in the sheath. The lowest grade insulation material is magnesia. This is usually found in sheaths assembled from mineral-insulated metal cable. Magnesia has the unfortunate property of absorbing moisture which leads to low values of insulation resistance and problems with moisture-induced hysteresis.

5.3 RESISTANCE MEASUREMENT

In order to realise the full potential of resistance thermometers, we must know how to measure resistance. To achieve an accuracy in temperature measurement of ±1°C, the resistance must be measured to better than 0.4%. Even an apparently ordinary temperature requirement is a non-trivial resistance measurement. Fortunately resistance measurement is a well-developed science and for most thermometry measurements the errors are not only well known but also reasonably simple to avoid.

In this section we give an overview of resistance measurement as it relates to platinum thermometry. It will provide the basis for understanding some of the sources of error and also for a critical assessment of the suitability of instruments for temperature measurement.

5.3.1 General principles

Ohm's law (equation (5.1)) suggests that resistance can be measured very simply by measuring the voltage across the resistance, and the current through it, and then calculating the ratio:

$$R = V/I. \tag{5.11}$$

However, electrical current is not easily measured or defined except in terms of a voltage and a second, known, resistance. In practice resistances are measured by comparison with other resistances, thus eliminating the need to know or measure the current accurately. There are two basic methods.

Potentiometric methods

Figure 5.4 shows a general potentiometric resistance measurement. The terminology *potentiometric* is derived from the fact that this was the circuit used to measure resistance when potentiometers were used to measure voltage. Nowadays high-impedance voltmeters replace the potentiometers. To measure a resistance in this way, a standard resistor and two good voltage measurements are required. A current is passed through both the standard resistor R_s and the unknown resistance $R(t)$. Since the current through the resistances is the same, the voltage across the resistances is in the ratio of the resistances:

$$R(t) = \frac{V_t}{V_s} R_s. \tag{5.12}$$

The essential features of the circuit are that the voltmeter draws no current from the circuit and, when properly assembled, the lead resistances as drawn in Figure 5.4 have no effect on the resistance measurement. This technique is particularly attractive to digital-multimeter manufacturers as most digital meters are based on integrated circuits which measure voltage ratios. Unfortunately few manufacturers of hand-held thermometers use this opportunity to provide a true four-lead resistance measurement.

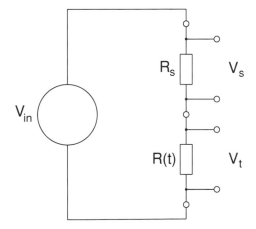

Figure 5.4
A potentiometric resistance measurement. When the circuit is properly assembled the effects of lead resistances are eliminated.

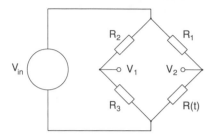

Figure 5.5
The Wheatstone bridge eliminates the
need to measure voltages accurately.

Bridge methods

The second group of resistance measurements is based on the Wheatstone bridge as shown
in Figure 5.5. A bridge measurement compares the output voltage from two voltage
dividers, one of which includes the resistance thermometer. The output voltage of the
bridge is

$$V_{out} = V_1 - V_2 = \frac{R_2 R(t) - R_3 R_1}{(R_2 + R_3)(R_1 + R(t))} V_{in}. \tag{5.13}$$

There are two modes of operation. In the balanced mode one of the bridge resistors
is adjusted until the output voltage is zero, then the unknown resistance is determined as

$$R(t) = \frac{R_3}{R_2} R_1. \tag{5.14}$$

When this relation is satisfied, the voltages from the two arms of the bridge are
equal, and the bridge is said to be balanced. The thermometer resistance can then be
determined in terms of three well-defined resistances. The advantage of this technique
is that the voltmeter only has to detect a null, greatly easing its accuracy requirements.
One of the disadvantages is that it is not as simple to eliminate the lead resistance of the
resistors. The accuracy demands on the variable resistor are also quite high, generally
making this option expensive. However, we shall see later that the balanced mode of
operation is ideally suited to high-accuracy a.c. measurements.

In the second mode of operation of the Wheatstone bridge, the variable resistor is
adjusted so that the bridge is balanced at one temperature, say the ice point, i.e. from
equation (5.14), $R(0°C) = R_3 R_1/R_2$. Then the output voltage becomes the measure of
temperature:

$$V_{out} = \frac{R(t) - R(0°C)}{(R_1 + R(0°C))(R_1 + R(t))} R_1 V_{in}. \tag{5.15}$$

Since the adjustment range for the variable resistor is only $\pm 0.1\%$ this bridge is much
cheaper than the balanced bridge. Now if R_1 is also large relative to $R(t)$ the output
voltage is approximately

$$V_{out} \approx \frac{V_{in}}{R_1} R(0°C) \, \alpha t \tag{5.16}$$

so that the output voltage is approximately proportional to temperature. This again is a common method for instruments measuring to about $\pm 1°$C. One disadvantage of this mode of operation is the additional nonlinearity introduced by the temperature dependence of the denominator of equation (5.15).

The main disadvantage of both bridge methods is the difficulty in eliminating the errors caused by the resistances of the lead wires.

5.3.2 Two-, three- and four-lead measurements

One of the most significant errors in resistance thermometry is caused by the resistance of the lead wires connecting the sensing element to the resistance-measuring instrument. Because the thermometer is usually several metres or more from the measuring instrument, and the lead resistances may be indistinguishable from the resistance of the PRT, significant errors can be introduced. For a two-lead measurement with a resistance R_L in each lead wire the error in the temperature measurement is approximately

$$\Delta T_m = \frac{2R_L}{\alpha R(0°C)}.$$
(5.17)

The error in a two-lead measurement of a 100 Ω thermometer with about 1 Ω total lead resistance is about 2.5°C. This is typical of most hand-held platinum thermometers. While some of this error can be compensated by treating the thermometer as a 101 Ω thermometer with a slightly reduced temperature coefficient, the measurement has no immunity to changes in the lead resistance. Such changes may be due to the temperature dependence of the wire leads, deterioration of plug and socket contacts, or deterioration of the cable as strands of the wire break. Most two-lead measurements are limited to accuracies of about $\pm 0.3°$C.

The ideal solution to the lead-resistance problem is to measure resistance by a four-lead method such as the potentiometric measurement of Figure 5.4. Because there is (almost) no current flowing in the leads to the voltmeter, there is no voltage drop due to the resistances in those leads, and therefore no error. If the voltmeter is not ideal and draws current from the circuit then there will be two sources of error: one due to the lead resistances, and one due to the inequality of the current in the two resistors. Note that the resistance of a four-lead resistor is well-defined, being the resistance between the points where the two pairs of leads meet.

Unfortunately the four-lead measurement principle is not easily incorporated into bridge instruments. Figures 5.6 and 5.7 show two common industrial approaches to the problem. In the first a three-lead resistance thermometer is used, so one lead resistance is introduced into each arm of the bridge. If the bridge is balanced (equation 5.14), if $R_3 = R(t)$, and if the lead resistances are equal, then there is no error. On the other hand, if any one of these three conditions is not satisfied there will be a small error.

The pseudo-four-lead bridge shown in Figure 5.7 is a very similar approach. A set of dummy lead wires is inserted into the other arm of the bridge to make the bridge balance correctly. As with the three-lead design, the bridge is sensitive to changes in one lead resistance that do not occur in the other, and to both lead resistances when the bridge is unbalanced. Both designs therefore provide no immunity to deterioration in the cables or plugs and sockets. The main advantage is that they can accommodate long cable lengths

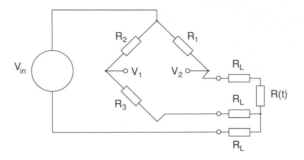

Figure 5.6
The three-lead bridge. By introducing one lead resistance into each arm of the bridge the lead-resistance error is reduced.

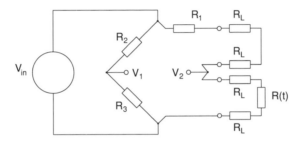

Figure 5.7
The pseudo-four-lead bridge. A set of dummy leads is used to reduce the lead-resistance error.

so long as the bridge is balanced. For this reason three-lead designs are very common in industrial temperature controllers where the leads are long and the control action seeks a bridge balance. Overall, the accuracy of three- and pseudo-four-lead measurements is only marginally better than that of two-lead measurements. True four-lead measurements should always be preferred.

5.3.3 D.C. resistance measurement

Potentiometric and bridge methods both rely on voltage measurements to establish either voltage ratios or the equality of voltages. Any extraneous voltages in these measurements can therefore cause errors. In d.c. systems there are three main sources of these extraneous voltages: thermoelectric effects, as with thermocouples; amplifier offset voltages and currents; and electrolytic effects, as in batteries.

Thermoelectric voltages are generated in conductors by temperature gradients. The voltage generated is the product of the temperature gradient and the thermoelectric constant of the wire (the Seebeck coefficient), this property being different for different materials. In an ideal resistance measurement all the lead wires have the same temperature profile, so the voltage generated in one lead will be equal to the voltage generated in all other leads. The meter will then measure the correct voltage difference across the PRT. However, if the materials differ, for example the lead-wire changes from platinum to copper, and the temperature profile across the leads is different, then the thermoelectric voltages will not balance. This leads to an error which appears to depend on the relative temperatures of the platinum–copper junctions. Since the difference in the Seebeck coefficients for platinum and copper is about 7 $\mu V/^\circ C$, the error in a typical measurement of a resistance with a 1 mA measuring current corresponds to about $0.02^\circ C$ error per degree Celsius difference in the junction temperatures. Thermoelectric effects are particularly troublesome at exposed instrument terminals which are subject to heating by convection and radiation.

In an ideal voltmeter the reading is zero when both of the input connections are held at zero potential. Any non-zero reading under these conditions measures the input *offset voltage* of the meter. The offset voltage is additive for all voltage measurements so will affect both the voltage ratio in potentiometric systems and the null measurement in bridge systems. For most modern electronic meters the offset voltage may be between 0.1 and 40 μV, causing errors of up to $0.1^\circ C$. In practice it is the temperature dependence of offset voltages that limits the performance of d.c. instruments. Accurate d.c. instruments are usually restricted to temperature-controlled laboratories.

When thermometers are operated in wet environments there is the possibility of electrolytic activity. This occurs if there is any moisture connecting the lead-wires to any earthed metal in the vicinity. The metal and the lead wires will behave as a small battery and cause currents to flow along the lead-wires to the meter or through the PRT resistance, in either case causing a significant and generally unpredictable error. In a wet environment all effort must be made to ensure that there is no electrical connection between the leads and the outside world other than through the measuring instrument itself. Errors due to electrolytic effects are normally seen as very noisy and erratic readings.

Overall the combination of thermoelectric effects and offset voltages limit simple d.c. measurements to accuracies of about $\pm 0.02^\circ C$.

5.3.4 A.C. resistance measurement

All of the d.c. errors described above are independent of the measuring current. Thus by reversing the current systematically and averaging the readings, all of these errors should disappear. This is the principle behind a.c. resistance measurement although, in practice, the measuring current or voltage is usually not chopped d.c. but true sinusoidal a.c.

The use of a.c. techniques has additional benefits. Firstly, the offset voltage of many meters and amplifiers varies erratically, exhibiting so-called *1/f noise*. By going to frequencies above a few tens of hertz the $1/f$ noise is all but eliminated. Secondly, at frequencies above a few tens of hertz, transformers can be used to establish extremely accurate ratios of a.c. voltage.

Figure 5.8 shows a simplified diagram of an a.c. resistance bridge. One arm of a conventional Wheatstone bridge is replaced by a variable-ratio transformer. In use the transformer turns-ratio is varied to obtain a bridge balance as determined by a null detector so that

$$R(t) = \frac{N_2}{N_1} R_s. \tag{5.18}$$

With the best a.c. bridges multistage transformers are used to obtain an effective number of turns exceeding 100 000 000 so that resistances can be measured with a

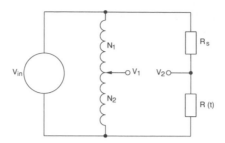

Figure 5.8
A simple schematic diagram of an a.c. resistance bridge. One of the arms of the Wheatstone bridge is replaced by a variable ratio transformer.

precision corresponding to 25 μK. With these designs it is absolutely essential that the lead resistances are eliminated.

Overall the temperature accuracy of a.c. techniques is limited by the performance of the thermometers, although there are a number of errors peculiar to a.c. measurement that users will need to be aware of (see Section 5.4.11). The bridges employing a.c. techniques are also expensive, in part because of the transformers used and in part because of the extra electronic circuitry required to balance the bridge automatically. Manual-balancing a.c. bridges are much cheaper but can be frustrating to use.

5.4 ERRORS IN RESISTANCE THERMOMETRY

The errors in resistance thermometry fall naturally into four main groups:

- thermal contact errors due to immersion, settling, lag and radiation which are much as discussed in Section 4.4;
- sensor errors due to mechanical, chemical, electrical and thermal effects;
- signal transmission errors due to lead resistances, electromagnetic interference, and thermoelectric effects;
- signal processing errors due to imperfect standard resistors and linearisation.

Most of the errors are easily recognised from simple tests and fall naturally into the general measurement model given in Figure 2.10. It is relatively easy to identify the causes of errors and to separate the various functional elements in the PRT's construction, and it is this clarity of operation which has allowed the PRT to develop into such a reliable and accurate thermometer.

5.4.1 Immersion errors

The immersion considerations for platinum resistance thermometers are relatively straight-forward and follow the general guide given in Section 4.4.1. The main concern with PRTs is that the sensing element is not small, so that extra immersion is required. Also, except for flexible-film types, PRTs are generally too bulky for surface temperature measurement.

Ice points can be difficult, particularly with stainless steel sheathed PRTs: the combination of the thermally conductive sheath and the self-heating can make it difficult to realise the ice point to better than $\pm 0.01°C$. It is important that the ice is very fine and well packed. For measurements requiring accuracies better than $\pm 0.01°C$ a triple point should be used in preference.

Standard PRTs also have quite demanding immersion requirements simply because of the high-precision required. At the zinc point, for example, the SPRT is required to measure the temperature to about 0.0001% (0.5 mK in 400°C) which requires immersion of about 14 diameters. Additionally the open structure of the assembly and transparent sheath make the effective length of the sensing element longer than just the length of the coil of wire. (See Section 5.4.3 on radiation errors.)

5.4.2 Lag and settling errors

The errors due to the response time of PRTs follow the general guide given in Sections 4.4.2–4.4.5. Time-constants for PRTs vary considerably: 0.2 seconds for film types; 2–6 seconds for larger fully and partially supported types; 5 seconds for standard PRTs; and 5–20 seconds for stainless steel sheathed assemblies. Additionally many of the larger sheathed assemblies exhibit a second and longer time-constant. Thus 95% of the settling may occur very quickly in perhaps 20 seconds while the remaining 5% of the error takes minutes to die away. A simple experiment, such as withdrawing and reinserting the thermometer, will normally reveal any problems.

5.4.3 Radiation errors

The most common situations in which radiation errors affect measurements made with PRTs are covered in Section 4.4.6. For glass or quartz sheathed PRTs, however, the considerations go beyond those for other thermometers.

The sheath of the long-stem PRTs provides a transparent 'light pipe' along which radiation can carry heat to and from the sensing element. Thus the PRT is not only in thermal contact with the medium immediately surrounding the platinum element but also in radiative contact with whatever it 'sees' down the sheath. At low temperatures this error will cause temperature readings to be high: room lights, for example, will heat an SPRT in a water triple point by a few tenths of millikelvin. At high temperatures the error will cause readings to be low, by an amount in excess of 30 mK at the aluminium point ($\sim 660°C$) and 5 mK at the zinc point ($\sim 420°C$).

The error can be substantially reduced by roughening the lower part of the thermometer sheath from just above the sensor for about 20 cm. This can be done either by sandblasting or by coating with graphite paint. Note that the thermometer still 'sees' the lower portion of the sheath, so long-stem PRTs have slightly more demanding immersion requirements than other thermometers.

5.4.4 Self-heating

Because a current is passed through the sensing element to measure its resistance, the element dissipates heat which in turn causes the temperature of the element to increase.

This self-heating error is very simply modelled as the power dissipated divided by the dissipation constant h. Thus the error in the temperature measurement is

$$\Delta T_m = R(t)I^2/h \tag{5.19}$$

where $R(t)$ is the resistance of the sensing element and I is the sensing current. The dissipation constant h is normally expressed in milliwatts per degree Celsius. The dissipation constant may also be expressed as the self-heating coefficient $s = 1/h$. Hence

$$\Delta T_m = sR(t)I^2. \tag{5.20}$$

The self-heating coefficient is normally given in kelvins per milliwatt.

Example 5.1   Self-heating of a sheathed PRT

Estimate the self-heating of a 100 Ω stainless steel sheathed PRT at 80°C in a water bath, operated at a sensing current of 1 mA. The manufacturer's specification for the dissipation constant is 30 mW/°C in water moving at 1 m/s.

From equation (5.2), the resistance of the element at 80°C is about 130 Ω, hence by applying equation (5.19) we obtain

$$\Delta T = \frac{130 \times (0.001)^2 \times 1000}{30} \text{K}$$

$$= 4.3 \text{ mK}.$$

The factor of 1000 in the numerator converts the power unit from watts to milliwatts.

Exercise 5.1

A bare 100 Ω detector element is used to measure air temperature near 40°C. The manufacturer's specification for the dissipation constant in still air is 1.3 mW/°C. Estimate the self-heating when the sensing current is (a) 1 mA; (b) 2.5 mA.

The range of typical values for h is wide, varying from 1 mW/°C for very small film elements in still air to 1000 mW/°C for large wire-wound elements in moving water. Table 5.1 shows the ranges of values of h and the self-heating error for different sensing elements in air and water.

Because the error increases as the square of the current, the current is probably the most significant factor in self-heating. For example, the errors in Table 5.1 are given for a 1 mA sensing current, and for typical applications the error is quite tolerable. However, for most PRT elements the sensing current may be as large as 10 mA, for which the errors would be 100 times greater, and the error then becomes a problem in almost every situation. As a rule most PRTs are operated at power dissipations of less than 1 mW; for a 100 Ω sensor 1 to 2 mA is a typical sensing current.

One of the problems with the self-heating error is that it is highly dependent on the immediate environment of the thermometer. The sheathing of elements may increase the

Table 5.1. The typical range of dissipation constants for unsheathed platinum resistance elements. The error is calculated for 100 Ω elements and 1 mA sensing current.

Condition	Dissipation constant (mW/K)	Self-heating coefficient (K/mW)	Error (mK)
Still air	1-10	0.1-1	10-100
Still water	2-400	0.0025-0.5	0.25-50
Moving water	10-1000	0.001-0.01	0.1-10

error by as much as a factor of five times, and use in air by as much as 100 times. Clearly it is not possible to improve the accuracy of a measurement significantly by applying a correction based on the manufacturer's estimate of the dissipation constant. The specification is indicative only.

Corrections for self-heating can be made by altering the sensing current and making a second measurement. The pair of results can then be used with equation (5.20) to calculate the zero-current reading (Exercise 5.2). For a pair of readings T_1 and T_2, made with currents I_1 and I_2, the zero-current reading is

$$T_0 = T_1 - \frac{I_1^2}{I_1^2 - I_2^2}(T_1 - T_2). \tag{5.21}$$

The correction formulae for common ratios of I_1 and I_2 are tabulated below.

I_2	T_0
$\sqrt{2}I_1$	$T_1 - (T_2 - T_1)$
$I_1/\sqrt{2}$	$T_2 - (T_1 - T_2)$
$2I_1$	$T_1 - (T_2 - T_1)/3$
$I_1/2$	$T_2 - (T_1 - T_2)/3$

Exercise 5.2

(a) Use equation (5.20) to derive equation (5.21).

(b) Assuming the uncertainties in T_1 and T_2 are σ_T, show that the uncertainty in the corrected temperature T_0 is

$$\sigma_{T_0} = \frac{(I_1^4 + I_2^4)^{1/2}}{|I_2^2 - I_1^2|}\sigma_T.$$

In making the correction it is assumed that the temperature T_0 does not change. This is the case when PRTs are used in fixed points; indeed the ITS-90 scale is defined entirely in terms of the zero-current resistance of SPRTs. For most SPRTs the self-heating effect is between 0.3 mK and 2 mK, depending on the fixed point. Unfortunately, in practice there are few other situations where the temperature is sufficiently stable to allow accurate corrections to be applied.

5.4.5 Mechanical shock and vibration

Vibration and mechanical shock are two of the main contributors to long-term drift in PRTs. Any acceleration of the thermometer will cause unsupported wire to flex against the supports or substrate. The flexing in turn causes work-hardening and an increase in the ice-point resistance of the thermometer. The example given earlier of standard thermometers drifting 100 mK due to vibration highlights the importance of choosing the appropriate PRT element for use in a high-vibration environment.

While most manufacturers specify the shock and vibration that the PRTs will withstand, the specifications are usually for a once-only event. No PRT will withstand indefinite exposure. In all cases an environment with excessive vibration will cause the ice-point resistance to increase, so regular checks are necessary to detect the problem before an element fatigues and fails completely. Thermometers should also be mounted so as to isolate the source of the vibration.

5.4.6 Thermal expansion effects

In industrial PRTs deformation of the wire due to various thermally driven mechanical effects is the single greatest source of error and uncertainty. There are two main effects, both caused by the differential expansion of the platinum wire and the substrate: firstly, elastic deformation, which gives rise to hysteresis; and secondly, plastic deformation and work-hardening, which give rise to drift.

All materials change their dimensions with temperature. For platinum wire this change is about 9 ppm (0.009 mm per m) for every degree change in temperature. Similarly all the materials used as substrates for PRTs expand or contract with temperature. Ideally the substrate should expand and contract at exactly the same rate as the platinum. This would ensure that there would be no strain on the wire. The two most common substrates, glass and alumina ceramic, very nearly satisfy this requirement.

Glasses designed to support platinum thermometers usually have a coefficient of expansion between 8 and 10 ppm/°C, which for most purposes is a good match. One of the problems with glass is that the coefficient of expansion increases by a factor of three or more above the softening temperature of the glass. The softening temperature is typically 400°C to 500°C but for some glasses it is as low as 250°C.

Ceramic alumina substrates also have expansion coefficients of about 8–10 ppm/°C. The structure of ceramics is quite different from that of glass: they are not a uniform solid but a mass of very small crystals bonded together. Alumina crystals have several forms, each of which has a different coefficient of expansion. Further, the crystals have different coefficients of expansion depending on alignment. For individual alumina crystals the coefficient varies between about +13 and −5 ppm/°C, both extremes being very different from that of platinum. In general the finer the raw alumina used to make the ceramic, the more uniform the coefficient of expansion. The porosity of the ceramic also decreases. The final coefficient of expansion is also process-dependent. Thus although alumina is better than glass in respect of electrical resistivity and purity, it is porous and may have a non-uniform and slightly unpredictable coefficient of expansion.

Elastic deformation and hysteresis

For both types of substrate there will be some differential thermal expansion, typically 1 ppm/°C or less for the better substrates. As the wire is stretched the length of the wire increases. The dimensional changes are not permanent deformations since the wire is elastic for small strains, and as soon as the strain is released the wire returns to its original shape. Because the lattice is distorted as the wire is stretched, a differential expansion coefficient of 1 ppm/°C results in increases or decreases of about 5 ppm/°C in the temperature coefficient of platinum. Since the temperature coefficient is about 4000 ppm/°C, the error introduced is usually within ±0.12% of the temperature change.

In most elements the substrate will be unable to maintain the strain on the wire; this allows the wire to relax and slip against the substrate. A thermometer undergoing stretching on the way to high temperatures will first relax and undergo compression as it returns to low-temperatures. This gives rise to hysteresis as shown in Figure 5.9. Some PRTs also exhibit relaxation with time, with a relaxation period as long as several hours. The relaxation can also give rise to erratic behaviour if it occurs in fits and starts. Above 250°C most PRTs exhibit relaxation as the strain is removed by annealing.

The typical hysteresis in the fully supported PRT of Figure 5.9 ranges between ±0.02% and ±0.05% and is very dependent on the range. For the best partially supported PRTs the maximum strain that the substrate (alumina powder) will support is low, so the hysteresis may be as low as 0.0002%, almost as good as SPRTs. Curiously hysteresis is also quite low (0.01%) in some thick-film elements because the platinum is bonded to the substrate and is not as free to relax.

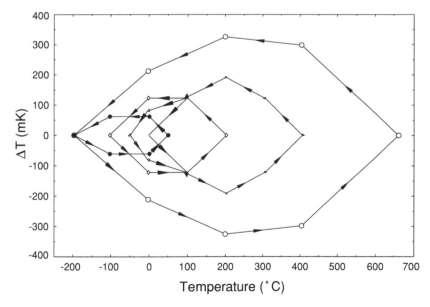

Figure 5.9
Hysteresis in a fully supported industrial PRT. The hysteresis error is proportional to the temperature range covered.

Plastic deformation and drift

Metals including platinum cannot be stretched indefinitely; once the strain exceeds about 0.1% the metal yields, and the deformation is said to be plastic. Superficially this would not seem to be a problem; resistance elements with a differential thermal expansion of about 1 ppm/°C would have to be cycled about 1000°C to reach the required strain levels. However, the platinum wire is not supported uniformly along its length. For example, platinum wire supported on an alumina insulator, which is microscopically rough, may be supported by only a few percent of its surface area. Thus very small localised areas of the platinum wire are subject to very high strain and undergo plastic deformation on every cycle.

Plastic deformation has two detrimental effects. Firstly, it permanently changes the dimensions of the wire. Secondly, the deformation introduces defects into the wire as the crystal structure is deformed and fractured. Both effects increase the ice-point resistance of the wire and can be distinguished by determining the temperature coefficient. If the ice-point resistance increase is associated with a decrease in temperature coefficient, then by Mathiessen's rule (Section 5.1.3) the increase in resistance is probably due to defects. In this case the increase in ice-point resistance can be removed by annealing. Usually both effects are present, so that annealing will not completely restore the thermometer to its original condition.

Because the rate of plastic deformation endured by the wire is closely related to the amount of hysteresis, the amount of hysteresis is a good indicator of the likely future drift of the thermometer. Experience shows that in the absence of contamination few thermometers have an annual drift that exceeds the hysteresis.

5.4.7 Other thermal effects

Thermal expansion affects not only the sensing element but also the lead wires, and lead-wire expansion is potentially much more damaging. For both stainless steel and quartz sheathed PRTs, differential thermal expansion can easily cause the lead-wires to be strained beyond their yield point. High-temperature SPRTs are probably the extreme example; the thermal expansion of quartz is close to zero so the differential expansion is about 10 ppm/°C. Over 800 mm of sheath and a 960°C cycle to the silver point, the leads expand nearly 8 mm! It is absolutely essential that the leads are allowed to move freely to prevent tangles and breaks.

Stainless steel sheaths are a little more forgiving in that the coefficient of expansion of steel is about 16 ppm/°C, so that the differential expansion is less ($\sim$ 7 ppm/°C). They are not suitable for temperatures much above 400°C. As a general rule any industrial thermometer used above 400°C with frequent cycling has a limited life, since the thermal expansion is sufficient to fatigue the lead wires, so they eventually break.

At high temperatures the thermal energy (lattice vibrations) is sufficient to cause atoms to form dislocations and other defects. The equilibrium concentration of defects, η, usually grows exponentially with increasing temperature according to

$$\eta = \eta_0 \exp(-E_d/T) \tag{5.22}$$

where η_0 is a constant and E_d is related to the energy required to create the defect. At temperatures above 600°C the thermal defects make a significant (>1 mK) contribution to the resistance. In order to get repeatable results above 600°C time must be allowed for the defect concentration to come to equilibrium. Similarly, the high-temperature defect concentration is sufficient to upset the resistance at lower temperatures, if they are allowed to remain. Therefore standard thermometers used at high temperatures must be cooled slowly to allow the defects to anneal out of the metal. To cool a thermometer from 960°C to 450°C, for example, requires in excess of six hours to ensure that the defects are not quenched in. The sheaths of high-temperature PRTs also become very fragile with exposure to temperatures above 600°C, especially when they are hot.

5.4.8 Contamination

At temperatures above 250°C platinum thermometers become progressively more susceptible to contamination. The effect of the contaminants is to increase the impurities in the metal and hence increase the resistance. If the level of impurities is high the resulting departures from the resistance tables can be in excess of several degrees, effectively destroying the thermometer. The damage is irreparable since, unlike crystal defects, the impurities cannot be removed by annealing.

Probably the most common cause of contamination is the migration of iron, manganese and chromium from stainless steel and inconel® sheaths. An overnight exposure of an unprotected ceramic element at 500°C can easily cause several degrees error. The migration of contaminants can be reduced by heat treating the sheaths in air or oxygen before the thermometer is assembled. This builds a layer of metal oxide which is relatively impervious to metal atoms. This allows intermittent exposure to 450°C.

Above 450°C ceramic elements require additional protection. The main weakness of the ceramic elements is that the ceramic is porous, particularly where the lead wires are cemented into the substrate. Glass substrates on the other hand are very effective at blocking the migration of impurities. Some manufacturers supply glass-encapsulated partially supported elements, which have the advantages of both ceramic and glass types. In some cases glass inner sheaths are used to protect ceramic elements.

Glass and glass-encapsulated elements are prone to contamination from within the element itself. Above the softening point of the glass, typically 400°C to 500°C, the metallic constituents of the glass are able to move readily. Therefore glass elements should never be used above the softening point. Unfortunately few manufacturers are prepared to supply information on the softening points of their glasses, some of which are usable up to 600°C. In principle the onset of the softening point can be detected by comparing a.c. and d.c. resistance measurements of the elements, but this is rarely practical.

Above 500°C the only robust strategy for preventing contamination is to use quartz (fused silica) sheaths. Very high-purity alumina may also be usable if there are no other metallic contaminants in the environment. For high-temperature SPRTs even a quartz sheath is insufficient. Above 800°C the only reliable protection for the PRTs is a platinum sheath. When used in the silver point, the platinum sheath (0.2 mm thick!) is mounted in the graphite well of the fixed point and protected from mechanical damage by a second quartz well.

All contamination causes an increase in the ice-point resistance of the PRTs. When significant, it also causes changes in the curvature of the resistance–temperature curve. Any probe exhibiting a large ice-point change that cannot be removed by annealing has usually been contaminated and should be discarded as unreliable.

5.4.9 Compensation and assessment of drift

The cumulative effect of work-hardening, contamination and plastic deformation is to increase the resistance of the PRT element. By following Mathiessen's rule (equation (5.3))
and assuming only linear dependence of resistance with temperature, we can use the
change in ice-point resistance to assess the likely temperature errors. The drift-affected
resistance is

$$R'(t) = (R(0°C) + \Delta R_d)(1 + \alpha t) + \Delta R_i \tag{5.23}$$

where ΔR_d is the change in resistance due to dimensional changes and ΔR_i is the change
induced by impurities. The error in temperature determined from measurements of $R'(t)$
depends on how the temperature is calculated. There are three methods:

Method 1   *Assume no change in ice-point resistance.* Using the original value for
the ice-point resistance, assuming a linear resistance–temperature relationship, the temperature is calculated as

$$t' = \frac{1}{\alpha}(W'(t) - 1), \tag{5.24}$$

where

$$W'(t) = R'(t)/R(0°C) \tag{5.25}$$

for which the error is

$$\Delta t_m = t' - t = \frac{\Delta R_d + \Delta R_i}{\alpha R(0°C)} + \frac{\Delta R_d}{R(0°C)}t. \tag{5.26}$$

The first term of equation (5.26) is a constant-temperature error due to the ice-point shift,
while the second describes the effect on the temperature coefficient.
   Method 2   *Use the most recent value of the ice-point resistance.* The error indicated
by equation (5.26) can be reduced considerably by using the most recent value of the
ice-point resistance to calculate the temperature, i.e. let

$$W'(t) = R'(t)/R'(0°C). \tag{5.27}$$

Then the error is

$$\Delta t_m = -\frac{\Delta R_i}{R'(0°C)}t. \tag{5.28}$$

For PRTs operated below 250°C this method reduces the errors considerably. For thermometers used above 250°C it becomes increasingly likely that the ice-point shift is due
to contamination, and a variation on this method is advantageous.
   Method 3   *Subtract the ice-point shift from the reading.* By using a value for $W'(t)$
of

$$W'(t) = \frac{R'(t) - \Delta R}{R(0°C)} \tag{5.29}$$

where $\Delta R = \Delta R_i + \Delta R_d$ is the total ice point shift, the effect of impurities is minimised.
The error in this case is

$$\Delta t_m = +\frac{\Delta R_d}{R(0°C)}t, \tag{5.30}$$

which depends only on dimensional changes induced by plastic deformation.

An example will demonstrate an error assessment based on these equations.

Example 5.2   The assessment of drift errors

Assess the likely errors caused by a 0.1 Ω shift in the ice-point resistance of a 100 Ω PRT. Evaluate the two extremes of the likely error at 100°C and 500°C, by substituting 0.1 Ω for ΔR_d and ΔR_i in the above equation. Use $\alpha R(0°C) = 0.4 \ \Omega/°C$.

The results are summarised in Table 5.2.

Table 5.2.   The temperature error due to a 0.1% change in ice point resistance versus the three methods of calculating $W'(t) = R(t)/R(0°C)$.

	$t = 100°C$		$t = 500°C$	
	min ($\Delta R_i = 0.1$)	max ($\Delta R_d = 0.1$)	min ($\Delta R_i = 0.1$)	max ($\Delta R_d = 0.1$)
Method 1: (equation (5.26))	+0.25	+0.35	+0.25	+0.75
Method 2: (equation (5.28))	−0.1	0.0	−0.5	0.0
Method 3: (equation (5.30))	0.0	+0.1	0.0	+0.5

Table 5.2 confirms that at high temperatures ($> 250°C$) Method 3 is probably the best since it is insensitive to impurity-induced resistance changes. At low-temperatures both Method 2 and Method 3 are suitable. Note that a combination of Methods 2 and 3 would produce error limits with a zero mean. Method 2 is the simplest and also reduces any error due to the inaccuracy of the standard resistors (Section 5.4.15).

Exercise 5.3   Uncertainty assessment for drift

Carry out a Type B uncertainty assessment for drift in a measurement carried out at 300°C: (a) assuming that Method 3 is used; (b) assuming that Method 1 is used. The ice-point shift is 0.1 Ω.

5.4.10 Leakage effects

Accurate resistance measurements require all of the measuring current to pass through the PRT element. This is relatively easy at low temperatures, where insulators have a very high resistance. However, at high temperatures even the very best insulators break down and form a short circuit around the sensing element. Moisture is the other main cause of leakage effects, particularly at lower temperatures where there is insufficient heat to drive the water out of the assembly.

The effect of any leakage resistance on the measurement is well modelled by a leakage resistance in parallel with the sensing resistance. The total resistance of the assembly $R'(t)$ is

$$R'(t) = \frac{R(t)R_{\text{ins}}}{(R(t) + R_{\text{ins}})} \qquad (5.31)$$

where $R(t)$ is the resistance of the PRT alone and R_{ins} is the resistance of the insulation; ideally R_{ins} is infinite. For large values of the insulation resistance, equation (5.31) is well approximated by

$$R'(t) = R(t) \left(1 - \frac{R(t)}{R_{ins}} \right).$$ (5.32)

Using a value for alpha for the PRT of 4000 ppm/°C we can then estimate the temperature error (in °C) due to poor insulation as

$$\Delta t_m \approx -(250 + t) R(t)/R_{ins}.$$ (5.33)

This also allows us to calculate the minimum insulation resistance for a given maximum temperature error:

$$R_{ins,min} = -\frac{(250 + t) R(t)}{\Delta t_{m,max}}.$$ (5.34)

Example 5.3 *Calculation of errors due to insulation resistance*

Calculate the error due to a leakage resistance of 10 MΩ on a 100 Ω sensor at 0°C.

Substitution into equation (5.33) yields

$$\Delta t_m = \frac{250 \times 100}{10\,000\,000} K$$

$$= 2.5 \text{ mK}.$$

Exercise 5.4

Calculate the minimum insulation resistance which ensures that the leakage error is less than 0.01°C at 400°C. The sensor is a 100 Ω PRT.

Specifications vary but most PRT standards require an insulation resistance at 0°C that exceeds 100 to 1000 MΩ; this is usually measured at 100 V d.c. The reason why the resistance of the insulator has to be so high at 0°C is that it does not remain high as the temperature increases. In fact, insulators behave very differently from metals and their resistance decreases rapidly with increasing temperature according to the exponential relationship

$$R_{ins} = R_0 \exp(E_d/T).$$ (5.35)

The similarity of this equation to equation (5.22) for the defect concentration in metals is not a coincidence; it is the thermally generated defects in insulators that aid electrical conduction.

As the temperature rating of the thermometer increases, the purity requirements on the insulation also increase. It also becomes advantageous to use elements with a low ice-point resistance. This is why high-temperature SPRTs have ice-point resistances of 0.25 Ω and use only high-purity quartz for the sheaths and substrates.

The main cause of leakage errors at low temperatures is moisture. The combination of soluble impurities and moisture in a thermometer assembly can, in extreme cases, cause errors of several degrees.

Moisture is also a major cause of hysteresis in thermometer assemblies. The hysteresis effects depend on how the thermometer is constructed. The parts of the assembly that are particularly prone to leakage resistances are: the connections between the element and lead wires, the connections between the lead wires and the flexible cable, and if a ceramic element is used, the element itself. When a thermometer is left unused for a period the water diffuses evenly throughout the assembly. This even distribution will correspond to a particular value for the leakage error. When the thermometer is used, the distribution of the moisture changes as it diffuses to the cooler parts of the thermometer. Thus the leakage error will vary strongly with use of the thermometer. Ceramic elements are particularly susceptible because they are porous and allow the water access to the platinum winding, which has a large surface area.

Moisture is a problem for almost all industrial assemblies because it is almost impossible to make a reliable, low-cost and airtight seal on steel sheathed thermometers. As a thermometer is cycled, the air within the assembly expands and contracts, and with time moisture is drawn into the assembly. This is a serious problem for the thermometers with magnesia insulation, which has a strong affinity for water.

Some thermometer manufacturers pack the sensing element in thermally conducting grease to prevent the ingress of moisture into the ceramic elements and to improve the thermal response times of the thermometers. However, this can seriously damage the ceramic elements. Because ceramic is porous the grease will gradually invade the pores and cause the wire to be stressed as the ceramic swells. With thick film elements the grease will strip the thin outer layer of ceramic. Thermal grease should be used only on working thermometers exposed to very wet environments and not cycled to temperatures above 100°C where the viscosity of the grease is low. Thermometers using glass elements are better suited to these applications.

5.4.11 A.C. leakage effects

Alternating current measurements are different from d.c. measurements because energy may be dissipated by the alternating electromagnetic fields around the conductors (resistors, lead wires etc.). To make a high-quality a.c. measurement we must consider not only the conductors in a circuit, but also their placement and the materials between them. The main concern is with the substrate materials and insulators.

Glass elements exhibit the largest a.c. leakage effects in thermometry. These effects also occur in many other substrates, including quartz and alumina. With glasses, however, the effects are more pronounced because they occur at low-temperatures within the operating range of industrial PRTs (250°C upward). As a general rule, glass elements should not be used on a.c. systems unless it can be proved that a.c. leakage is absent.

Glass elements used at temperatures near the softening point (400°C to 500°C) become highly conductive to a.c. as the metallic ions become mobile. The d.c. conductivity also rises, although usually well past the softening point and beyond temperatures where the element would normally be used. The problem with a.c. leakage is that it may begin 100°C or more below the softening point, well within the normal operating range of the element. The effect has, for example, been observed near 100°C, with a resulting error of about 10°C.

Other situations where a.c. leakage effects are important are generally restricted to high-accuracy applications. For example, PTFE insulated leads and cables should always be used in preference to PVC, to prevent errors of a few millikelvins. The effect is also known to afflict high-temperature SPRTs used near the silver point. The silver point is sufficiently close to the softening point of quartz for errors of several millikelvins to be apparent if too high a frequency is used.

Tests to expose a.c. leakage errors exploit the frequency dependence. The effect is absent at d.c. (zero frequency) and gets progressively worse as the frequency increases. It is for this reason that resistance thermometry bridges operate at very low frequencies, typically 25 and 75 Hz, and 30 and 90 Hz, depending on the local mains supply frequency. The simplest test which exposes a.c. leakage is to change the carrier frequency of the bridge, and some thermometry bridges have this facility.

5.4.12 Electromagnetic interference

Electromagnetic interference (EMI) is any unwanted voltage or current that originates outside the measurement circuit. Causes of EMI include electric motors, transformers, power cables, radio and TV transmissions, leakage currents from electric heaters, and ground loops.

It is commonly believed that EMI due to magnetic fields can be reduced by metal screens. However, a screen would have to be several metres thick to have a direct effect on the field at d.c. and the low frequencies used in resistance thermometry. There are two basic techniques for reducing magnetic EMI. Firstly, separate the EMI source and the thermometer as much as possible. This exploits the fact that the coupling between source and the thermometer falls off as $(distance)^3$. Secondly, ensure that all lead wires are kept close together. Twisted-pair and coaxial cables are very effective in reducing the loop area exposed to magnetic fields. Some examples are shown in Figure 5.10.

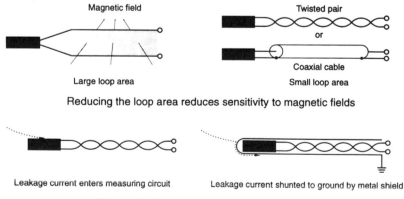

Magnetic field   Twisted pair

or

Coaxial cable

Large loop area   Small loop area

Reducing the loop area reduces sensitivity to magnetic fields

Leakage current enters measuring circuit   Leakage current shunted to ground by metal shield

Using a shield to intercept leakage currents

Figure 5.10
Examples of measurement practices that are, left: susceptible to EMI, and right: relatively immune to EMI.

The main benefit of screens in low-frequency instruments is that they can be used to eliminate the effects of leakage currents and ground loops. A common example of leakage currents affecting resistance thermometry occurs in electric furnaces, where the heaters are wound on ceramic. At high temperatures the ceramic will conduct very slightly, allowing small currents to flow into an unscreened thermometer assembly. Surrounding the thermometer with an earthed metal screen intercepts the leakage current and shunts it harmlessly to ground.

Ground-loop effects are very similar to the leakage current problem except that the currents are induced by differing ground voltages or magnetic fields. The solution is also the same: surround the thermometer by an earthed screen which intercepts the current. For screens to be effective there must be high insulation resistance between the screen and the thermometer and lead wires.

5.4.13 Lead resistance errors

The errors due to lead resistances were discussed in Section 5.3.2. In most measurements the errors can be estimated on the basis of the measurement technique, estimates of the lead resistance, and knowledge of the thermometer resistance.

For a two-lead measurement the temperature error is

$$\Delta T_m \approx 500 \; R_L/R(t) \tag{5.36}$$

where R_L is the resistance in one lead and $R(t)$ is the thermometer resistance. The error can be large: for example 0.5 Ω lead resistance in each lead of a 100 Ω thermometer gives rise to an error of approximately 2.5°C.

For ideal three-lead and pseudo-four-lead measurements the errors are less and depend largely on the difference in lead resistances, which are characterised by the uncertainty U_{R_L},

$$U_{T_m} \approx 250 \; U_{R_L}/R(t). \tag{5.37}$$

In true four-lead resistance measurements the errors should be negligible.

For all measurement techniques, a simple check will expose any susceptibility to lead-resistance errors. Simply insert a small resistance successively into each of the leads. Then with estimates of the lead resistances it is relatively easy to estimate the error and uncertainty. This check is necessary where an instrument is used with excessively long lead wires or there are doubts about the instrument's sensitivity to lead resistances.

Example 5.4 Assessing errors due to lead resistances

A three-lead resistance thermometer indicator is to be connected to a remote 100 Ω thermometer probe. The lead resistances are all measured and found to be within 7 $\Omega \pm 1$ Ω (95% CL). Estimate the expected error and uncertainty due to the lead resistances.

(1) *The error* A 1 Ω resistor is successively inserted into each of the three leads to the thermometer and the changes in reading are

lead 1: $\Delta T = 2.6°C$

Continued on page 182

Continued from page 181

lead 2: $\Delta T = -4.1°C$

lead 3: $\Delta T = 0.1°C$

The changes ΔT suggest that the instrument does not compensate correctly for lead resistance. In a good three-lead or pseudo-four-lead resistance measurement the sum of the changes should be zero. In a true four-lead measurement each of the changes should be zero. For this example there would be an error of

$$\Delta T_m = (2.6 - 4.1 + 0.1) \times 7°C$$

$$= -9.8°C.$$

Or equivalently, the correction for lead resistance error is $+9.8°C$.

(2) *The uncertainty*   The simplest method is to substitute the values directly into equation (5.37). Hence

$$U_{T_m} = \pm 2.5°C \ (95\% \ CL).$$

This is the uncertainty in the lead resistance correction.

5.4.14 Thermoelectric effects

Thermoelectric effects as discussed in Section 5.3.3 generally affect only high-accuracy d.c. platinum thermometry. The few microvolts generated by thermoelectric effects are not significant compared to the 0.4 mV/°C output voltage of most resistance thermometers. The problem is serious only in extreme cases where, for example, a lead wire has been replaced by a dissimilar metal, or connection terminals are exposed to high temperature gradients.

When voltage errors are expected or known, their influence on the measurement can be assessed as

$$\Delta T_m = \frac{250 \ V_e}{I R(0°C)} \tag{5.38}$$

where I is the sensing current through the PRT, V_e is the error voltage, and $R(0°C)$ is the ice-point resistance of the PRT. The typical error for a 1 mA sensing current and a 100 Ω PRT is about 2.5 mK/μV.

5.4.15 Reference resistor stability and accuracy

Ultimately the accuracy of a resistance thermometer depends on the accuracy of one or three reference resistances. For example, the balance equation for the Wheatstone bridge (equation (5.14)) is

$$R(t) = \frac{R_3}{R_2} R_1.$$

Any changes in the values of R_1, R_2 and R_3 will be interpreted incorrectly as changes in the value of $R(t)$. For small changes, the perceived change in $R(t)$, namely $\Delta R(t)$, is

$$\frac{\Delta R(t)}{R(t)} = \frac{\Delta R_1}{R_1} - \frac{\Delta R_2}{R_2} + \frac{\Delta R_3}{R_3} \tag{5.39}$$

where ΔR_1, ΔR_2 and ΔR_3 are the changes in each of the reference resistors. This equation provides us with the information to assess the stability and accuracy of the bridge. By expressing the percentage changes in resistance in terms of the temperature coefficients of the resistors, β_i, we obtain

$$\Delta T_m = \frac{1}{\alpha}(\beta_1 - \beta_2 + \beta_3)\Delta T_a \tag{5.40}$$

where ΔT_m is the apparent change in the measured temperature, α is the temperature coefficient of the PRT (≈ 4000 ppm/°C), and ΔT_a is the change in the ambient temperature.

Example 5.5   Temperature stability of a bridge

A platinum thermometer bridge is required to indicate temperature to an accuracy of ±0.1°C. The bridge will be exposed to ambient temperatures between 10°C and 40°C. Estimate the maximum temperature coefficients of the reference resistors.

By matching the temperature coefficients of R_1 and R_2 we need consider only the temperature coefficient of R_3. Rearranging equation (5.40) we obtain the maximum acceptable value for β_3:

$$\beta_3 < \alpha \frac{\Delta T_m}{\Delta T_a}.$$

Now substituting $\Delta T_m = 0.1$°C, $\Delta T_a = 30$°C and using $\alpha = 4000$ ppm/°C we find

$$\beta_3 < 13 \text{ ppm/}°C.$$

Since we must also accommodate some mismatch between R_1 and R_2, resistors with temperature coefficients of less than 10 ppm/°C would be appropriate.

Typical temperature coefficients are 50–200 ppm/°C for ordinary resistors and 0.2–15 ppm/°C for precision resistors. For temperature accuracies of 0.01°C or better, it is usually necessary to restrict the ambient temperature range, or to control the precision resistors with a thermostat.

Equation (5.39) can also be modified (see equation (2.27)) to estimate the uncertainty in temperature caused by uncertainties in the reference resistors:

$$U_{T_m} = \frac{1}{\alpha}\left[\left(\frac{U_{R_1}}{R_1}\right)^2 + \left(\frac{U_{R_2}}{R_2}\right)^2 + \left(\frac{U_{R_3}}{R_3}\right)^2\right]^{1/2}. \tag{5.41}$$

Example 5.6   Estimating the accuracy of a bridge

A simple Wheatstone bridge is assembled using resistors with a 0.01% (100 ppm) tolerance. Estimate the accuracy of the bridge.

Substituting the values directly into equation (5.41) and using $\alpha = 4000$ ppm/°C we obtain

$$U_{T_m} = (100^2 + 100^2 + 100^2)^{1/2}/4000°C$$

$$= 0.043°C.$$

5.4.16 Linearisation

All direct-reading platinum thermometers include some form of linearisation in their electronic systems. As discussed in Section 4.6, the linearisation is required to convert the non-linear response of the platinum thermometer into a signal that is directly proportional to temperature. Because the resistance–temperature characteristic for platinum is so very nearly linear, the linearisation is relatively simply achieved by comparison with other thermometers. Indeed most of the residual error after linearisation is due to small departures of the sensing element from the standard tables, and would typically be less than 0.1°C over a 200°C range.

5.5 CHOICE AND USE OF RESISTANCE THERMOMETERS

5.5.1 Choosing a thermometer

If a reference thermometer is required for any temperature below 200°C, then PRTs should be the first choice. Although the initial cost of a PRT may be higher than that of a liquid-in-glass thermometer, the maintenance and recalibration costs of PRTs are much less, and they are less fragile. If accuracies of 0.1°C or better are required, then PRTs should be the only choice.

At temperatures above 200°C the limitations of PRTs begin to affect their suitability; in particular, whether or not the PRT will be subject to vibration or regular cycling. Figure 5.11 summarises the best temperature range and accuracy that can be expected from the three main types of PRTs.

There are four main factors to consider in the choice of industrial platinum thermometers.

- *Accuracy* The accuracy of calibrated PRTs is between $\pm1°C$ and ±1 mK, depending on the construction and the required temperature range. A good rule-of-thumb is that the cost of the thermometer is inversely proportional to the required accuracy. A ±1 mK system will cost about 1000 times more than a $\pm1°C$ system, with most of the cost in the bridge or display unit.

  The accuracy of industrial PRTs is also strongly dependent on the temperature range: it is below 0.005% of the range for partially supported PRTs and below 0.1% for fully supported PRTs. For the highest-accuracy applications PRTs can be selected for low hysteresis by cycling them, for example between 100°C (boiling

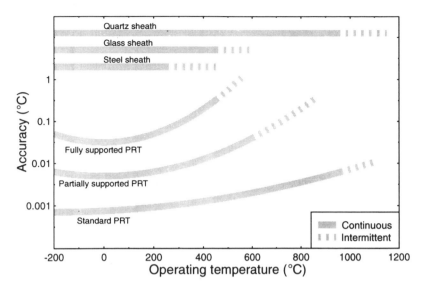

Figure 5.11
Approximate accuracy and range that can be achieved with fully supported, partially supported and standard PRTs.

water) and $-196°C$ (liquid nitrogen). Measuring the change in triple-point or ice-point resistance between exposures to the high and low temperatures will reveal the amount of hysteresis. The best partially supported PRTs have less than 0.0002% hysteresis.

D.C. instruments are suitable for accuracies between $\pm0.02°C$ and $\pm1°C$. A.C. bridges are necessary for accuracies better than $\pm0.02°C$.

- *Temperature range*   As the temperature range increases, the lower-grade PRTs are excluded, and demands on the quality of the environment and sheath increase.

  — *Above 250°C*   The environment should be free of contaminants — ceramic elements in stainless steel sheaths are suitable only for intermittent use. Fully supported elements should not be exposed to regular (e.g. daily) cycling.

  — *Above 450°C*   Silica or quartz sheaths only should be used. Fully supported and partially supported elements have a limited life at these temperatures due to the eventual fatigue and failure of lead wires.

  — *Above 650°C*   Only high-temperature SPRTs survive readily, although some of the best of the partially supported PRTs will survive intermittent use to 850°C.

  To obtain the best accuracy from reference PRTs it is worth restricting their use to narrow ranges in order to limit the hysteresis and drift. Depending on the accuracy required, one PRT per 200°C range is a reasonable guide.

- *Environment*   The major environmental considerations are vibration and mechanical shock. If either of these are present, fully supported elements should be used. Partially supported elements may be suitable if the vibration is small or if the assembly can be decoupled from the source of the vibration. In a very wet or humid environment, glass elements should be used to prevent excessive leakage and moisture-induced hysteresis.

- *Construction* Most manufacturers of PRT elements also assemble and sell sheathed PRTs. Since sheathing is required for most applications, elements should be purchased sheathed. As can be seen from the earlier discussion on construction, the construction of a reliable sheathed PRT is not trivial and is best left to the experts. The best techniques are also often proprietary information. Remember also that a calibration laboratory may be unwilling to certify a thermometer which has obviously not been manufactured using well-established techniques.

Except for the very lowest-accuracy applications, the main cost of a platinum thermometer is the electronic display unit or bridge. A good-quality PRT is partially supported, has four leads, and has a good seal where the cable joins the sheath. The cable should have a braided screen which is connected to the sheath if it is metal. In some applications PTFE insulated cable may be advantageous; it exhibits less a.c. loss and withstands temperatures of up to 200°C. The length of the sheath should be chosen according to the application and temperature range. As a guide, the minimum sheath length should be about 200 mm plus 100 mm per hundred degrees of duty above 200°C. For example, a minimum length for duty at 400°C is 400 mm.

5.5.2 Care and maintenance

Platinum resistance thermometers are relatively easy to care for. PRTs have a long life so long as they are not exposed to vibration, temperature cycling and potentially contaminating environments.

All PRTs should be checked regularly at the ice point or triple point, since a change in the ice-point resistance will expose almost all signs of faulty behaviour or misuse. A decrease in ice-point resistance is normally an indicator of excessive leakage due to moisture. With steel sheathed PRTs this can be checked very easily by measuring the insulation resistance between the sheath and element. The moisture can be removed by drying the thermometer in a drying oven for a day or so. The assembly should not be heated above the maximum temperature rating of the head and leads, typically 65°C for PVC cable. PTFE insulated assemblies may be dried at 100°C.

An increase in ice-point resistance caused by work-hardening can be removed by annealing. This is accomplished by heating the thermometer to 400–450°C for several hours. This should be repeated until the ice-point resistance of the PRT stabilises at a single value. In principle any PRT can be annealed. However, fully supported PRTs and metal-sheathed PRTs may be damaged by the exposure to 450°C and should be annealed only when necessary.

All SPRTs and partially supported PRTs mounted in silica sheaths should be annealed regularly (e.g. annually). This should be repeated until the triple-point resistance of the PRT has stabilised. For SPRTs not used above 700°C, this value should be stable to better than 1 mK for periods of years. In some cases, where very severe mechanical shock is known to have caused a large resistance shift, the annealing temperature may need to be increased to 600°C or up to the working temperature limit, whichever is lower. If the triple-point resistance increases on annealing, the thermometer may be being damaged by the treatment.

PRTs that exhibit large permanent ice-point resistance shifts should be treated with suspicion. If the permanent shift exceeds 0.1% the PRT should be discarded as unreliable.

5.5.3 Use

Restrictions on the use of PRTs are also straightforward. The main restrictions are associated with the temperature range.

- *Below 250°C* There are few restrictions on the PRTs except that they should be exposed to minimal vibration. Unsheathed ceramic PRTs should never be exposed to water or any other fluid or grease which wets the ceramic. Fluid penetrating the pores of the ceramic will cause it to swell and strain the wire. In water, for example, a fully supported wirewound ceramic element will drift as much as 10 mK per hour and be very noisy. Condensation on unsheathed elements at low temperatures can be a similar problem. Because the seal at the head of sheathed PRTs is rarely perfect, PRTs should be mounted so that the head is not exposed to moisture or high humidity.
- *Above 250°C* The environment should be free of contaminating agents; for example, ceramic elements in stainless steel sheaths are suitable for intermittent use only. Fully supported elements should not be exposed to regular (e.g. daily) cycling.
- *Above 450°C* Quartz sheaths only should be used. Fully and partially supported elements may have a limited life at these temperatures due to eventual fatigue and failure of the lead wires.
- *Above 600°C* Only high-temperature SPRTs survive readily. Some of the best of the partially supported PRTs will survive intermittent use to 850°C.

SPRTs used at temperatures above 450°C should be cooled slowly to allow thermally generated defects to anneal and to prevent strain caused by uneven cooling (see Section 3.5).

5.6 CALIBRATION OF RESISTANCE THERMOMETERS

Platinum resistance thermometers have been a part of all the temperature scales since 1927. Consequently there has been a great deal of research on calibration and interpolation equations, with over a dozen recommended at various times for various applications. In this book we describe only two basic forms of calibration equation, depending on the accuracy required. The ITS-90 formulation which is suited to SPRTs and the best of the partially supported PRTs is discussed in Chapter 3. In this section we recommend a formulation based on the Callendar–van Dusen (CVD) equation, which is simpler and suited to lower-accuracy applications. Both recommendations apply to reference thermometers that are calibrated in terms of resistance. Direct-reading thermometers should be calibrated according to the procedure outlined in Section 4.7.

As with all high-accuracy calibrations, a high level of expertise is required of personnel involved in the calibration of platinum resistance thermometers. In particular, good algebraic and computing skills are required to handle both the ITS-90 formulation and the simpler CVD equation. Additionally, for the CVD equation an understanding of least-squares fitting is essential.

5.6.1 Calibration equations

For all of the earlier temperature scales the Callendar–van Dusen (CVD) equation was the accepted interpolation equation for platinum resistance thermometers. It is also the defining function for all the industrial PRTS. The general form of the equation is

$$W(t) = 1 + At + Bt^2 + Ct^3(t - 100). \tag{5.42}$$

where C is zero above $0°C$, and $W(t) = R(t)/R(0°C)$. Note that ITS-90 uses the triple-point ratio rather than the ice-point ratio.

The CVD equation is very much simpler than the ITS-90 formulation and is well suited to least-squares fitting. There might seem to be a problem with the C parameter having two values, but in practice it is not a problem. Thermometers calibrated for low-temperature work are not usually used above $0°C$, so equation (5.42) is used with $C \neq 0$. If thermometers used above room temperature are rarely used below $-40°C$, then the C term can be ignored, the resulting error being less than $-0.01°C$ at $-40°C$.

At temperatures above 150 to $200°C$ the simple quadratic equation given as equation (5.42) with $C = 0$ may prove to be inadequate. Most partially supported elements exhibit a t^3 dependence which begins to dominate the residual errors when the temperature range gets large. For large ranges above $0°C$ the equation may be extended to

$$W(t) = 1 + At + Bt^2 + Dt^3. \tag{5.43}$$

Use of this equation is equivalent to the deviation-function approach described in Section 4.6.

After the thermometer has been calibrated the temperature can be calculated from the measured resistance, and the calibration constants by successive approximation. For the cubic equation the temperature is calculated by repeated application of

$$t_n = \frac{W(t) - 1}{A + Bt_{n-1} + Dt_{n-1}^2}. \tag{5.44}$$

This gives an improved estimate t_n, based on the previous estimate t_{n-1}. With repeated application of equation (5.44) the estimate of the temperature improves steadily to an accuracy greater than can be achieved by solving the equation directly; the conventional technique involving square roots can suffer from round-off errors.

The recursion relation for the CVD equation is

$$t_n = \frac{W(t) - 1}{A + Bt_{n-1} + Ct_{n-1}^2(t_{n-1} - 100)}. \tag{5.45}$$

When $W(t)$ is a lot different from 1, five or six iterations of equations (5.44) or (5.45) may be necessary before t_n converges to the correct value.

Exercise 5.5

Apply the recursion equation, equation (5.44), to finding the temperature reading of a PRT with $W(t) = 2.6$. The calibration constants for the PRT are $A = 4 \times 10^{-3}$, $B = -6 \times 10^{-7}$, $D = 0$.

5.6.2 Calibration at fixed points

There are two basic methods for calibrating PRTs, namely calibration at fixed points as described in this section, and calibration by least squares, as described in the next section.

Calibration by direct comparison with fixed points has been described in Chapter 3. This technique is the more accurate but is subject to some serious restrictions when applied to industrial PRTs.

The method has a number of advantages:

- It provides a very accurate determination of the calibration constants.

- Since all measurements are normally corrected for self-heating errors (Section 5.4.4), the effects of self-heating are eliminated from the calibration.

- Relatively few points are required: typically two or three, depending on which ITS-90 interpolation equation is used.

- Within $-38°C$ to $420°C$, the ITS-90 provides five fixed points at approximately -38, 30, 157, 232, and $420°C$. This is sufficient choice for most calibration ranges.

Disadvantages include:

- By using the same number of fixed points as unknown constants, no additional information is made available on the likely uncertainty in the calibration. For standard PRTs which are always calibrated at the fixed points there is enough generic knowledge available to make a good Type B assessment. This is not true for industrial PRTs, which differ considerably between grades and manufacturers.

- In many cases the user of the calibrated thermometer does not have access to bridges with the facility to change the measuring current. All measurements made with the thermometer will therefore be subject to self-heating errors of between 5 and 30 mK which the user is unable to assess. This will make the full accuracy of the thermometer unrealisable and introduce a serious systematic error. Again this is not a serious problem for SPRTs because the self-heating error is usually less than 2 mK and users of SPRTs use bridges which allow the current to be changed.

An assessment of the uncertainty in a thermometer's readings is essential if a calibration is to satisfy the requirements described in Chapter 4. For industrial PRTs it is therefore necessary to make more measurements than is required simply for determination of the calibration constants. Least squares provides the best means for analysing the results.

5.6.3 Calibration by least squares

In a least squares calibration the data are acquired through intercomparison with an SPRT. There are several choices of calibration equation, but for most applications equations (5.42) or (5.43) are more than adequate. The measurements required are: one measurement of the triple-point or ice-point resistance, and a number of measurements of $W(t)$ distributed evenly over the calibration range. The exact number of points required depends on the number of parameters to be fitted; about four points per unknown parameter is sufficient.

Once all the measurements have been made the values of A and B are determined by the method of least squares described in Section 2.10.2. For the quadratic equation

$$W(t) = 1 + At + Bt^2 \qquad (5.46)$$

where $W(t) = R(t)/R(0°C)$ or $R(t)/R(0.01°C)$ as appropriate. The best values of A and B are found by following the principles described in Section 2.10.2 on least squares, so that

$$\begin{pmatrix} A \\ B \end{pmatrix} = \begin{pmatrix} \Sigma t_i^2 & \Sigma t_i^3 \\ \Sigma t_i^3 & \Sigma t_i^4 \end{pmatrix}^{-1} \begin{pmatrix} \Sigma t_i[W(t_i) - 1] \\ \Sigma t_i^2[W(t_i) - 1] \end{pmatrix}. \qquad (5.47)$$

The error-of-fit (equation (2.42)) which describes how well the equation describes the measurements, and the uncertainties in the values of A and B (equation (2.43)), can then be computed.

The advantages of this method include the following:

- It provides an assessment of the uncertainty in temperatures measured by the PRT, the error-of-fit. Just as a low value for the error-of-fit shows that the thermometer has the expected resistance–temperature relationship, so a high value for the error-of-fit is indicative of a faulty thermometer. The fault may be excessive hysteresis and relaxation, contamination, or an excess of moisture.
- The calibration is carried out under the same conditions as those in which the thermometer will be used. This ensures that the relationship determined is realisable by the user of the thermometer. Such conditions might be: 1 mA sensing current and 200 mm immersion in a stirred fluid bath. The effects of self-heating will then be the same (or very similar) in use as in calibration.
- It can be applied to any calibration range. For most calibrations only the quadratic form of the CVD equation is required. For very wide-range calibrations a cubic term may need to be added.
- It provides an assessment of the uncertainties in the fitted values, A and B. This is useful in determining the precision for reporting the values.

Disadvantages include the following:

- It requires more calibration points. However, unlike measurements at fixed points the measurements are more amenable to automation and the sensing current is constant.
- The conditions under which the thermometer may be used with full accuracy are restrictive, although they are probably less restrictive than for a fixed point calibration.

Both methods described above (here and in Section 5.6.2) are quite complicated. While this would traditionally have been considered an impracticality, this is no longer the case. The availability and power of even the lowest-cost computers now mean that algebraic complexity is no longer an issue, at least for laboratory applications.

5.6.4 A calibration procedure

The calibration procedure for a platinum resistance thermometer follows closely the outline given in Section 4.3.6. In this section we highlight additional features relevant to PRTs.

Step 1 — Initiate record-keeping

Resistance bridges are normally calibrated independently of the thermometer. Direct-reading thermometers follow Example 4.7. Otherwise proceed as for Section 4.3.6.

Step 2 — General visual inspection

As for Section 4.3.6.

Step 3 — Conditioning and adjustment

SPRTs and partially supported PRTs in silica sheaths are appropriately constructed to withstand duty at 450°C and may therefore benefit from periodic annealing to relieve accumulated strain in the wire. The procedure should be repeated until the triple-point (or ice-point) resistance becomes stable. If the ice point increases steadily on annealing, then the PRT may have been contaminated.

Step 4 — Generic checks

- *Detailed inspection*   As for Example 4.7.
- *Insulation resistance*   The insulation resistance of metal-sheathed resistance thermometers should be checked to confirm that there is no build-up of moisture in the insulation.
- *Ice-point or triple-point resistance*   Measurement of one of these will confirm that the lead wires are intact, and indicate whether the thermometer has been exposed to damaging environments. For industrial PRTs the resistance should be within 0.2% of the nominal resistance. This allows for 0.1% on initial tolerance plus a further 0.1% shift due to drift. Brand-new thermometers should be within 0.1%. Resistance values that are high are indicative of contamination or exposure to vibration and shock. Low-resistance values are indicative of moisture build-up in the insulation, and occasionally short-circuited lead wires. The resistance should be measured before and after the calibration to enable assessment of the stability of the thermometer.
- *Hysteresis assessment*   There are two situations depending on the expected usage of the thermometer.
  (i) *The measured temperature will always be approached from room temperature* This follows the rationale of Example 2.6 for reducing the uncertainty due to hysteresis. To assess the likely uncertainty due to hysteresis a number of additional points must be included in the intercomparison on return from the highest (or lowest) temperatures. Once the width of the hysteresis loop has been determined the uncertainty (95% CL) is estimated as half of the loop width.
  (ii) *The measured temperature may be approached from either direction* In this case the intercomparison must cover the required calibration range in both directions to avoid biasing the intercomparison data. If the data for both directions are included in the least squares fit then the uncertainty due to hysteresis will be included in the error-of-fit and no additional uncertainty need be included in the total.
- *Self-heating assessment*   This applies only to thermometers that are calibrated at non-zero current. The self-heating effect in resistance thermometers can depend on the environment in which they are used. Because the self-heating error in use will be different from the self-heating in calibration, the additional uncertainty must be

included in the assessment of total uncertainty. For all 100 Ω sheathed PRTs operated at 1 mA the variation in self-heating when used in oil baths, water baths, ice points and triple points is usually less than ±2 mK. If the thermometer is unsheathed or to be exposed to an environment very different from the calibration environment then a subsidiary experiment must be designed to enable estimation of the likely change in self-heating.

Step 5 — The intercomparison

For SPRTs and the very best of the partially supported PRTs a fixed point comparison (Section 5.6.2) is appropriate. Otherwise the least-squares fit approach (Section 5.6.3) should be adopted. For most PRTs the simple quadratic version of the Calendar–van Dusen equation is adequate so that a minimum of 8 points is required. These should be distributed evenly over the calibration range. The intercomparison must also duplicate the expected usage in respect of hysteresis. If the thermometer is expected to be used so that the measured temperature is always approached from room temperature (to reduce the hysteresis) then the intercomparison must be carried out in the same way. If the thermometer usage is not expected to be controlled and significant hysteresis is expected, the intercomparison must cover the expected temperature range in both directions. This doubles the number of calibration points in the intercomparison.

Step 6 — Analysis

The analysis comprises the determination of the constants in the calibration equations. We demonstrate the analysis for an industrial PRT in Example 5.7 below.

Step 7 — Uncertainties

As for Section 4.3.6, and Example 5.7 below. For SPRTs the assessment of uncertainty is based entirely on generic history and would typically be ±2 mK (95% CL). For all other PRTs additional information specific to the thermometer under test is required.

Step 8 — Complete records

As for Section 4.3.6.

Example 5.7   Calibration analysis for a platinum resistance thermometer

A fully supported steel sheathed PRT is calibrated over the range −10 to 180°C. The calibration data and the results of a least-squares fit analysis are summarised in Table 5.3 while a hysteresis check is summarised in Table 5.4. Discuss these results and prepare a calibration certificate for your (imaginary) firm, Calvin, de Gries & Co. The uncertainty in the SPRTs is 2.0 mK (95% CL) and the uncertainty due to bath non-uniformity is 1.0 mK (95% CL).

Part 1 — The resistance–temperature relationship

The results of the least squares fit are included in Table 5.3. The pre-calibration triple-point resistance has been used to determine the resistance ratio and the data have been fitted to the simple quadratic version of the Callendar–van Dusen equation (5.42) to determine the values for A and B. A check of the residual errors in the fit (Table 5.3) shows that there is a slight pattern in the signs of the errors, which suggests a small cubic or S-shaped nonlinearity (Section 4.6).

Continued on page 193

— Continued from page 192 —

Table 5.3. Summary of initial readings and intercomparison.

Insulation resistance:			Greater than 1000 MΩ	
Pre-calibration triple-point resistance:			100.0384 Ω	

Reading number	Measured resistance	Measured temperature	Fitted temperature	Residual error
1	96.1462	−9.9482	−9.9443	−0.0039
2	100.7751	1.9065	1.8858	0.0207
3	105.4064	13.7761	13.7633	0.0128
4	110.0186	25.6411	25.6332	0.0079
5	114.6116	37.5020	37.4952	0.0068
6	119.1831	49.3438	49.3433	0.0005
7	123.7829	61.3051	61.3070	−0.0019
8	128.3179	73.1410	73.1442	−0.0032
9	132.8424	84.9942	84.9960	−0.0018
10	137.3440	96.8296	96.8298	−0.0002
11	141.8415	108.6959	108.6953	0.0006
12	146.3178	120.5399	120.5473	−0.0074
13	150.7780	132.3917	132.3993	−0.0076
14	155.2155	144.2333	144.2337	−0.0004
15	159.6404	156.0842	156.0773	0.0069
16	164.0506	167.9291	167.9246	0.0045
17	168.4467	179.7764	179.7773	−0.0009

Standard error-of-fit	= 0.0030 Ω
Standard deviation of residual errors	= 0.0078°C
$R(0.01°C)$	= 100.0384 Ω
Fitted parameters: A	= $3.90670294 \times 10^{-3} \pm 2.8 \times 10^{-5}/°C$
B	= $-5.72849136 \times 10^{-7} \pm 2.0 \times 10^{-7}/°C^{-2}$

Table 5.4. Uncertainty information and summary for hysteresis check.

Reading number	Measured resistance	Measured temperature	Predicted temperature	Residual error
1	168.4530	179.7825	179.7944	−0.0119
2	164.0582	167.9242	167.9451	−0.0209
3	159.6502	156.0905	156.1036	−0.0131
4	155.2254	144.2419	144.2601	−0.0182
5	150.7847	132.3985	132.4171	−0.0186
6	146.3264	120.5462	120.5701	−0.0239
7	141.8509	108.6992	108.7201	−0.0209
8	137.3530	96.8396	96.8535	−0.0139
9	132.8498	85.0018	85.0154	−0.0136
10	128.3241	73.1561	73.1604	−0.0043
11	123.7871	61.3033	61.3179	−0.0146
12	119.1884	49.3544	49.3570	−0.0026
13	114.6226	37.5073	37.5237	−0.0164
14	110.0309	25.6426	25.6649	−0.0223
15	105.4197	13.7823	13.7974	−0.0151
16	100.7901	1.9104	1.9242	−0.0138
17	96.1490	−9.9482	−9.9372	−0.0110

Mean hysteresis error	= −0.0150°C
Post-calibration triple-point resistance	= 100.0438 Ω

— Continued on page 194 —

Continued from page 193

Part 2 — Determining the total uncertainty

The contributing factors are as follows:

Reference thermometer   The uncertainty in the reading of the reference thermometer, determined from its certificate, is

$$U_{ref} = 2.0 \text{ mK (95\% CL)}.$$

Calibration medium   The uncertainty in the uniformity of the calibration bath (from bath commissioning tests) is

$$U_{bath} = 1.0 \text{ mK (95\% CL)}.$$

Uncertainty in the fit   This is determined from the standard deviation of the residual errors

$$s_{fit} = 7.8 \text{ mK } (v = 15).$$

From Student's t-distribution (Table 2.2) the uncertainty is determined as

$$U_{fit} = 16.6 \text{ mK (95\% CL)}.$$

Hysteresis   In this example the average hysteresis error is -15 mK. The half-width of the loop is used as an estimator of the uncertainty, hence

$$U_{hys} = 7.5 \text{ mK (95\% CL)}.$$

It is noted that the change in triple-point resistance of -5.4 mΩ (-14 mK) also shows the effect of hysteresis.

  This example highlights the worst of the problems with hysteresis. The jagged appearance of the hysteresis loop (Figure 5.12) is due to relaxation.

Drift   In the absence of specific information on the observed drift for the thermometer, the half-width of the hysteresis curve is a good indicator of the expected drift:

$$U_{drift} = 7.5 \text{ mK (95\% CL)}.$$

Self-heating   The PRT is steel sheathed, nominally 100 Ω, and operated at a measuring current of 1 mA. Therefore the type B assessment recommended in the procedure gives an estimate of the uncertainty as

$$U_{sh} = 2 \text{ mK (95\% CL)}.$$

The *total uncertainty* is the sum of these terms in quadrature:

$$U_{total} = 20 \text{ mK (95\% CL)}.$$

Figure 5.13 shows a completed certificate for the thermometer.

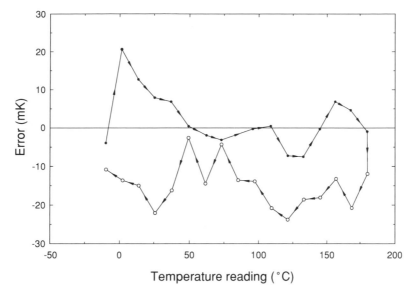

Figure 5.12
A graphical summary of the PRT calibration data for Example 5.7. The ascending sequence of measurements (filled circles) is used to determine the coefficients in the Callendar equation while the descencing sequence of measurements (open circles) is used to determine the amount of hysteresis. The jagged appearance of the curves is due to relaxation associated with the hysteresis.

Exercise 5.6

Rework Example 5.7 assuming that the thermometer will be used in a manner such that the measured temperature may be approached from above or below. You will need to include the data from both Tables 5.3 and 5.4 in the least-squares fit, and do not need to include the uncertainty due to hysteresis in the total uncertainty.

5.7 OTHER RESISTANCE THERMOMETERS

5.7.1 Thermistors

Thermistors are semiconducting ceramic resistors made from various metal oxides. They have one outstanding advantage over all other resistance thermometers, namely very high sensitivity. It is not difficult to build thermistor thermometers with sensitivities of 50 mV/°C or more, more than 100 times that of most platinum thermometers and more than 1000 times that of most thermocouples. They are also very small and fast to respond.

There are two main classes of thermistor, namely PTC or positive temperature co-efficient thermistors, and NTC or negative temperature coefficient, the latter being the most suitable for thermometry. An approximate equation relating resistance of the NTC thermistor to temperature is

$$R(T) = A \exp(B/T), \tag{5.48}$$

CALVIN, DeGRIES AND CO

1 Traceability Place, P O Box 31-310, Lower Hutt; New Zealand
Telephone (64) 4 566-6919   Fax (64) 4 569-0003

CALIBRATION CERTIFICATE

Report No.:   T92-2003.

Client: ACME Thermometer Co, 100 Celsius Avenue, P O Box 27-315,
                            Wellington, New Zealand

Description of Thermometer: A stainless steel sheathed platinum resistance thermometer
                            manufactured by ACME, serial number GRT10.

Date of Calibration: 13 to 16 January 1992.

Method: The thermometer was compared with standard thermometers held
                            by this laboratory. The temperature scale used is ITS-90.

Conditions: The thermometer was immersed in a stirred bath to a minimum
                            depth of 200 mm.   The sensing current for all resistance
                            measurements was 1 mA.

Results: The temperature, $t°C$, was related to the thermometer resistance,
                            $R(t°C)$, and the resistance at the triple point of water by the
                            equation

$$\frac{R(t°C)}{R(0.01°C)} = 1 + At + Bt^2$$

The constants $R(0.01°C)$, A and B were found to be

$$R(0.01°C) = 100.0384 \text{ ohm}$$
$$A = 3.9067 \times 10^{-3} \ (°C)^{-1}$$
$$B = -5.728 \times 10^{-7} \ (°C)^{-2}$$

Note $R(0.01°C)$ should be measured with the user's instrument and the value obtained used in the
        equation.

Accuracy: The uncertainty in temperatures measured with the thermometer over the range $-10°C$
                to $180°C$ and determined using the above constants is estimated to be $\pm0.02°C$ at the
                95% confidence level.

Checked: _____   Signed: _____
         W Thomson   R Hooke

This report may only be reproduced in full.

Figure 5.13

A completed certificate for an industrial platinum resistance thermometer, based on the information
given in Example 5.7.

which is of the same form as that for the leakage resistance of insulators, equation (5.35). Values of $R(T)$ range from less than $100\ \Omega$ to more than $100\ M\Omega$, depending on the temperature and the values of A and B. For convenience equation (5.48) is usually written

$$R(T) = R(T_0) \exp \left(\frac{B}{T} - \frac{B}{T_0} \right), \tag{5.49}$$

where T_0 is $290.15\ K$ ($25.0^\circ C$) or $273.15\ K$ ($0^\circ C$). The resistance typically varies by a factor of $100\,000$ or more over the $-100^\circ C$ to $150^\circ C$ operating range. The temperature coefficient of thermistors is approximately

$$\alpha = \frac{-B}{T^2} \tag{5.50}$$

with typical values between $-3\%/^\circ C$ and $-6\%/^\circ C$.

The main disadvantages of thermistors include the extreme non-linearity of the resistance with temperature and instability with time and cycling. The best thermistors are glass-encapsulated or epoxy-encapsulated bead types and are now available with an interchangeability of $0.1^\circ C$. Their long-term stability approaches a few millikelvins per year.

Equation (5.49) is a satisfactory calibration equation for only very narrow ranges ($10^\circ C$) or low accuracy applications. A better equation is

$$\frac{1}{T} = a_0 + a_1 \log(R) + a_2 \log^2(R) + a_3 \log^3(R) \tag{5.51}$$

which will fit most thermistor responses over ranges of $100^\circ C$ or more to within a few millikelvins.

The high sensitivity and fast response of thermistors make them ideally suited to precision temperature control and differential temperature measurement where resolutions in excess of $5\ \mu K$ can be obtained. They are also attractive for simple hand-held thermometers because the sensitivity and high resistance make them relatively immune to lead resistance errors. Thermistors are available in a wide variety of sheathed assemblies including air–temperature, surface–temperature, veterinary and hypodermic probes.

5.7.2 Copper and nickel resistance thermometers

Platinum is not the only metal used for resistance thermometry, although it is the most widely used. Other metals include copper, nickel and nickel–iron, as well as the rhodium–iron thermometer used for cryogenic thermometry (Chapter 3).

The main attraction of copper resistance thermometers is their very high linearity, within $0.1^\circ C$ over ranges less than $200^\circ C$. The disadvantages are their low-resistance, typically $10\ \Omega$ at $25^\circ C$, and their susceptibility to corrosion. Typical operating ranges are from $-80^\circ C$ to $260^\circ C$. The temperature coefficient, $\alpha = 4.27 \times 10^{-3}/^\circ C$, is marginally higher than that for platinum.

Nickel resistance thermometers are chosen principally for their low cost and high sensitivity. They are also subject to greater standardisation than copper thermometers. The DIN 43760 standard defines a nickel thermometer for the range -60 to $180^\circ C$

with a resistance–temperature relationship similar to the CVD equation (equation (5.42)) although C is differently defined:

$$R(T) = R_0[1 + At + Bt^2 + Ct^4] \tag{5.52}$$

where $R_0 = 100\ \Omega$, $A = 5.450 \times 10^{-3}/°C$, $B = 6.65 \times 10^{-6}/°C^2$, $C = 2.605 \times 10^{-11}/°C^4$.

The α value for nickel is $6.18 \times 10^{-3}/°C$, nearly twice that of platinum. The non-linearity of nickel thermometers is about three times that of platinum.

Nickel–iron resistance thermometers are used for their high sensitivity and resistance, for example, in air-conditioning systems. A typical nickel–iron thermometer has a resistance of $100\ \Omega$ at $21.1°C$ ($70°F$), and an α value marginally less than that of nickel. They have a useful temperature range of $-20°C$ to $150°C$.

FURTHER READING

Supplementary Information for the International Temperature Scale of 1990, BIPM (1990).
Platinum Resistance Thermometer Calibrations, B. W. Mangum, NBS Special Publication 250-22, US Department of Commerce (1987).
These are two key references for anyone working with standard PRTs.

Techniques for Approximating the International Temperature Scale of 1990, BIPM (1990).
Temperature, its Measurement and Control in Science and Industry, American Institute of Physics, New York, Vol. 5 (1982), Vol. 6 (1992).
The BIPM Booklet in particular provides a very good summary of work with industrial PRTs. The proceedings from the ten-yearly Temperature symposia detail most of the research on PRTs and their application to difficult measurements.

6

Liquid-in-glass Thermometry

6.1 INTRODUCTION

Liquid-in-glass thermometers were one of the earliest forms of thermometer and their use dominated temperature measurement for at least 200 years. They have had a profound effect on the development of thermometry and in popular opinion they are the only 'real' thermometer! Liquid-in-glass thermometers have been developed to fill nearly every niche in temperature measurement from $-190°C$ to $600°C$, including the measurement of temperature differences to a millikelvin. The popularity of these thermometers continues in spite of the fragile nature of glass and depends largely on the chemical inertness of glass, as well as the self-contained nature of the thermometer.

The trend is, however, to move away from liquid-in-glass thermometers. For higher precision, platinum resistance thermometry gives superior performance and is readily accessible with modern resistance-bridge designs and high-quality probes, both of which are available commercially. At the lower-quality end of the market many simple battery-operated thermometers are available and these have a similar convenience in use, without the disadvantages of glass. Liquid-in-glass thermometers are still specified in many test procedures world-wide, and they represent a cost-effective solution in situations where only a few temperature measurements are made.

We examine the construction and modelling of a modern liquid-in-glass thermometer and show that it is not as simple as it first appears. Various sources of error need to be taken into account in order to achieve effective procedures for the calibration and use of a liquid-in-glass thermometer.

6.2 GENERAL DESCRIPTION

Many types of liquid-in-glass thermometers have been developed but only a few are suitable for holding and transferring the temperature scale. Solid-stem glass thermometers are a suitable type and are considered in this chapter. Typical examples are illustrated in Figures 6.1 and 6.2. We concentrate on mercury-in-glass thermometers because they perform better than thermometers containing organic liquids. Many of the principles will, of course, apply to other types of liquid-in-glass thermometers.

There are four main parts to the thermometer:

- *Bulb*   A thin glass container holding the bulk of the liquid. The glass should be an approved type and correctly annealed.

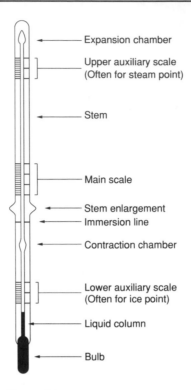

Figure 6.1
The main features of a solid-stem glass
thermometer. The thermometer may
have an enlargement in the stem or
an attachment at the end of the stem
to assist in the positioning of the
thermometer.

- *Stem* A glass capillary tube. Again of approved glass type which may differ from
  that of the bulb. The bore may be gas-filled or vacuous.

- *Liquid* Usually mercury or an organic liquid (see Table 6.1 for commonly used
  liquids).

- *Markings* Usually etched or printed onto the stem. The markings include the various
  scales as well as other information.

Table 6.1. Working range of some thermometric liquids and their apparent thermal expansion coefficient in thermometer glasses around room temperature.

Liquid	Typical apparent expansion coefficient ($°C^{-1}$)	Possible temperature range
Mercury	0.000 16	−35 to 510°C
Ethanol	0.001 04	−80 to 60°C
Pentane	0.001 45	−200 to 30°C
Toluene	0.001 03	−80 to 100°C

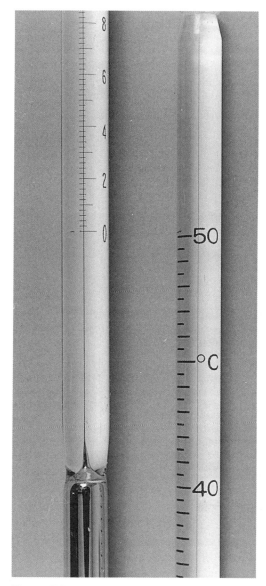

Figure 6.2
Calibration marks are usually scratched on at both ends
of a thermometer's scale to locate the ruling of the scale.
Left: A good quality thermometer. The calibration mark
is immediately alongside the 0°C mark. Right: A general-
purpose thermometer. Here the calibration mark is about
$\frac{1}{4}$ scale division above the 50°C mark. Since it is a cheaper
thermometer the manufacturer is content to locate the
scale within the $\frac{1}{4}$ scale division, and this would vary from
thermometer to thermometer in the same batch. Indeed
the total length of the scale varies somewhat for this type
of thermometer. Readings could be expected to be accurate to about one scale division, 0.5°C in this instance.

Figure 6.1 illustrates the main parts of a liquid-in-glass thermometer along with a nomenclature for other features commonly found. The purpose of most of these features will be met later. Variations on all the features will be found with real thermometers.

The operation of liquid-in-glass thermometers is based on the expansion of the liquid with temperature, that is the liquid acts as a transducer to convert thermal energy into a mechanical form. As the liquid in the bulb becomes hotter it expands and is forced up the capillary stem. The temperature of the bulb is indicated by the position of the top of the mercury column with respect to the marked scale.

The equation which best describes the expansion of the liquid volume is

$$V = V_0(1 + \alpha t + \beta t^2), \tag{6.1}$$

where V_0 is the volume of the liquid at $0°C$ and α and β are the coefficients of thermal expansion of the liquid.

For mercury

$$\alpha = 1.8 \times 10^{-4}°C^{-1}$$

and

$$\beta = 5 \times 10^{-8}°C^{-2}.$$

Equation (6.1) is the ideal equation for a liquid-in-glass thermometer. In practice there are several factors which modify the ideal behaviour because of the way in which the thermometers are constructed.

Because the glass of a liquid-in-glass thermometer also expands it is the apparent expansion coefficient due to the differential expansion of the liquid with respect to the glass that is of interest. Glass used in a typical thermometer has a value of $\alpha = 2 \times 10^{-5} °C^{-1}$, about 10% that of mercury. Typical apparent expansion coefficients of various liquids are given in Table 6.1. Hence both the glass and liquid act as temperature transducers and thus justify the description 'liquid-in-glass'.

The liquid also serves as the temperature indicator in the stem and consequently may not be at the same temperature as the liquid in the bulb. Fortunately this effect is small for mercury where the bulb volume is 6250 times the volume of the mercury in a $1°C$ length of the capillary stem. Organic liquids have higher expansion coefficients resulting in less liquid volume and give rise to greater non-linearity.

The bore in the stem, which is the signal transmission path, needs to be smooth and uniform. An allowed departure from uniformity is a *contraction chamber* which, by taking up a volume of the expanding liquid, allows the overall length of the thermometer to be kept a reasonable size. The chamber shape must be very smooth to prevent bubbles of gas being trapped. An auxiliary scale is usually added for the ice point if a contraction chamber is used.

The marked scale is required to allow the user to measure the column length as a temperature. For a well-made thermometer the change in length is proportional to the change in volume and hence to the temperature, as per equation (6.1). In order to make the scale, the manufacturer first places 'calibration' marks on the thermometer stem, as shown in Figure 6.2. Depending on the range and accuracy, more than two calibration marks may be used, thus dividing the thermometer stem into segments. A ruling engine is then used to rule a scale between each pair of marks, with careful alignments between the adjacent segments if they occur. The scale rulings will be spaced to approximate equation (6.1) to the accuracy expected for the thermometer type.

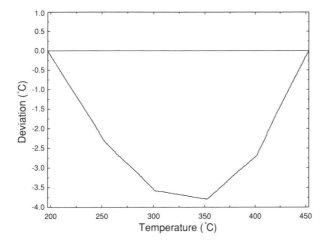

Figure 6.3
The deviations of markings on a high-temperature thermometer
scale from length linearity. The scale has been ruled in five
segments to approximate a curve. The positions of the scale
markings were measured with an automatic laser length-bench
as if they were a ruler. The length of the scale and deviations
from linearity have been expressed in equivalent temperatures.

For example, consider Figure 6.3. This illustrates the achieved scale rulings for a
wide-range high-temperature thermometer. The spacings have been represented by temperatures for ease of interpretation. Five segments have been ruled from the six calibration
marks. Each segment is very linear, i.e. equally spaced steps. However, between segments
the spacings differ. This is an example of a segmented scale similar to those discussed
in Section 4.6. Even though the scale markings have been applied to a precision better
than 1/20 of a scale division, the segmented approximation to equation (6.1) will give
rise to errors greater than this.

Construction details are covered by performance standards published by several organisations. The international ones by ISO do not appear to be greatly followed and instead
thermometers follow the requirements of the British Standards Institute (BSI), the American Society for Testing and Materials (ASTM), or the Institute of Petroleum (IP). Some
of their documentary standards are given in the references at the end of the chapter.

6.3 ERRORS IN LIQUID-IN-GLASS THERMOMETRY

Correcting for errors in liquid-in-glass thermometers can become a complex process.
Even for a moderate accuracy three correction terms may be needed with possibly one
or two subsidiary measurements.

For an overview of the problems consider the measurement model of Figure 2.10.
The liquid-in-glass thermometer is a very compact measuring instrument and hence it is
difficult to isolate the various components of the model because of strong interactions.
Figure 6.4 shows a possible representation. A consequence is that it is difficult to isolate
the causes of error and analyse them.

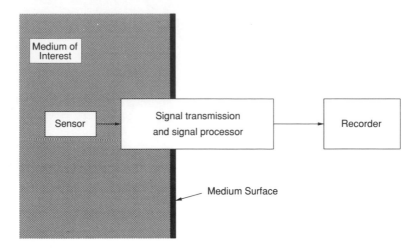

Figure 6.4
A simple measurement model for a liquid-in-glass thermometer.

Ideally the bulb is the sensor. However, because there is also liquid in the contraction chamber and bore, the sensor is not entirely isolated from the temperature profile of the thermometer's signal transmission path and signal processor, i.e. the stem.

The bulb has a thin glass wall which couples the sensor thermally to the physical system and the glass is a major cause of the problems. Because the glass is delicate any mechanical stress affects the readings. The mechanical properties of glass also tend to change with time and temperature.

The other major source of problems is the multiple function of the liquid in the stem, which acts as part sensor, a transmission medium and an indicator. Both the bore and marked scale are intimately related and require high mechanical precision in their construction. As seen above, some of the bore errors are corrected by the way the scale markings are applied. However, only a full calibration can determine how successful this was.

Most sources of error therefore can be seen to arise from thermal and mechanical effects. Acoustic or vibrational effects may make the reading of the meniscus difficult, and in extremes cause effects similar to those for mechanical stress.

Nuclear and magnetic effects need not be considered unless there is a major source of these nearby.

Electrical effects can generally be ignored except where microwaves are used for heating, in which case mercury should not be used.

Glass is inert chemically and only a few substances that are known to attack glass need to be avoided, e.g. hydrofluoric acid. The chemical consequences of any breakage allowing mercury or organic liquids to enter the system under investigation needs to be carefully evaluated, e.g. in food products.

Radiant energy can directly heat the bulb to cause measurement errors. Radiation may induce a chemical change, usually polymerisation, in organic liquids.

6.3.1 Time-constant effects

Thermal lag errors for liquid-in-glass thermometers are as discussed in Sections 4.4.3 and 4.4.4 and may readily be determined for a mercury-in-glass thermometer. The time-constant is determined almost entirely by the diameter of the bulb since heat must be conducted from its outside to its centre. A typical bulb of diameter 5 mm has a relatively short time-constant. The length of the bulb is then determined by the sensitivity required of the thermometer, given that there is a minimum useful diameter for the capillary bore. See Section 6.3.5 for factors affecting the bore diameter.

Table 6.2 gives the $1/e$ time-constants in various media for a 5 mm diameter bulb. Time-constants for other diameters can be estimated by scaling the time in proportion to the diameter. The table clearly indicates that the thermometer is best used with flowing (or stirred) fluids.

6.3.2 Thermal capacity effects

Glass thermometers are bulky and can have considerable mass, especially if a high-precision is called for. The high thermal mass or heat capacity can upset measurements, making high-precision measurements difficult. Inappropriate use of liquid-in-glass thermometers occurs where the thermometer is too massive to achieve the precision required. Pre-heating the thermometer can alleviate the worst of the problem. For higher precision and low mass choose a platinum resistance thermometer or thermistor (see Chapter 5).

Simple estimates of the heat requirements are made by measuring the volume of thermometer immersed, and assuming 2 J are required to raise 1 cm^3 of the thermometer volume (glass or mercury) by 1°C. Section 4.4.2 covers methods for estimating the thermal capacity error.

6.3.3 Pressure effects

The volume of the bulb is sensitive to both external and internal pressure, a pressure coefficient of about 0.1°C per atmosphere being typical for the mercury thermometer. One atmosphere of pressure corresponds to 760 mm height of mercury. Thus the height of the mercury column in the stem can cause a significantly different reading between horizontal and vertical positions (see Figure 6.5). For this reason precision thermometers are calibrated and used vertically. Pressure variations in the fluid being

Table 6.2. Time-constants for a mercury-in-glass thermometer with a 5 mm diameter bulb.

Medium	Still	0.5 m/s flow	Infinite flow velocity
Water	10 s	2.4 s	2.2 s
Oil	40 s	4.8 s	2.2 s
Air	190 s	71 s	2.2 s

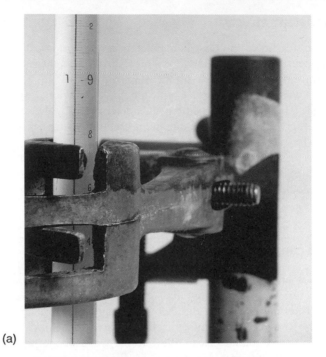

(a)

(b)

Figure 6.5
The pressure effect can be quite large for a long thermometer. Shown here is a bomb calorimeter thermometer with a mercury column about 400 mm long: (a) the vertical reading (as usual in calibration and use) is 19.374°C; (b) the horizontal reading is 19.451°C. The difference of 0.077°C is mostly due to the internal pressure change in the bulb.

measured may also affect the position of the meniscus, making the thermometer difficult to read.

Pressure variations may also be due to directional mechanical forces, such as caused by resting the thermometer on its bulb. In fact this can give rise to long-term problems as the bulb may be so stressed that it does not relax back to its original shape. Liquid-in-glass thermometers should be supported by their stems and not the bulb, and the bulb should be protected against any knocks. Because thermometers appear to be ideal stirring rods, some lower-quality thermometers are made with thick bulbs to allow their use as stirring rods, but stirring should not be done with a precision thermometer!

6.3.4 Bulb hysteresis and drift

A liquid-in-glass thermometer can be considered as having two moving parts: the thermometric liquid and the glass bulb. The liquid volume follows the temperature change rapidly but the glass does not. The result is that the bulb volume, and hence the reading, depends on its thermal history. This is one of the major sources of error that needs to be accounted for in liquid-in-glass thermometry. An accuracy of 0.05°C implies a volume change less than 1 part in 100 000. Volume changes in the bulb are classified in two ways:

- slow changes with time; and
- changes due to temperature.

Not all glasses are suitable for thermometric use: Table 6.3 lists some of the glasses approved by the British Standards Institute. Note that several types have colour stripes in the stem to identify them. A good thermometer should indicate the make of glass so that its generic history can be assumed. The bulb glass is not necessarily the same as the stem glass. All glass does change with time and temperature, thus altering the basic calibration of a thermometer.

Glass is not a stable substance because it is essentially a very viscous liquid and not a crystalline solid; by comparison, quartz would give considerably less hysteresis and drift. A properly annealed glass bulb will steadily shrink in volume with time. This volume change is called a *secular change* and can be corrected for in use. Thermometers are made with an initial ice-point reading just below zero to allow for this rise over the lifetime of the thermometer. All readings will be affected similarly by the volume change so that after the initial calibration any change in the ice-point reading is applied as a correction.

When heated, the bulb expands with the temperature rise, but once the temperature falls the bulb does not contract immediately. For a good thermometer glass, it takes three days for the bulb to relax back close to its original volume. Poor glasses may take weeks if they come back to the original volume at all. A good glass has as much as a 0.01°C change for each 10°C rise in temperature due to the initial expansion, so that the observed hysteresis is about 0.1% of the temperature change — about ten times the hysteresis observed in platinum resistance thermometers. If accuracies of this level are wanted, then careful procedures are needed to ensure that the thermometer is carefully conditioned before and after a reading. You may need to wait three days just to recheck the ice point and be sure your readings were accurate! This period of time is usually of no concern for a calibration laboratory and indeed some national standards laboratories

Table 6.3. Thermometer glasses.

Glass	Identification stripe(s) or approved abbreviation	Normal maximum working temperature (°C)
Normal glass, made by Whitefriars Glass Ltd	Single blue stripe	350
Normal glass, Dial, made by Plowden and Thompson Ltd	Double blue stripe	350
Normal glass, Schott-N16, made by Jenaer Glaswerk Schott and Genossen, Mainz	Single red stripe	350
Normal glass, 7560, made by Corning Glass Co.	CN	350
Corning borosilicate glass, made by Corning Glass Co.	CB	450
Thermometric glass, Schott 2954, made by Jenaer Glaswerk Schott and Genossen, Mainz	Single black stripe	460
Borosilicate glass, made by Whitefriars Glass Ltd	Single white stripe	460
Corning glass, 1720, made by Corning Glass Co.	C1720	600
Schott-Supremax R8409, made by Jenaer Glaswerk Schott and Genossen, Mainz	SPX8409	600

Note. The maximum temperatures given in the last column of the table are a guide to normal practice. The performance of a thermometer depends greatly on the stabilising heat treatment which it has been given during manufacture, and a well-made thermometer of 'normal glass' may be satisfactory for many purposes at temperatures as high as 400°C. On the other hand, for the best accuracy it may be preferred to use one of the borosilicate glasses for temperatures lower than 350°C. In general the lower the maximum temperature of use in relation to the approved temperature of the glass the better will be the 'stability of zero' of the thermometer.

have quite time-consuming procedures to obtain the best out of a thermometer. If you use such high-precision thermometers, follow the conditioning recommendation of the calibration laboratory in order to achieve the claimed accuracy.

A consequence of the calibration methods used to cope with the bulb hysteresis is that the liquid-in-glass thermometer is essentially an instrument for measuring increasing temperatures, i.e. immediately after use at a high temperature the calibration will not apply to subsequent lower temperatures. A variety of methods have been developed to help overcome this limitation, e.g. taking ice points immediately after each reading and adjusting the calibration value according to the shift in the zero. However, if you really need to achieve such accuracy, then use a more suitable type of thermometer such as a platinum resistance thermometer.

6.3.5 Bore non-uniformity effects

The bore needs to be smooth and uniform with the cross-sectional area not varying more than 10% from the average, or not more than 5% for high-precision applications such as calorimetric thermometers. During production the bore is checked by introducing a known amount of mercury and measuring the length of the mercury column at various positions along the bore. Thus changes in the diameter of the bore can be recognised by a change in the length of the column. After manufacture thermometer bore errors cannot be assessed so directly and instead possible errors are controlled by visual inspection. If irregularities in the bore smoothness or uniformity, such as in Figure 6.6, are noticeable, then the thermometer should be discarded.

Where an expansion or contraction chamber is added, it needs to be sufficiently far away from the scale to ensure that the bore is uniform over the scale region. Expansion chambers are safety devices to prevent permanent damage occurring if the thermometer is accidentally overheated. Contraction chambers, on the other hand, transmit the signal to allow a shorter length for high-temperature thermometers. They need to be well shaped to avoid breaks or bubbles occurring in the liquid column. In use the contraction chamber must be at the same temperature as the bulb.

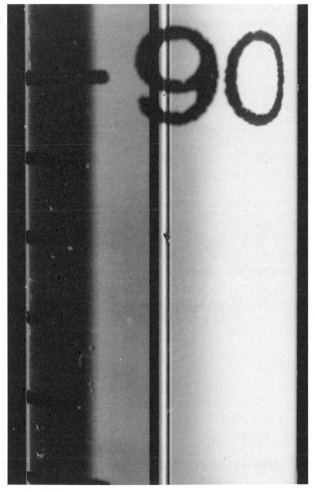

Figure 6.6
Capillary distortion in an inexpensive thermometer. The small inclusion appears to be a piece of foreign material embedded into the glass. To locate such faults requires careful visual examination.

The choice of the bore diameter is a compromise involving several error effects. A large-diameter bore requires a larger-volume bulb to achieve a given resolution, thus increasing the thermal capacity. A small-diameter bore not only becomes difficult to read but also suffers from *stiction* — the mercury moving in fits and starts due to the surface tension between the mercury and the bore wall. Stiction should be kept less than 1/5 of a scale division. In use, follow procedures of Section 6.3.7 to minimise stiction effects. Small-diameters bores also cause the mercury column to break readily and become difficult to rejoin (see Section 6.3.6).

6.3.6 Separated columns

A common problem in use is for a part of the thermometric liquid in the stem to become separated from the main volume. While this will show as an ice-point shift it is still important to make a simple visual check when using the thermometer.

With organic liquids the problem may be harder to identify because liquid adheres to the surface of the capillary tube and may not be visible. Spirit thermometers need to be held vertically to allow the thermometric liquid to drain down. Warm the top of the thermometer to prevent condensation of any vapour. Allow time for drainage of the liquid in the thermometer if the temperature is lowered quickly; approximately 3 minutes per centimetre. Cool the bulb first in order to keep the viscosity of the liquid in the stem low for better drainage.

For mercury the separation is usually visible. Two causes can be identified: boil-off (Figure 6.7) and mechanical separation (Figure 6.8).

To help retard the boil-off of mercury vapour at high temperatures (e.g. above 150°C) the capillary tube is filled with an inert gas when manufactured. Usually dry nitrogen under pressure is used to prevent oxidation of the mercury. The expansion chamber must be kept cooler than the bulb to prevent a high pressure build-up.

Mechanical separation of the liquid column is, unfortunately, a common occurrence, particularly after shipment. A gas fill will also help prevent this separation but, conversely, the gas makes it more difficult to rejoin. There is the risk of trapped gas bubbles in the bulb or chambers and careful inspection is needed to locate them. A vacuum in the capillary tube will give rise to more breaks but they are easily rejoined.

With care it is often possible to rejoin the column and still have a viable thermometer. However, it must be realised that attempts to join a broken column could also result in the thermometer being discarded if the procedure is not successful. Breaks that occur only when the thermometer is heated often require the thermometer to be discarded.

Various procedures for joining a broken mercury column can be tried. The ones below are given in order of preference.

- Lightly tap the thermometer while holding it vertically. This may work for a vacuous capillary stem.

- Apply centrifugal force, but avoid a flicking action, and be careful to avoid striking anything. This can be best done by holding the bulb alongside the thumb protecting it with the fingers, and with the stem along the arm. Raise the arm above the head and bring it down quickly to alongside the leg.

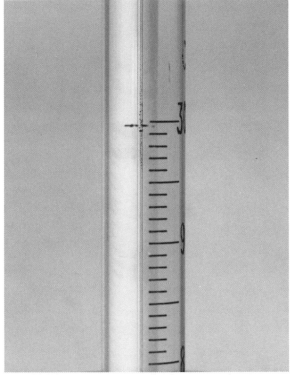

Figure 6.7
A 300°C thermometer showing small globules of mercury
in the bore around the 300°C mark. These have evaporated
off the mercury column at 250°C and condensed up the
capillary tube. (*Note.* the calibration scratch at 300°C is 1/3
scale division lower than the 300°C mark.).

- If both the above are unsuccessful, a cooling method can be tried. This method
  relies on sufficient cooling of the bulb for all the mercury to contract into the bulb,
  leaving none in the stem. The first two methods may also need to be applied to assist
  movement of the mercury. The column should be rejoined when it has warmed to
  room temperature. Carry out the warming slowly so that all the mercury is at the
  same temperature. More than one try may be needed. Three cooling mediums readily
  available are:

(1) salt, ice and water (to −18°C);
(2) 'dry ice', i.e. solid CO_2 (−78°C);
(3) liquid nitrogen (−196°C).

The last two refrigerants require more care as they could freeze the mercury. An
excessive cooling rate may stress the glass. *Cold burns to the user could also occur.*

- A more drastic method of rejoining is to apply heat to allow the rejoining to occur
  in the expansion chamber at the top. This method should be one of last resort!
  Do not use it on high-temperature thermometers in case the bulb breaks releasing
  mercury vapour. Much care is needed to avoid overheating, and ideally a temperature-
  controlled bath should be used. This method sometimes takes the thermometer well

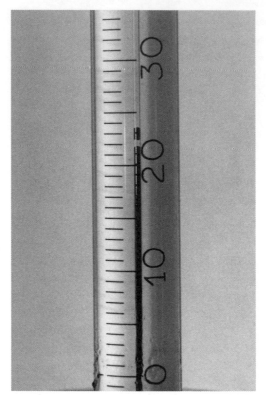

Figure 6.8
A typical break in the mercury column of a thermometer.

above the temperatures at which it was annealed, causing irreversible changes in the thermometer.

If the broken column has been successfully rejoined, then an ice-point (or other reference point) check should be made. If the reading is the same as obtained previously, then the join can be considered completely successful and the thermometer ready for use. (It is essential to keep written records here.) However, a significant ice-point shift indicates that the join was not successful and that the thermometer should be discarded. If the ice-point shift is within that specified for the thermometer type, then treat the thermometer with suspicion until there is evidence of long-term stability, i.e. no significant ice-point changes after use.

6.3.7 Errors in reading

Thermometers have an analogue display which requires some care in reading. The basic principle is to line up the indicator (the liquid meniscus) with the scale in a systematic and repeatable manner.

The problem of lining up two objects with the eye gives rise to *parallax* errors. Parallax is the amount of apparent displacement of the objects when the position of the

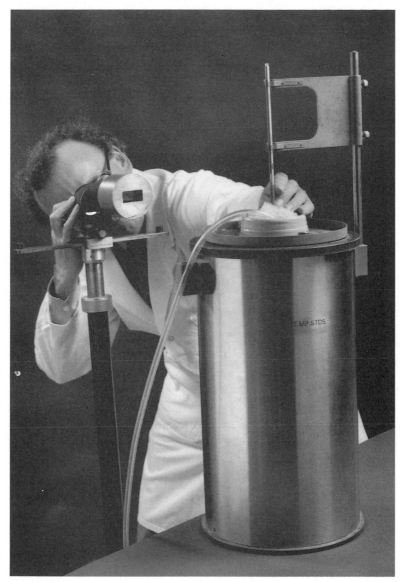

Figure 6.9
A simple monocular telescope with a close-up lens being used to read the ice point on a mercury-in-glass thermometer. Note that (i) the monocular is on a heavy stand to free the user's hands, (ii) its height is set on the same level as the top of the mercury to avoid parallax errors, (iii) the distance for viewing is about an arm's length to allow the user to adjust the thermometer for position, clear any fogging or obstacle away from the scale and to be able to tap the thermometer just before the reading is made.

eye is changed. For a solid-stem thermometer the scale is located 2–4 mm in front of the meniscus and can give rise to significant parallax errors. To avoid the error in reading a scale the eye has to be in the same relationship to the scale and indicator every time.

For liquid-in-glass thermometers this usually means having the eye at the same height as the meniscus.

A convenient way to do this is to have a mounted telescope set to the correct height. The telescope acts as an optical aid to read the distant thermometer and to ease interpolation of the scale.

There are two disadvantages. The smaller field of view means it is harder to know where you are on a thermometer, e.g. does the 2 mark correspond to 22°C or 32°C? The optical magnification also increases the parallax effect. Placing a horizontal slot in front of the telescope will help reduce the parallax error (see Figure 6.9).

Many general-purpose thermometers have coarse markings and variable bores and it is advisable to estimate only whether the meniscus is on a scale marking or between markings, i.e. readings are made to no better than half a scale division. This is the best policy for the general user, so that if readings to 0.1°C are needed, then a thermometer divided to 0.1°C or 0.2°C should be used.

For precision thermometers, such as those used for calibration, the scale markings are fine and the bore uniform (see Figure 6.10). Interpolating readings between scale divisions may give an improvement in accuracy. Estimates to 1/20 of a scale division can be readily made with practice but the accuracy will not be as great as this. The limiting factor is that even skilled observers are not accurate; they tend to favour certain numbers in making estimates.

In the authors' experience, a 99% confidence level of about one-fifth scale division is possible. For this accuracy care should be exercised in reading, as follows:

- Read the column when it is rising, if possible.
- Tap the thermometer lightly before reading to prevent stiction. A small artist's brush is useful and it can also serve to clear any frosting or fog on the scale.
- Use a telescope or other optical aid to read the scale. This allows easier reading as well as eliminating most of the parallax errors. See Figure 6.9 for more details.
- Take at least two independent readings. For each reading re-check the scale markings to ensure they have been interpreted correctly, e.g. is the scale divided to 0.1°C or 0.2°C?
- Remember to divide the interval from the centres of the graduation lines, not their edges. For mercury the top of the meniscus is to be used as the indicator and for spirits the bottom, i.e. the flattest part of the meniscus. See Figure 6.11 for more details.

6.3.8 Immersion errors

We have already mentioned that problems are expected if all the liquid in a liquid-in-glass thermometer is not at the same temperature as the bulb. Because the scale has to be read visually, liquid-in-glass thermometers are used at various immersion depths, which results in different parts of the thermometric liquid being at different temperatures. In addition, the clutter around an experimental set-up often necessitates the thermometer being placed in a non-ideal position.

Three distinct immersion conditions are recognised for a liquid-in-glass thermometer and each requires a different error treatment. Figure 6.12 illustrates the three conditions.

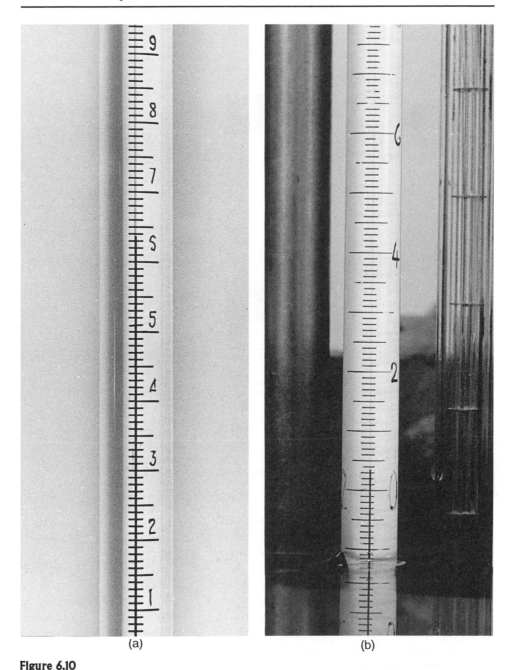

(a)   (b)

Figure 6.10

Difference in quality between two thermometers marked to 0.1°C. Both had been purchased to allow readings to be interpolated down to a few hundredths of a degree. Note that on both the number is not complete, e.g. the 2 displayed in the photographs corresponds to 22°C. (a) Thermometer markings are thick (about 1/5 scale division) and appear uneven; 25.1°C marking is bent and would indicate in this instance that the thermometer is not suitable for the purpose it was bought. Also the numbers are not well formed; there appear to be two 5s and no 6. (b) Markings are fine (1/10 scale division) and appear even. Interpolation between marks would be meaningful.

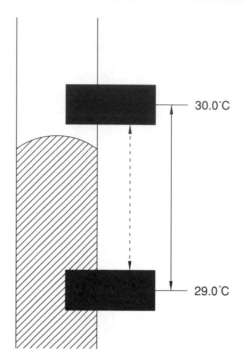

Figure 6.11
Diagram of mercury meniscus in a bore and the scale markings. The tendency is for an observer to subdivide the interval between the inside edges of the scale markings (the dotted line). The interval should be divided between the centres of the marks (solid line). In this instance the first method would give 29.9°C, whereas the correct method gives 29.8°C.

Other types of immersion errors arising from heat transfer, such as those covered in Section 4.4.1, are minor in comparison due to the low thermal conductivity of glass.

Complete immersion   If the complete bulb and stem are immersed at the same temperature the thermometer is completely immersed (obvious!). This condition is not common and should be avoided, especially at higher temperatures. High-pressure build-up in the thermometer may cause it to rupture, spreading deadly mercury vapour throughout the laboratory. Laboratories where there is a danger of mercury exposure to high temperatures should be kept well ventilated. In other words *do not* put your mercury thermometer completely inside your oven to measure the temperature.

Thermometers designed for complete immersion are rare and are not covered here. Special calibration procedures are also needed to handle pressure changes.

Total immersion   Total immersion applies to the situation where all the thermometric liquid, i.e. all the mercury in the bulb, the contraction chamber and the stem, is at the temperature of interest. The remaining portion of the stem will have a temperature gradient to room temperature (approximately). Especially at high temperatures the expansion chamber should be close to room temperature. A very small part of the mercury column may be outside the region of interest, to allow visual readings to be made. The error

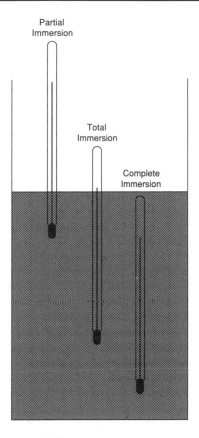

Figure 6.12
Three types of immersion conditions
used for liquid-in-glass thermometers.
The preferred immersion condition
will be given by the thermometer
specification and is usually marked as
a line or distance on the stem.

introduced by this can be estimated by the procedures given below for partial-immersion
thermometers. Obviously the thermometer will have to be moved if a range of temperatures is to be measured. Total immersion thermometers are generally calibrated at total
immersion and therefore do not need additional corrections.

Partial immersion One way around the problem of scale visibility and the need to
move the thermometer is to immerse the thermometer to some fixed depth so that most,
but not all, of the mercury is at the temperature of interest. The part of the mercury
column not immersed is referred to as the *emergent column*. Corrections will be needed
to compensate for the error arising from the emergent column not being at the same
temperature as the bulb. Many thermometers are designed and calibrated for partial
immersion and are marked accordingly with an immersion depth or an immersion line
(see Figure 6.1).

A partial-immersion thermometer is not properly defined unless the temperature profile
of the emergent column is also specified. Usually an average *stem temperature* is quoted

to represent the temperature profile of the emergent column. Thermometer specifications may define the expected stem temperature for a set of test temperatures but they do not usually define stem temperatures for all possible readings. Such partial-immersion thermometers should only be used in tests which specify their use. Many ASTM and IP tests are of this nature.

A measure of the stem temperature is required if the accuracy of the thermometer reading is to be assessed. The traditional way to measure the stem temperature is with a *Faden thermometer*. These are mercury-in-glass thermometers with a very long bulb, with various bulb lengths available. The bulb is mounted alongside the part of the stem containing the emergent column with the bottom of the bulb in the fluid. An average temperature is obtained as indicated by the Faden thermometer. Other ways of measuring the temperature profile are to use thermocouples along the length of the thermometer or even several small mercury-in-glass thermometers. The stem temperature may be calculated as a simple average but strictly it should be a length-weighted average.

Because the measured stem temperature may not be the same as that given on the calibration certificate it is necessary to make corrections for the difference.

For partial-immersion thermometers the true temperature reading t is given by:

$$t = t_i + N(t_2 - t_1)k \tag{6.2}$$

where t_i = indicated temperature

N = length of emergent column expressed in degrees, as determined by the thermometer scale

t_2 = the mean temperature of the emergent column when calibrated (i.e. the stem temperature on a certificate for partial immersion or the thermometer reading for a total-immersion certificate)

t_1 = the mean temperature of the emergent column in use

k = the coefficient of apparent expansion of the thermometric liquid used, in the glass of which the thermometer stem is made. See Table 6.1 for suitable values to use for normal temperature ranges.

The use of equation (6.2) with typical k values from Table 6.1 is estimated to give a 10% accuracy (95% CL) for the correction term. Consequently the correction is a major source of uncertainty for large-temperature differences.

Figure 6.13 gives a chart derived from equation (6.2) for mercury thermometers which enables the stem correction to be determined graphically. You should become familiar enough with it to make quick estimates in order to determine whether the immersion condition error is significant and therefore needs correction.

Thermometers are usually calibrated at their stated immersion conditions and the actual stem temperatures during calibration are measured and quoted on the certificate. In many applications a thermometer is used for a standard test method (such as specified by the ASTM or IP). For these instances the expected stem temperature is specified and there is no requirement for the stem temperature to be measured. The user will, however, need to adjust the certificate correction terms to the immersion conditions of the specification in order to see that the thermometer corrections meet the appropriate quality criteria of, say, the ASTM or IP standard.

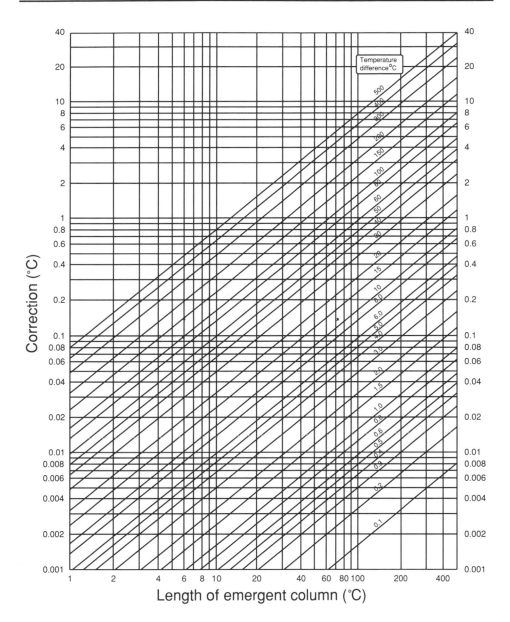

Figure 6.13
Chart of stem exposure corrections for mercury-in-glass thermometers with $k = 0.00016°C^{-1}$.

Using the chart (Figure 6.13) for different immersion conditions

Total-immersion thermometer used in partial immersion Measure the total length of the emergent column in degrees Celsius. If part of the stem is undivided, its length in degrees is estimated by comparison with the divided scale. From the length of the emergent column at the top or bottom of the chart, follow a line vertically until it reaches the diagonal line corresponding to the difference between the thermometer reading and the mean temperature of the emergent column. The height of this intersection, measured on the vertical scale, gives the correction to be applied. It is positive when the stem temperature is lower than the thermometer reading and negative when it is higher.

Example 6.1   Stem correction for total-immersion thermometer used in partial immersion

A high-temperature total-immersion thermometer is used in partial immersion to measure the temperature of an oil bath. The reading is 375.0°C after applying certificate corrections and allowing for any ice-point shift. The mean temperature of the emergent column is 75°C and the length of the emergent column is 150°C. Calculate the true temperature (a) using equation (6.2), (b) using the chart of Figure 6.13.

(a)   *Using equation (6.2)*

$$t = 375.0 + 150(375 - 75)1.6 \times 10^{-4}°C$$
$$= 375.0 + 7.2°C$$
$$= 382.2°C.$$

This is the temperature that the thermometer would indicate if it were used in total immersion. Note that the calculation should be iterated using 382.2°C in place of 375°C in order to obtain a better accuracy.

(b)   *Using the chart*
Find the vertical line corresponding to an emergent-column length of 150°C. Move up the line until it intersects with the angled line corresponding to the 300°C temperature difference (300°C = 375 − 75°C). Now move horizontally from the intercept to the vertical scale and read the correction of 7.2°C. Since the mercury in the column is cooler than it was in calibration, the thermometer will be reading low. Hence 7.2°C must be added to the reading.

Exercise 6.1

A total-immersion thermometer is used in partial immersion to measure the temperature of a cold bath. The indicated temperature is −31.50°C after all certificate corrections have been applied. The emergent column length is 10°C, and the mean stem temperature is 15°C. Calculate the true temperature. (Answer −31.57°C)

Change in stem temperature of partial-immersion thermometer   When a partial-immersion thermometer is used at the immersion for which it has been calibrated, but with a different stem temperature, follow the same procedure as in correcting the readings of a total-immersion thermometer. But, in evaluating the temperature difference, use the stem temperature given on the calibration certificate instead of the thermometer reading.

Example 6.2   Stem correction for partial-immersion thermometer

A partial-immersion thermometer indicates a temperature of 375°C after the certificate corrections have been applied. In calibration the stem temperature was 70°C and in use it is 85°C. The emergent column length is 150°C. Calculate the true temperature (a) using equation (6.2), (b) using the chart of Figure 6.13.

Continued on page 221

Continued from page 220

(a) *Using equation (6.2)*

$$t = 375.0 + 150(70 - 85)1.6 \times 10^{-4}°C$$
$$= 375.0 - 0.36°C$$
$$= 374.6°C.$$

(b) *Using the chart*

Find the vertical line corresponding to 150°C emergent-column length. Move up the line until it intersects the 15°C temperature-difference line. Now move horizontally to read 0.36°C off the horizontal scale. Since the stem of the thermometer is hotter in use than it was in calibration, the thermometer reading is high. Hence the correction is −0.36°C.

Exercise 6.2

A partial-immersion thermometer indicates a temperature of −31.50°C after the certificate corrections have been applied. In calibration the stem temperature was 20°C and in use it is 15°C. The emergent-column length is 10°C. Calculate the true temperature.

(Answer −31.49°C)

It is instructive to compare the results of Examples 6.1 and 6.2. The stem correction for the total-immersion thermometer used in partial immersion is about 20 times that for the partial-immersion thermometer. It is important that thermometers are used in immersion conditions as close as possible to their calibration conditions so that the stem corrections and the accompanying $\sim 10\%$ uncertainty are kept small. Example 2.11 in Section 2.10.1 discusses the uncertainty in the stem correction.

Change in both temperature and length of emergent column   To convert from one condition of partial immersion to another, when both the lengths and temperatures of the emergent columns are different, it is best to find the correction to full immersion for each condition. The difference between the two corrections then gives the correction to apply to convert from one condition of partial immersion to the other.

Example 6.3   A difficult stem-correction calculation

A partial-immersion thermometer indicates a temperature of 200.0°C after the certificate corrections have been applied. In use it has an emergent column length of 100°C, with a mean stem temperature of 70°C. In calibration it had an emergent column length of 50°C and a mean stem temperature of 90°C. Calculate the true temperature.

(a) *Correction to total immersion for calibration*

$$\Delta t = 50(200 - 90)1.6 \times 10^{-4}°C$$
$$= +0.88°C.$$

(b) *Correction to total immersion for use*

$$\Delta t = 100(200 - 70)1.6 \times 10^{-4}°C$$
$$= 2.08°C.$$

Continued on page 222

Continued from page 221

Since the correction in use is greater than when calibrated, the thermometer is reading low. Hence the correction must be positive, i.e.

$$\Delta t = 2.08 - 0.88 = +1.2°C.$$

Hence the true temperature is 201.2°C.

Perhaps more important than the size of the correction is the sign, and it pays to double-check it. Get a colleague to go over the calculation independently. As a guide, if the thermometer is in a hot medium then the average temperature of the emergent mercury is lower, and so a positive correction is needed.

6.3.9 Scale errors

Scale errors are a major source of error in liquid-in-glass thermometry and the calibration is the main means of dealing with them. Because it is difficult to isolate the effect of the bore and scale, some bore errors will be taken into account by the treatment of scale errors.

Three sources of scale errors can be distinguished:

- misalignment of the scale;
- the method of linearisation; and
- ruling irregularities of the ruling machine.

A common feature is that discontinuities are introduced. (Note that by a discontinuity we mean a rapid change in the thermometer error, which usually occurs within one scale division but possibly over several scale divisions.)

Misalignment errors

Section 6.2 has discussed how a scale is manufactured. If the ruling machine is not aligned to the calibration marks then several types of error occur. Firstly, there can be a zero-shift error due to a simple linear displacement. This error can be handled in the same way as the ice-point correction. Secondly, the overall length of the scale may be shorter or longer than it should be. This error will show as a systematic change in the error along the scale. A third error type will occur if two or more segments have to be ruled. Incorrect alignment between segments will create a discontinuity: a scale division larger or smaller than its neighbours. Visual examination can identify the worst of these errors but otherwise they will have to be handled by the statistical treatment of errors (see Section 6.5.2).

Linearisation errors

A main function of the scale is to compensate for the non-linearity in the expansion of the liquid. This was highlighted in Figure 6.3, which showed the length linearity of a five-segment thermometer scale. In the context of the measurement model of Figure 6.4 the signal-processing function performed by the scale is a form of segmented linearisation (see Section 4.6).

The segmented linearisation as applied to liquid-in-glass thermometers is not ideal for two reasons. Firstly, it usually results in a very jagged error curve which makes interpolation difficult. Secondly, if the scale is placed on the stem correctly, only the ends of each scale segment have no error. Everywhere else on the thermometer scale there will be errors, all with the same sign. Thus all the readings on the thermometer are subject to a systematic error. Some manufacturers use non-linear ruling engines to rule the scales on their precision thermometers to overcome both of these problems.

As the linearisation errors have been designed into the thermometer and are unknown to the user they can be handled only by a statistical treatment of calibration errors. Some of the errors will be periodic and exhibit discontinuities.

Ruling errors

The ruling machine used in creating the scale illustrated by Figure 6.3 created the desired scale with a high precision. However, not all ruling machines will always achieve this precision. Consider the scale markings in a thermometer whose scale is illustrated in Figure 6.14. The curve suggests that the scale may have been ruled in 12 segments of 50 scale divisions each. No sign of the calibration marks appears on the thermometer and they may have been covered by the scale markings. However, it is clear that the ruling machine has caused distortions in the scale. The resulting errors will be periodic and discontinuous as well as random. The worst of the errors may be eliminated by visual

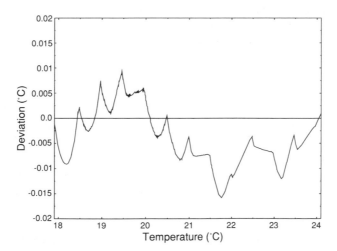

Figure 6.14

The deviations from length linearity of a calorimeter thermometer scale. The scale appears to have been ruled in 12 segments. The ruling machine has created distortions. The positions of the scale markings were measured with an automatic laser length-bench as if they were a ruler. The length of the scale and deviations from linearity have been expressed in equivalent temperatures.

examination but a statistical treatment will be needed to cope with the expected error distribution.

The specifications for this type of thermometer require a calibration at every 50 scale divisions to enable an accuracy of 1/5 scale division. In the case of Figure 6.14, this would give an unrepresentative view of the likely error for the thermometer.

6.4 CHOICE AND USE OF LIQUID-IN-GLASS THERMOMETERS

Considerable care is needed in the selection and use of liquid-in-glass thermometers. Many of the errors can be reduced to manageable levels if we do not require either high-precision or extremes of the usable range. The recommendations made here therefore concentrate on the non-specialised use of liquid-in-glass thermometers. Consult the references at the end of the chapter for details on more specialised thermometers.

In many cases the choice and use of a liquid-in-glass thermometer may be completely covered by the documentary standard for the test to be conducted, e.g. many ASTM and IP tests are of this nature. In such a situation, follow the documentary instructions and use this book as an aid to understanding the requirements. Many of the tests are for situations where it is difficult to measure the true temperature of the physical system of interest. Therefore careful adherence to the test specification is needed for repeatability and consistency between different laboratories.

6.4.1 Choice of thermometer

Choice of a liquid-in-glass thermometer depends on various factors with the main concern being its ability to measure the required temperature with sufficient accuracy. Like most probe thermometers they are best suited to measurement of flowing or stirred fluids. In some applications their long time-constant and high thermal capacity may be disadvantageous. The main advantages of liquid-in-glass thermometers are that they are fully self-contained and have lower initial cost, chemical inertness, low susceptibility to electrical interference, and low thermal conductivity. Their chemical inertness makes them ideal in chemical laboratories using a lot of glassware; there are not many thermometers that would survive having acid accidentally spilt over them. Against this is their fragile nature, the risk of mercury and glass contamination, the need for line-of-sight viewing, and the more complex procedures for high-precision.

A bewildering number of types of liquid-in-glass thermometers are available. Where possible choose those that are made to a recognised specification. In particular, determine what dimensional tolerances are suitable for your application since most thermometers have to be replaced eventually. The dimensional variations amongst thermometers of the same type may require your apparatus to be rebuilt or a change in procedures. Thermometers with controlled dimensions will be more expensive to purchase.

6.4.2 Range and type

For best performance solid-stem mercury-in-glass thermometers should be restricted to operation over the maximum range $-38°C$ to $250°C$. Outside of this range choose a

Table 6.4. Summary of requirements for ASTM precision thermometers.

ASTM thermometer number	Range (°C)	Maximum length (mm)	Graduation at each (°C)	Maximum error (°C)
62C	−38 to +2	384	0.1	0.1
63C	−8 to +32	384	0.1	0.1
64C	−0.5 to + 0.5 and 25 to 55	384	0.1	0.1
65C	−0.5 to +0.5 and 50 to 80	384	0.1	0.1
66C	−0.5 to +0.5 and 75 to 105	384	0.1	0.1
67C	−0.5 to +0.5 and 95 to 155	384	0.2	0.2
68C	−0.5 to +0.5 and 145 to 205	384	0.2	0.2
69C	−0.5 to +0.5 and 195 to 305	384	0.5	0.5
70C	−0.5 to +0.5 and 295 to 405	384	0.5	0.5

different type of thermometer, e.g. a platinum resistance thermometer, which may allow you to do away with glass thermometry altogether. The purchase should be guided by a specification as published by a recognised standards body, e.g. ASTM, BS, IP, or ISO. Thus you will have a generic history on which to base procedures, as well as having a basis for negotiation with suppliers if they fail to meet specification. Beware that some type numbers may be the same, yet may refer to different thermometers, such as in the IP and ASTM ranges. Make sure the specification body is referred to, e.g. an order for a 16C thermometer could result in either an ASTM 10C, the equivalent of IP 16C, or an IP 61C, the equivalent of ASTM 16C.

Your choice of thermometer will most probably be a compromise between the best range, scale division and length for your purpose. If you need good precision, then the thermometer range will be constrained so as to avoid extremely long and unwieldy thermometers. Table 6.4 gives the specification for precision thermometers based on the compromise as seen by the ASTM. The best precision for ASTM liquid-in-glass thermometers is around 0.1°C with the thermometers supplied being accurate to one scale division. If higher resolution is required, say for reference thermometers with a resolution of 0.01°C, then there may be scale errors of several scale divisions. Table 6.5 shows how a British Standard sees the necessary compromises. As a rule, choose thermometers subdivided to the accuracy you wish to achieve and do not rely on visual interpolation to increase the accuracy. If you find you are relying heavily on interpolation, then a better choice of thermometer should be made.

Tables 6.4 and 6.5 have thermometers with an ice point either in the main scale or as an auxiliary scale. The ice point is a very convenient way to check on the on-going performance of a thermometer, and without it more expensive time-consuming procedures may be needed.

6.4.3 Acceptance

Once a thermometer has been received from a supplier you should subject it to a thorough visual inspection, preferably with a magnification up to 20×. Figure 6.15 gives a checklist of main physical dimensions that need to be checked. In principle, the supplier should have carried out this inspection but unless it was written into the purchase contract, with evidence supplied, then it is unlikely that a detailed inspection was made.

Table 6.5. Details of British Standard secondary reference thermometers (reproduced by permission of BSI, from BS1900).

Designation mark	Range (°C)	Maximum overall length	Gradua- tion at each (°C)	Maximum error at any point (°C)	Maximum permitted interval error in an interval* (°C)
SR1/30C	−80 to +30	405	0.5	1.0	1.0/2.0
SR2/2C	−40 to +2	455	0.1	0.3	0.3/5
SR3/20C	−20 to +20	405	0.1	0.2	0.2/5
SR4/1C	−11 to +1	485	0.02	0.1	0.1/2
SR4/11C	−1 to +11	485	0.02	0.1	0.1/2
SR5/20C	−0.5 to +0.5 and 9.5 to 20.5	485	0.02	0.1	0.1/2
SR5/30C	−0.5 to +0.5 and 19.5 to 30.5	485	0.02	0.1	0.1/2
SR5/40C	−0.5 to +0.5 and 29.5 to 40.5	495	0.02	0.1	0.1/2
SR5/50C	−0.5 to +0.5 and 39.5 to 50.5	485	0.02	0.1	0.1/2
SR5/60C	−0.5 to +0.5 and 49.5 to 60.5	485	0.02	0.1	0.1/2
SR5/70C	−0.5 to +0.5 and 59.5 to 70.5	485	0.02	0.15	0.15/2
SR5/80C	−0.5 to +0.5 and 69.5 to 80.5	485	0.02	0.15	0.15/2
SR5/90C	−0.5 to +0.5 and 79.5 to 90.5	485	0.02	0.15	0.15/2
SR5/100C	−0.5 to +0.5 and 89.5 to 100.5	485	0.02	0.15	0.15/2
SR6/18C	−1 to +18	485	0.05	0.1	0.1/3
SR6/34C	−0.5 to +0.5 and 16 to 34	485	0.05	0.1	0.1/3
SR6/51C	−0.5 to +0.5 and 33 to 51	485	0.05	0.1	0.1/3
SR6/68C	−0.5 to +0.5 and 50 to 68	485	0.05	0.15	0.15/3
SR6/85C	−0.5 to +0.5 and 57 to 85	485	0.05	0.15	0.15/3
SR6/102C	−0.5 to +0.5 and 84 to 102	485	0.05	0.15	0.15/3
SR7/51C	−1 to +51 and 99 to 101	505	0.1	0.2	0.2/10
SR7/101C	−1 to +1 and 49 to 101	505	0.1	0.2	0.2/10
SR8/151C	−1 to +1 and 99 to 151	540	0.1	0.2	0.2/10
SR8/201C	−1 to +1 and 149 to 201	540	0.1	0.3	0.3/10
SR8/251C	−1 to +1 and 199 to 251	540	0.1	0.5	0.5/10
SR9/202C	−2 to +2 and 98 to 202	540	0.2	0.4	0.4/20
SR10/302C	98 to 102 and 198 to 302	540	0.2	1.0	1.0/20
SR11/452C	98 to 102 and 198 to 452	590	0.5	1.5	1.5/25
SR12A/505C	95 to 505	590	1.0	2.0	2.0/50

*Expressed in the form maximum permitted interval error/interval, the interval error being the algebraic differ- ence between the errors at opposite ends of the interval. For example, 0.2°C/5°C is written as 0.2/5 and means that the change of error in any interval of 5°C does not exceed 0.2°C.

Suppliers may request a premium for carrying out such a preselection, but the cost will usually repay itself in reducing the time wasted on a bad thermometer and the cost of a calibration.

The points to watch for are:

- breaks and bubbles in the mercury which may be repairable (Section 6.3.6);
- any foreign matter in the capillary;
- distortions in the capillary or scale;
- the existence of fine scale marking, less than one-fifth of a division;
- the scale markings to be at multiples of 1, 2 or 5;

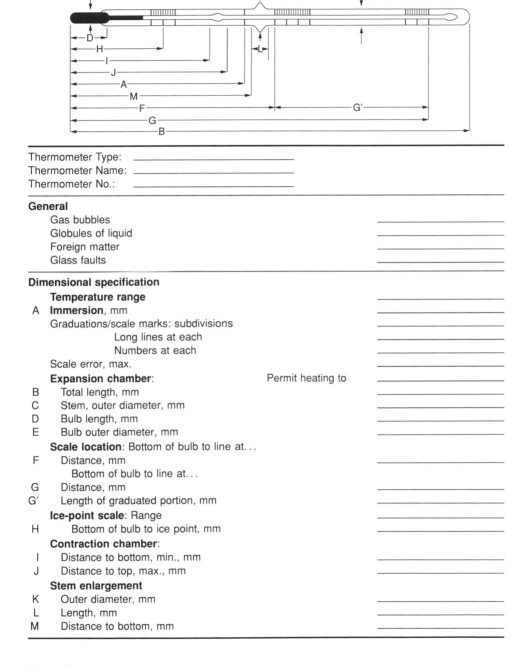

Thermometer Type: _____
Thermometer Name: _____
Thermometer No.: _____

General
    Gas bubbles _____
    Globules of liquid _____
    Foreign matter _____
    Glass faults _____

Dimensional specification
    Temperature range _____
A **Immersion**, mm _____
    Graduations/scale marks: subdivisions _____
             Long lines at each _____
             Numbers at each _____
    Scale error, max. _____
    Expansion chamber: Permit heating to _____
B Total length, mm _____
C Stem, outer diameter, mm _____
D Bulb length, mm _____
E Bulb outer diameter, mm _____
    Scale location: Bottom of bulb to line at. . .
F Distance, mm _____
     Bottom of bulb to line at. . .
G Distance, mm _____
G′ Length of graduated portion, mm _____
    Ice-point scale: Range _____
H   Bottom of bulb to ice point, mm _____
    Contraction chamber:
I Distance to bottom, min., mm _____
J Distance to top, max., mm _____
    Stem enlargement
K Outer diameter, mm _____
L Length, mm _____
M Distance to bottom, mm _____

Figure 6.15
A checklist to record the result of the visual and dimensional inspection of a thermometer when first obtained. This is important for ASTM and IP thermometers, where the correct physical size is essential for replacement.

- the dimensions to match the specifications;
- the required markings to be on the stem, e.g. possible markings include:
  - temperature scale
  - immersion condition
  - immersion line
  - gas fill or vacuum
  - bulb glass
  - serial number
  - vendor's name
  - specification body
  - type number.

See Section 6.4.4 for adding your own markings or serial number.

If the thermometer is mechanically sound then an ice-point calibration should be made and recorded. Check this value with the value supplied (if at all!) by the supplier. Any large difference, e.g. greater than one-fifth scale division, could indicate a fault or mishandling of the thermometer.

Any special purpose thermometer not purchased to a recognised specification should undergo a thermal cycling test. This will also apply to specified thermometers used over 300°C and those you are suspicious about. Warm the thermometer to the maximum temperature you propose to use it for (providing it is within the range of the thermometer!). Leave for an appropriate period, e.g. the expected time in use, and let it cool to room temperature. An ice point taken straight away should not be out by more than a scale division, except in the case of a high-precision or high-temperature thermometer. After three days the ice point should have relaxed back to within one-fifth of a scale division of its original value . Reject the thermometer if it does not stabilise, or subject it to more thermal cycling tests. Note that this test is designed to eliminate bad thermometers and does not give an assurance of good behaviour in the future.

6.4.4 Etching

Manufacturers usually supply thermometers with serial numbers on them. If there is no serial number or if other required markings are absent it is necessary to put them on the thermometer so that an accurate record of the thermometer may be kept. Marking tools with tungsten carbide tips can be used to scratch markings directly on to the thermometer stem, and while this may endanger the thermometer if not done carefully, it is the preferred method for personal safety. The alternative method is to etch the glass stem with hydrofluoric acid. *This should **not** be attempted if your laboratory does not have safety procedures and facilities for handling dangerous chemicals. Hydrofluoric acid is extremely dangerous and should not be allowed to touch the skin.*

If hydrofluoric etching is used, the following procedure should be carried out in a fume cupboard with the operator's hands protected by rubber gloves. The thermometer is first degreased with a solvent such as white spirits. The top 50 mm are then dipped in a bath of melted wax (microcrystalline) which is maintained at such a temperature that a thin transparent layer of wax is left on the thermometer. When the wax has set, the required marking is made on the thermometer stem by cutting through the layer of wax with a stylus. The wax chips produced are brushed off the stem with a soft brush.

The hydrofluoric acid is then painted on the wax-covered stem and left for 4 minutes. The acid is removed by washing with water and the wax removed by remelting and wiping off. The etched markings can then be filled with black engraving filler, while the thermometer is still hot.

6.4.5 Use of thermometer

The main constraint in the use of a liquid-in-glass thermometer is the need to read it visually. Any apparatus employing liquid-in-glass thermometers needs to be designed to allow this, with the main variable available to the designer being the depth of immersion, unless specially shaped thermometers are used. Methods to cope with different immersions are given in Section 6.3.8. If you find you have to move the thermometer away from its proper position in order to read it then you should use a different type of thermometer or should redesign the apparatus.

Some general points in the use of the thermometer:

- *Do not* drop the thermometer; it causes irreversible changes!
- Hold the thermometer vertically by the stem and do not let it rest on its bulb.
- Keep the bulb protected and free from knocks.
- Keep the thermometer below its maximum indicated temperature.

6.4.6 Organic liquids

Thermometers with organic liquids have three possible uses:

- for measuring temperatures below $-38°C$;
- for situations where mercury is to be avoided; and
- for inexpensive thermometers.

The utility of spirit thermometers is limited because of the lower achievable accuracy, the high non-linearities and the volatile nature of the liquids. Organic-liquid thermometers are also difficult to read because of the very clear liquid and concave meniscus. However, the use of a suitable dye and wide bore can give them as good a readability as mercury. Follow the recommendations of Section 6.3.6 on Separated Columns and Section 6.4.7 on Storage in order to get the best result from organic-liquid thermometers.

6.4.7 Storage

Most mercury-in-glass thermometers can be stored horizontally on trays in cabinets, care being taken to avoid any weight or pressure on the bulbs (one reason for the horizontal position). Avoid vibration. Corrugated cardboard, or similar material, can be used as a liner for a tray to prevent the thermometers rolling.

Thermometers whose main range is below $0°C$ are better stored vertically, bulb down, in a cool place, but do not rest the thermometer on its bulb. This particularly applies to organic liquid thermometers, which also should be shielded from light sources as ultraviolet radiation can often degrade the liquid. If the top of the bore of a spirit thermometer is kept at a slightly higher temperature than the rest of the thermometer, then the volatile liquid will not condense in the expansion chamber.

6.4.8 Transport

It may be necessary to send your thermometer away periodically for calibration. Remember that there is often a considerable investment in a good thermometer, not just in the purchase price but in the calibration and a recorded history of its behaviour. Therefore, it is important that it survives shipment. Make an ice-point or similar reference point check before shipment.

The preferred method of transport of thermometers is by safe hand; otherwise use the most reliable delivery service.

Regardless of the method of despatch, thermometers should be adequately packed to ensure their safe arrival. The following procedure has been found to be highly reliable (see Figure 6.16):

- Use a wooden box with a lid secured by screws.
- Line the inside of the box with a flexible foam. The thickness and density of the foam must be sufficient to prevent the thermometer from coming into contact with the inside of the box while in transit.
- Support the thermometer firmly throughout its length to prevent vibration and sliding of the thermometer. The use of expanded polystyrene with slots cut by the hot wire

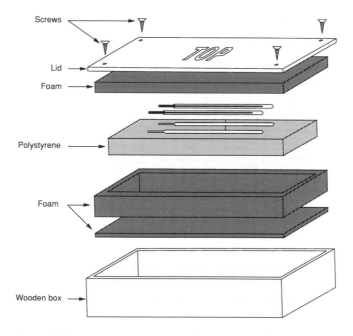

Figure 6.16
Outline diagram of a box suitable for transporting liquid-in-glass thermometers.

method is satisfactory. If thermometers are to be transported inside their protective tubes then it is essential that they are packed firmly into the tubes, supported from end to end and unable to move inside the protective tube. Firmly packed cotton-wool is suitable.

- Keep the thermometers separated within the box. Lengths of wooden dowelling may be used to fill up unused spaces in a multi-thermometer box.

- Place all thermometer bulbs at the same end of the box, and label the box to make clear which side is the top.

As air cargo is the most common means of transport for thermometers, a few aspects need to be noted. The transport of goods by air is covered by the International Air Transport Association (IATA) and its regulations need to be observed, especially those concerning restricted articles for which special documents have to be prepared. Mercury is a restricted article and the regulations require mercury to be properly sealed inside a non-breakable container. Clearly glass is not; therefore some form of sealing is required. If the thermometer tubes are used, and properly packed, they can be sealed with a suitable tape. Otherwise it will be necessary to wrap the wooden box or its insert in a strong plastic bag and seal it.

Spilt mercury inside an aircraft can be a direct danger to the aircraft itself because aluminium is used in its construction. Mercury removes the oxide coating off aluminium and thus allows the aluminium to burn slowly in air. Such damage is difficult to locate without a complete scan of the aircraft. A number of airlines refuse to ship mercury.

Two extracts from the IATA regulations are also of interest:

- Vibrations in commercial aircraft to which packages are exposed range from 5 mm amplitude at 7 cycles/s (corresponding to 1 g acceleration), to 0.05 mm amplitude at 200 cycles/s (corresponding to 8 g acceleration).

- Except as otherwise specified in these regulations, completed and filled packages shall be capable of withstanding one 1.2 m drop test on solid concrete or other equally hard surface in the position most likely to cause damage.

6.5 CALIBRATION

Calibration procedures for liquid-in-glass thermometers have to take into account larger discontinuities than is common for other thermometers. These discontinuities occur at fixed points on the temperature scale due to bore or scale irregularities and at variable temperatures due to mercury sticking in the bore. Indeed, as we have previously noted, visual inspection is important to identify and eliminate thermometers with obvious scale and bore irregularities. Also the reading procedure is designed to reduce the sticking of the column in the bore. The chance of a discontinuity occurring is not eliminated by these procedures, but the magnitude of the likely error is reduced to be comparable to other errors.

Over the long history of liquid-in-glass thermometry a variety of calibration practices has developed to cope with these problems. One aim was to obtain the best possible performance; at one time liquid-in-glass thermometers were the main reference thermometers used by national standards laboratories. As this is no longer the case a simpler approach can be taken. The procedure given in Section 6.5.4 follows directly from the main calibration principles as used in this text. Because the user will come across other procedures which may have to be followed, the background behind them is given below.

6.5.1 Generic type tests

Traditionally calibration procedures for liquid-in-glass thermometers have not used statistical techniques to evaluate the uncertainty of the calibration. However, quoted correction terms are often the simple average of at least two readings. Instead of a statistical emphasis, a variety of tests is carried out to show that the thermometer conforms to the standard specification for its manufacture. Among the tests required are usually the following:

- visual examination for defects;
- dimensional inspection of the thermometer;
- visual examination for scale clarity;
- thermal cycling to establish ice-point stability;
- verification of readings at selected points;
- restriction on the maximum error observed;
- restriction on the rate of change of error;

Tables 6.4 and 6.5 illustrate typical error restrictions. However, there is no objective assessment of uncertainties. A Type B assessment (see Section 2.7) of uncertainties based on the generic behaviour can be made by experts experienced with the thermometer type. Typically the reported uncertainty is 1/5 to 1/2 of a scale division.

The above verification tests and the error restrictions are usually based on data from intercomparisons with a reference thermometer, taken at set intervals of 50–100 scale divisions. A problem with this approach is how to determine the correction for an intermediate value and assign it an uncertainty. To meet this essential calibration requirement further Type B assessments would have to be made by the calibration laboratory or the user.

The main disadvantage of relying solely on Type B assessments is that an experienced person with a high expertise in liquid-in-glass thermometry is required for consistent results. It is not a suitable calibration method for beginners.

Example 6.4

A calibration certificate for a mercury-in-glass thermometer divided to 0.1°C gives a correction of +0.1°C at 20°C and a correction of −0.1°C at 30°C. The uncertainty is given as ±0.03°C. What is the correction and uncertainty at 25°C?

If we could assume a smooth function we could use a linear interpolation and assign, say, an additional uncertainty of 1/5 of a scale division. Thus the correction at 25°C is 0°C ±0.04°C. However, the change in correction value could be due to a single discontinuity at any point between 20°C and 30°C. That is, the correction could be out by a scale division. If this discontinuity size was typical of the thermometer, then a better uncertainty estimate is ±0.2°C. More data would be required to establish this.

6.5.2 The error distribution for discontinuities

Both the above problems, the need for interpolation and the need for skill, can be largely overcome by a statistical approach, i.e. a Type A assessment of the uncertainties. By following, say, the calibration procedures for direct-reading electronic thermometers (see Section 4.6) the correction and the uncertainty can be determined for any thermometer reading. If the quality inspections overlooked any faults, then this will show as a high uncertainty. Thus a poor thermometer can be detected and discarded even by a person not familiar with the type of liquid-in-glass thermometer. For a good thermometer an evaluation of the uncertainties by a Type A or B approach should give similar results, i.e. 1/5 to 1/2 a scale division.

One way of making a Type A assessment is to calibrate the thermometer at every scale marking. As there can be over 500 scale markings, this task is not practical. However, in length measurements where the process can be automated, checking all the scale markings on a ruler is the preferred method. An alternative is to sample a sufficient number of scale markings and statistically estimate the uncertainty. The validity of this approach depends on the type of distribution function (see Section 2.3) used to describe the errors.

The consideration of scale errors, Section 6.3.9, has shown the types of errors involved if it were possible to examine all scale markings. Several types of error distribution are clearly involved and the mixture and number of the various types of distributions cannot be known in advance. Discontinuities can also give rise to distributions that are skew and narrower than the Gaussian distribution. It is reasonable to assume that the distribution is Gaussian since it gives a slightly pessimistic estimate of the uncertainty. Thus total immersion thermometers can be treated statistically as direct-reading thermometers (see Section 4.6). Depending on the required precision a single correction or a deviation function may be used.

There is a proviso with this approach, namely the assumption that the scale on the thermometer is a reasonable match to ITS-90. This is generally true for total-immersion thermometers but may not be for partial-immersion thermometers. The scale on a partial-immersion thermometer is a function of two temperatures: those of the bulb and the stem. Unfortunately the stem temperature is usually not defined by specifications for every point on the thermometer. Indeed it is possible to manufacture a partial-immersion thermometer to meet its accuracy requirements at the designated stem temperatures without knowledge or consideration of intermediate stem temperatures. There are two solutions to this problem: one is to use the short-range calibration method described in the next section for the defined points, and the other is for the user to define a stem-temperature profile for the whole thermometer scale.

6.5.3 Suitable fitting functions

Selection of the intercomparison points depends in part on the functions to be used to fit the deviations. These functions in turn depend on the level of uncertainty required. For a liquid-in-glass calibration we are not necessarily interested in the least uncertainty for the specific thermometer. Instead we want a correction value to remove the bulk of the error and a statistical assessment of the resulting uncertainty.

The simplest approach is to take the average of the deviations and calculate the uncertainty from their variance. This corresponds to determining a single zero shift on the thermometer; the most common error which arises from a change in the bulb volume. The calculation is simple enough to be carried out on a statistical calculator and does not require a least-squares fit. Also the method is easily applied to existing calibration certificates where there is an ambiguity about the reported uncertainty or the interpolation method.

Example 6.5

An ASTM 63C thermometer divided to 0.1°C has the following values on its calibration certificate:

Temperature	Error
−7°C	+0.03°C
0°C	+0.05°C
10°C	+0.06°C
20°C	+0.02°C
30°C	+0.05°C

Determine a correction value and an uncertainty for all readings.

The average of the errors gives an overall correction of −0.04°C to be applied to any reading on the thermometer. From the five readings a variance is calculated to give an uncertainty at the 95% CL of ±0.05°C, i.e. 1/2 a scale division. Knowing that each reported reading is the average of two values would lower the uncertainty to ±0.04°C.

The advantage of the average is its simplicity in assigning a Type A uncertainty to any set of thermometer readings. A minimum of five calibration points is required. The main disadvantage is that the quoted uncertainty is high compared to the best possible accuracy. Achieving higher accuracy will, of course, involve more calibration points.

In more complex situations one way to preserve the simplicity of the average approach is to calibrate parts of the scale for the short ranges of interest. A correction and uncertainty is determined for each short range. Such a short-range calibration solves the problem of the scale definition on a partial-immersion thermometer. It also allows a lower uncertainty to be quoted for selected points provided there is no significant discontinuity over the range selected. Example 4.6 in Section 4.5 illustrates this approach. This is a common calibration approach for liquid-in-glass thermometers and can be used to provide the basis for a Type B uncertainty assessment for other readings on the thermometer.

Calibration equations based on higher-order polynomials may give a lower uncertainty than a simple correction. In practice the polynomial need not be higher than a cubic. High-temperature thermometers are examples of thermometers which may need quadratic or cubic calibration equations. There is little value in going to other forms of calibration equations. For example, to fit a function to segmented-scale errors (Figure 4.14) requires a complicated periodic function not easily handled by the least-squares method. Care is obviously needed to avoid sampling with the same period as the segmentation to ensure that the full distribution of errors is covered.

6.5.4 Liquid-in-glass calibration outline

A procedure for calibrating a liquid-in-glass thermometer is outlined here. The outline follows the step-by-step calibration procedure of Section 4.3.6 and includes the extra requirements for liquid-in-glass thermometers.

Step 1 — Initiate record-keeping

Include in the records any information as to why and how this liquid-in-glass thermometer is being used as this may point to very specific requirements, such as short-range calibration.

Step 2 — General visual inspection

Besides checking for any broken glass or loose mercury, examine the column for any breaks. Rejoin any breaks found as per Section 6.3.6. Remember to consult with the client if difficulties are expected or occur with the procedure.

The general visual appearance of the scale and bore should indicate if the requirements of Step 1 are likely to be met.

Update the records if any thermometer is discarded.

Step 3 — Conditioning and adjustment

Adjustments are not possible with the glass thermometers described here.

Thermally cycle any thermometer which is to be used over 300°C and check for stability at the ice point. Also thermally cycle any thermometer you suspect has not been annealed.

Give the thermometers three days at room temperature before remeasuring the ice point. Store carefully (see Section 6.4.7). Keep the organic liquid thermometers vertical with the top slightly warmer than the bulb to ensure that all the liquid drains before making measurements.

Very high-precision thermometers may need additional conditioning if required by the client, e.g. keeping the whole thermometer below 0°C before the ice-point reading.

Step 4 — Generic checks

Carry out a detailed visual check at 20× magnification. Reject any thermometer with bore or scale irregularities. The scale markings need to be clear and marked according to their documentary standard. Check that the quality is consistent with the client's requirements.

Ensure that any dimensional requirements are met. Figure 6.15 gives the more common dimensions that need to be controlled. Ideally these should have been checked before submission for calibration as wrong dimensions are a common reason for non-compliance with the standard specification.

Step 5 — The intercomparison

There are two essential features for a liquid-in-glass intercomparison:

- use increasing temperatures for measurements; and
- ensure visual access to the scale.

An overflow bath is ideal for a total-immersion thermometer as it allows viewing across the liquid surface and hence a very short emergent column exposed. See, for example, Figures 1.1 and 6.10(b). Otherwise a window is required to view the thermometer. Provide a means to physically lower the thermometers during the calibration in order to keep the meniscus at the same height as the viewing telescope. A firm clip will be needed to prevent the thermometer dropping into the bath but not so firm that it stresses the glass.

Fix a partial-immersion thermometer so that the telescope can be moved up and down without any blockage in the line of sight. Arrange for a stem-temperature measurement. Alternatively the partial-immersion thermometer could be calibrated as a total-immersion thermometer and the stem corrections applied later. This can be done only where the 10% uncertainty in the stem correction, equation (6.2), will not be the dominant contribution to the total uncertainty.

To keep the k-factor for the confidence limit low, choose 8 or 9 points for the calibration (see Section 2.6). Additional points will be needed if a polynomial fit is used, about four per parameter. If the number of scale segments on the thermometer can be identified, then try and choose four points per segment.

Step 6 — Analysis

Make any necessary corrections that may arise due to the stem temperature not being at its designated value. This may apply to a total-immersion thermometer if the column was too far above the bath liquid. Calculate the corrections and fit the results to the chosen deviation equation.

Step 7 — Uncertainty

If a fitting procedure was used, obtain the uncertainty from the standard deviation of the fit (Section 4.3.3).

Drift uncertainty is usually taken as zero because corrections for the ice-point shift should be applied to any reading. If there is no ice-point marking on the thermometer, then experience will be needed to assign an uncertainty.

The hysteresis uncertainty can be estimated from the change in ice-point values before and after the calibration. Assume a rectangular distribution.

For a partial-immersion thermometer a Type B assessment of the error in the stem temperature may be sufficient. Use equation (6.2) or Figure 6.13 to convert the stem temperature error to a temperature reading uncertainty, U_{stem}. If a stem-temperature correction is applied then include the uncertainty of the correction in the total uncertainty.

Step 8 — Complete records

Decide if the thermometer's performance warrants issue of a certificate of calibration. If it does, prepare the certificate. Report the ice-point value on the certificate. The actual value is needed to track any drift which also may indicate abuse of the thermometer. The ice point may be on an auxiliary scale and not the main scale.

A table of correction terms can be drawn up for the thermometer.

A completed certificate for the short-range style of calibration is shown in Figure 4.13.

A certificate for the deviation function style will look more like that in Figure 4.16.

FURTHER READING

ISO 386-1977   *Liquid-in-glass Laboratory Thermometer — Principles of Design, Construction and Use.*

ISO 651-1415   *Solid-stem Calorimeter Thermometers.*

ISO 653-1980   *Long Solid-stem Thermometers for Precision Use.*

ISO 654-1980   *Short solid-stem Thermometers for Precision Use.*

ISO 1770-1981   *Solid-stem General-Purpose Thermometer.*

The above ISO publications give documentary standards related to the liquid-in-glass thermometer.

E1-91   *Specification for ASTM Thermometers.*

E77-89   *Test Method for Inspection and Verification of liquid-in-glass Thermometers.*

The above standards related to liquid-in-glass thermometers are included by ASTM in their standards on Temperature Measurement Vol 14.03.

BS 593   *Laboratory Thermometers.*

BS 791   *Thermometers for Bomb Calorimeters.*

BS 1704   *General Thermometers.*

BS 1900   *Secondary Reference Thermometers.*

The above BSI publications present a series of documentary specifications for thermometers.

Liquid-in-glass Thermometer Calibration Service, J. A. Wise, Natl. Inst. Stand. Technol. Spec. Publ. 250-23, September (1988).

This covers calibration practice for liquid-in-glass thermometers.

7

Thermocouple Thermometry

7.1 INTRODUCTION

Thermocouples are the most widely used of all temperature sensors. Their basic simplicity and reliability have an obvious appeal for many industrial applications. However, when accuracies greater than normal industrial requirements are called for, their simplicity in use is lost and their reliability cannot be assumed.

For example, a major manufacturer of Type K thermocouple wire advises, 'Once a thermocouple has been used at a high temperature, however, it is not a good practice to use it later at a lower temperature'. Yet commercial hand-held electronic thermometers using Type K wire are sold for use over the range $-200°C$ to $1400°C$ and at an accuracy far exceeding that claimed by the wire manufacturer!

Such misuse of thermocouple wire stems in part from a lack of understanding of how a thermocouple works. Thermocouple literature often mistakenly states that the thermocouple junction is the source of the signal voltage, whereas in a well-designed system the junction does not contribute to the signal at all! Instead the signal is generated along the length of the thermocouple wire. This has a profound effect on the way traceability can be claimed.

Most of the basic principles for thermocouple thermometry were known around 1900 but were partly obscured by the development of the 'laws of thermoelectricity'. These laws give a working model for thermocouple practice, but they obscure the physical source of the thermoelectric potential. Therefore the model is unsuitable for analysing and avoiding the common errors in thermocouple practice even though the 'laws' are commonly used to explain the operation of thermocouples. Periodically the basic principles are rediscovered when large errors result from the use of thermocouples in new and unusual applications, but then the basic principles seem to be overlooked or forgotten since they are not well covered in textbooks.

7.2 GENERAL DESCRIPTION

Thermocouples look deceptively simple in that the basic transducer comprises two pieces of dissimilar wire joined together. However, without some knowledge of the underlying processes it is not possible to understand how they can be used as thermometers.

7.2.1 Thermoelectric effects

Thermoelectric effects were given this name because they involve both heat and electricity. Three different, but interrelated, thermoelectric effects can be identified. The *Seebeck*

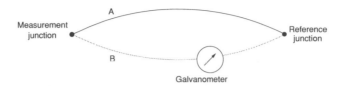

Figure 7.1
A basic thermoelectric circuit comprising two different wires, A
and B, two junctions and a galvanometer.

effect is the effect that is relevant for thermocouple thermometers, while the *Peltier* and
Thomson effects describe heat transport by an electric current.

The thermoelectric effects are observed when two dissimilar wires are joined together
to form an electrical circuit, and one junction of the wires is at a higher temperature than
the other. Thermoelectric effects cause an electric current to flow in the closed circuit, as
shown schematically in Figure 7.1. In this figure A and B are wires of different metals
or metal alloys and the galvanometer indicates that a current flows when the junctions
are at different temperatures.

The amount of current increases as the temperature difference between the junctions
increases. This phenomenon has both useful value (enabling the construction of thermocouple thermometers) and nuisance value (since unwanted thermoelectric effects can be
a major source of error in low-voltage measurements where you happen to have different
wire junctions at different temperatures). The two junctions are commonly labelled 'hot'
and 'cold', but a more useful terminology is to refer to the *measurement junction* and
the *reference junction*, as done in Figure 7.1.

A key concept throughout this chapter is that of an *isothermal* environment. This is
a region of space throughout which the temperature is the same. All the fixed points
described in Chapter 3 are examples of isothermal environments. A reference junction
at 0°C is often made isothermal using an ice point.

7.2.2 Seebeck effect

While the circuit of Figure 7.1 can serve as a self-contained thermometer, it is not
advisable for traceability because the current is dependent on the circuit resistance, which
itself depends on the chemical and physical condition of the wires and junctions. A
more suitable circuit for a thermocouple thermometer is shown in Figure 7.2 in which a
voltmeter is used instead of a galvanometer. The reference and measurement junctions
are each in an isothermal environment, each at a different temperature.

The open circuit voltage across the reference junction is the so-called *Seebeck voltage*
and increases as the temperature difference between the junctions increases. A typical
sensitivity is 40 μV/°C, which is well within the capability of modern voltmeters.

While the Seebeck effect is very easy to measure and demonstrate under the conditions
shown in Figure 7.2, it has taken physicists a long time to prove how the Seebeck effect
works. Part of the problem is that the Seebeck voltage is observable only in a complete
electrical circuit that involves at least two types of wires. Nevertheless, physicists can
show that the Seebeck effect occurs for any pair of points, which are not at the same
temperature, in any sample of electrically conducting wire. That is, while a Seebeck

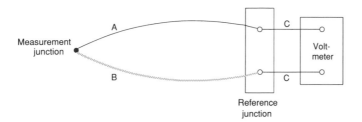

Figure 7.2
A circuit to measure the Seebeck potential, comprising two different wires, A and B, two junctions and a voltmeter. Copper wires (C) connect the reference junction to the meter.

voltage can be ascribed to a single metal wire, in practice it can be observed only with two different wires.

A simple explanation for the origin of the Seebeck voltage can be given by considering the metal wire as a lattice of fixed positive ions with the negative electrons free to move between them as a gas. At a fixed temperature and with no current flow the charges will balance, even though the electrons will be moving about under thermal motion. Once a temperature gradient is applied, some electrons move more than others, causing a redistribution of the electrons. A potential difference thus arises from the separation of the electrons and the fixed ions. This Seebeck voltage arising from a temperature gradient is a material property of the wire and does not depend on a junction or the presence of other wires in the circuit.

The thermocouple, which operates on the Seebeck effect, is therefore different from most other temperature sensors in that the output is not related to temperature directly but depends on the *temperature gradient*, i.e. on the difference in temperature along the thermocouple wire. Because of this dependence on temperature gradient we need to consider a short length of the wire. The increase in the Seebeck voltage, dE, along the short length of wire is proportional to the *Seebeck coefficient* of the wire, $s(T)$, and to the increase in temperature, dT, along the short length of wire:

$$dE = s(T)\,dT. \tag{7.1}$$

This is the basic equation with which to analyse a thermocouple circuit. Figure 7.3 shows the temperature dependence of the Seebeck coefficient, $s(T)$, for typical thermocouples.

We stress that dT arises from a temperature gradient and not a small change in overall temperature. To emphasise this, equation (7.1) can be rewritten for thermocouple wires as

$$dE = s(T, x)(dT/dx)dx, \tag{7.2}$$

where dT/dx is the temperature gradient along the wire and dx is a small length along the wire. The Seebeck coefficient $s(T, x)$ depends on the position along the wire, which in general is not uniform everywhere. The lack of uniformity may arise from deliberate change, such as connections in the circuit, or from inadvertent changes, such as stress in the wire or corrosion.

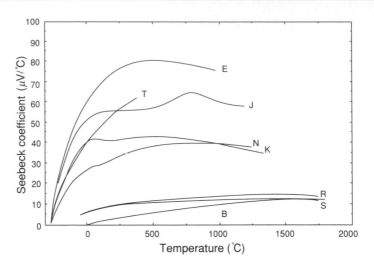

Figure 7.3
Seebeck coefficient for various thermocouples. The letters refer to the
thermocouple types of Table 7.I.

7.2.3 The thermoelectric basis of a thermocouple thermometer

The application of equation (7.2) to a circuit such as that in Figure 7.2 would imply
that a thermocouple would not make a practical thermometer. This is because detailed
information on the temperature profile and construction material would be needed, not
only for the wires but also for the myriad of components in the voltmeter. Obviously it
is not desirable to have the measuring instrument as a part of the sensor and therefore a
way of uncoupling the two is needed.

The application of two basic theorems to the circuit can show how to do this and
also how the circuit can perform as a thermometer. The theorems rely heavily on the
concepts of an isothermal environment (see Section 7.2.1) and *homogeneous* wires. The
word 'homogeneous' implies that every part of the wire is in an identical condition,
both physically and chemically. There is a wide variety of causes which can make the
wire inhomogeneous: some of these are a change in metallurgical structure due to work-
hardening, a change in metallurgical composition due to the migration of atoms, corrosion
of the wire, and a change in the wire diameter.

- **Theorem 1** If there is no temperature gradient, $dT/dx = 0$, i.e. an isothermal
  condition, then no Seebeck voltage is produced. (Proof by direct substitution in
  equation (7.2).)
- **Theorem 2** If a wire is homogeneous, i.e. $s(T, x) = s(T)$, then the Seebeck voltage
  depends on the end point temperatures of the wire. (Proof by the integration of
  equation (7.2).)

Note. The terms *isothermal* and *homogeneous* represent ideals which can only be approximated in practice. Theorem 1 can be considered the stronger theorem, simply because the
isothermal condition may be readily checked by measurement, whereas a measurement
of homogeneity is difficult.

The most complex part of the circuit in Figure 7.2 is the voltmeter and to remove its thermoelectric contribution to the measurement it should be maintained in an isothermal condition (Theorem 1). As the thermoelectric effects are a major error source in voltmeters, most voltmeters will be designed to minimise them. For example, brass input terminals are used on voltmeters to give a low thermoelectric effect with copper leads, but if used with other metal wires a good isothermal condition is essential. Heating one of the terminals with your fingers is a test to see how significant the thermoelectric effect is. Therefore, the main consideration will be to avoid draughts, to avoid locating the voltmeter near hot or cold objects, and to prevent rapid changes in the environmental temperature.

Junctions between wires are clearly inhomogeneous even if they are kept very small. Therefore, by Theorem 1, a junction should be kept isothermal to prevent any voltage being produced by the junction. Any other source of voltage, such as electrochemical activity or the existence of an electric current and circuit resistance, should also be avoided. Thus a golden rule for the use of thermocouples should be:

> ### THE JUNCTION SHOULD GENERATE NO VOLTAGE

Analysis of the circuit in Figure 7.2 now reduces to determining the thermoelectric effects for three wires: A and B for the thermocouple, and C for the instrument leads. Clearly from the above, any connections must be isothermal, which means keeping the heat capacity of the connection sufficiently high to ensure that there is no heat loading due to the use of heavy wires, particularly copper. The wires A, B and C should be kept as homogeneous as possible by avoiding any chemical or mechanical treatment of any part of the wire.

The voltage, E, produced by the circuit is the sum of three parts. Theorem 2 can be applied separately to each wire to give

$$E = E_A(T_M) - E_A(T_R) + E_C(T_R) - E_C(T_R) + E_B(T_R) - E_B(T_M), \qquad (7.3)$$

where for the wires A, B and C, E_A, E_B and E_C are the voltages for the end points T_M and T_R, which are the measurement- and reference-junction temperatures respectively. Note that the end point is not the position of the junction but the temperature of the isothermal region which includes the junction.

From equation (7.3) it can be seen that the net contribution of the instrument lead wire, wire C, is zero and hence under the above conditions the instrumentation for measuring a thermocouple sensor can be considered independent of the sensor.

Also from equation (7.3) the voltage output from the pair of wires (A and B) is related to the difference between the thermoelectric effect for the wires A and B. It is common to consider only a *relative Seebeck voltage*, E_{AB}, and *relative Seebeck coefficient*, S_{AB}, and thus equation (7.3) becomes

$$E = E_{AB}(T_M) - E_{AB}(T_R). \qquad (7.4)$$

A further simplification is made by choosing a single reference temperature for all thermocouples, namely $T_R = 0°C$, and setting $E_{AB}(0°C) = 0$,

$$E = E_{AB}(t_M) \qquad (7.5)$$

where the Celsius scale, t, is used to indicate that both the temperature and the voltage have an arbitrary zero point.

Equation (7.5) is the basic thermometer equation for a thermocouple which relates the voltage produced to the temperature being measured. However, the equation is based on several assumptions and the user needs to ensure that these do apply to the circuit in use. Any error analysis must use equations (7.1) or (7.2) to evaluate the error over the length of a thermocouple wire for the given temperature profile along the wire.

The assumptions made are:

- the reference junction is 0°C;
- the relative Seebeck coefficient is known;
- there are isothermal conditions for instruments and connecting wires;
- there are isothermal conditions for measurement and reference junctions;
- there are homogeneous wires connecting measurement and reference junctions.

In any real circuit these assumptions can only be approximated. The resulting error can be minimised by following good procedures for the construction and use of the thermocouples as outlined in this chapter.

7.2.4 Thermocouple tables

While equation (7.5) will apply to virtually any two pairs of wires, there are only a limited number of metals and alloys that have a history of reliability and are used for traceable measurements. Standard tables of the relative Seebeck voltage, $E_{AB}(t)$, have been approved internationally for a number of different wire pairs that are suitable for thermocouple thermometry (see Table 7.1 and Appendix D for detailed tables based on ITS-90). Manufacturers make thermocouple wires within a close tolerance to match these tables (see Table 7.2).

While the thermocouple types were originally based on the composition of the alloys, they are now based on mathematical functions which closely describe the thermoelectric behaviour of the alloy (see Appendix D). Manufacturers found problems conforming to an alloy composition because trace impurities in raw materials can change the Seebeck coefficient. Instead a more consistent product can be obtained by varying the alloy composition in order to match the tables. A manufacturer may also vary the alloy composition in order to make a superior wire, e.g. one more resistant to chemical attack. Thermocouple instruments may have the mathematical functions, or approximations to them, programmed into their microprocessors to convert the voltage reading directly to temperature.

Table 7.1 lists the type-lettering for pairs of thermocouple wires as well as for single thermoelements or 'legs' which can be combined in pairs to form thermocouples. The classification is based on the polarity of the separate legs: for example, a negative leg EN and a positive leg EP can be combined to form a Type E thermocouple. However, caution must be observed when considering 'mix and match' of thermocouple elements. In particular, it is very unwise to break up a pair supplied by a manufacturer and then to attempt to re-combine these legs with those from another pair. Normally the manufacturer has to ensure only that the pair conforms and thus mixing legs from different manufac-

Table 7.1. Compositions, trade names, and letter designations for standardised thermocouples.

Type	Materials
	Thermocouple combinations
B	*platinum*-30% rhodium/*platinum*-6% rhodium
E	*nickel*-chromium alloy/a *copper*-nickel alloy
J	iron/another slightly different *copper*-nickel alloy
K	*nickel*-chromium alloy/*nickel*-aluminium alloy
N	*nickel*-chromium alloy/*nickel*-silicon alloy
R	*platinum*-13% rhodium/platinum
S	*platinum*-10% rhodium/platinum
T	*copper*/a copper-nickel alloy
	Single-leg thermoelements
BN	*platinum*-nominal 6% rhodium
BP	*platinum*-nominal 30% rhodium
EN or TN	a *copper*-nickel alloy, constantan: Cupron[a], Advance[c], ThermoKanthal JN[b], nominally 55% Cu, 45% Ni
EP or KP	a *nickel*-chromium alloy: Chromel[d], Trophel[a], T-1[c], ThermoKanthal KP[b], nominally 90% Ni, 10% Cr
JN	A *copper*-nickel alloy similar to but usually not interchangeable with EN and TN
JP	iron: ThermoKanthal JP[b]; nominally 99.5% Fe
KN	a *nickel*-aluminium alloy: Alumel[d], Nial[d], T-2[c], ThermoKanthal KN[b], nominally 95% Ni, 2% Al, 2% Mn, 1% Si
NN	*nickel*-silicon alloy: nominally 95.5% Ni, 4.4% Si, 0.1% Mg
NP	*nickel*-chromium alloy: nominally 84.4% Ni, 14.2% Cr, 1.4% Si
RN, SN	high-purity platinum
RP	*platinum*-13% rhodium
SP	*platinum*-10% rhodium
TP	copper, usually electrolytic tough pitch

An *italicised* word indicates the primary constituent of an alloy and all compositions are expressed in percentages by weight. The use of trade names does not constitute an endorsement of any manufacturer's products. All materials manufactured in compliance with the established thermoelectric voltage standards are equally acceptable.

Registered trade marks: [a]Wilbur B Driver Co; [b]Kanthal Corp; [c]Driver-Harris Co; [d]Hoskins Manufacturing Co.

...N denotes the negative thermoelement of a given thermocouple type

...P denotes the positive thermoelement of a given thermocouple type

turers, or even the same manufacturer, could result in a thermocouple not conforming to the standard. For example, one well-known manufacturer makes seven different versions of Type K thermocouples, and mixing of legs from these pairs is not recommended.

The advantage of buying single 'legs' is that the user can then make a custom-designed thermocouple. While there are no international standards for single legs there may be national ones based on a thermocouple with platinum as one leg. Manufacturers can often supply single legs conforming to such a standard and this ensures that the performance of a thermocouple made from two single legs will conform to the standard.

By choosing a standard thermocouple type such as listed in the above tables the user will be able to draw on the generic history of that thermocouple type when designing the thermometer and procedures for its use. The choice of a thermocouple type suitable to become a standard is based on several factors, such as a high output voltage, good stability, the ability to withstand harsh environments, and of course a proven history of reliability. For some applications one of the developmental thermocouples may appear more suitable but the user will need to ascertain the advantages and disadvantages from

Table 7.2. Tolerance classes* for thermocouples (reference junction at 0°C).

	Class 1	Class 2	Class 3[†]
Tolerance values[‡] ($\pm$)	0.5°C or 0.4%	1°C or 0.75%	1°C or 1.5%
	Temperature limits for validity of tolerances		
Type T	−40°C to 350°C	−40°C to 350°C	−200°C to 40°C
Tolerance values[‡] ($\pm$)	1.5°C or 0.4%	2.5°C or 0.75%	2.5°C or 1.5%
	Temperature limits for validity of tolerances		
Type E	−40°C to 800°C	−40°C to 900°C	−200°C to 40°C
Type J	−40°C to 750°C	−40°C to 750°C	—
Type K	−40°C to 1000°C	−40°C to 1200°C	−200°C to 40°C
Type N	−40°C to 1000°C	−40°C to 1200°C	−200°C to 40°C
Tolerance values[‡] ($\pm$)	1°C plus 0.3% of $(t - 1100)$°C	1.5°C or 0.25%	4°C or 0.5%
	Temperature limits for validity of tolerances		
Type R or S	0°C to 1600°C	0°C to 1600°C	—
Type B	—	600°C to 1700°C	600°C to 1700°C

*These tolerances follow IEC584-2 except for Type N. Consult the latest version of IEC584-2 for the current status of the classes.

[†]Thermocouple materials are normally supplied to meet the manufacturing tolerances specified in the table for temperatures above −40°C. However, these materials may not fall within the manufacturing tolerances for low-temperatures given under Class 3 for Types T, E, K and N thermocouples if thermocouples are required to meet limits of Class 3, as well as those of Class 1 and Class 2. The purchaser shall state this, and selection of materials is usually required.

[‡]The tolerance is expressed either as a deviation in degrees Celsius or as a percentage of the actual temperature. The greater value applies.

the research literature and will also have to investigate suitable calibration methods to prove the performance.

7.2.5 Measurement model

A summary of the main features of a thermocouple measurement is given in Figure 7.4 for a 'typical' thermocouple. The figure is based on the measurement model of Figures 1.10 and 2.10.

There is a large variety of ways in which the thermocouple wires can be incorporated as a transducer into a sensor capable of measuring the temperature of the desired physical system. Whether the wire is very fine or very thick, it is necessary to ensure that the measurement junction is isothermal over the region of the junction. Because the transducer responds to temperature gradients, it has to be connected to two physical systems at different temperatures. As a result the transducer has to pass through a part of the environment over which the user may have little control or knowledge. It is essential to ensure homogeneity over this region or to have other provision for control over the possible errors. The reference junction must be held isothermal at a known temperature, firstly to provide this known fixed temperature, and secondly to help provide a signal interface which isolates the sensor from the instrumentation. The signal transmission leads from the reference junction to the instrument are often in a more controlled environment than that of other temperature sensors, especially if the reference junction is built into the instrument. If the instrument is a voltmeter then the data interpretation will require

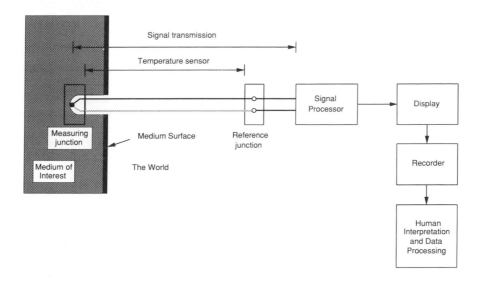

Figure 7.4
A model of a thermocouple measurement. The two boxes around the junctions indicate that they are isothermal and that no voltage is produced there.

extra information about the reference temperature and which thermocouple table to use, otherwise this information may be included in the instrument and the temperature may be displayed directly.

It must be emphasised that the model of Figure 7.4 is one of many possible variants, but the model should not be simplified to that of Figure 2.10, even though many instruments and probes that are sold would appear to approximate this quite well. The danger in the approach of Figure 2.10 is that the measurement junction is thought of as the sensor and the thermocouple wires are considered to be merely signal transmission lines. Both of these assumptions lead to serious measurement problems with thermocouples.

7.3 THERMOCOUPLE TYPES

The thermocouple types in Table 7.1 have been developed to meet perceived needs in temperature measurement. They do not meet all needs and hence there is continuing development of new types. In order to make an informed selection of a thermocouple type the user may need to acquire quite detailed knowledge about the properties of the materials involved. This section gives some of the basic starting information required and Tables 7.3 and 7.4 as well as Appendix D help summarise the information.

Three categories of thermocouple types are considered: *rare-metal* standard thermocouples, *base-metal* standard thermocouples and *non-standard* thermocouples. In practice the distinction between 'base' and 'rare' is that the rare metals contain platinum and the base metals contain nickel. As a consequence the rare-metal thermocouples are some 200 times more expensive than base-metal thermocouples, at current prices.

7.3.1 Rare-metal standard thermocouples

There are three standard types, B, R and S, as shown in Table 7.1. All involve platinum and its alloys with rhodium and hence they are relatively inert chemically. Their advantage is that inhomogeneity arises mainly from mechanical effects and can often be reversed by careful annealing. Pure platinum, however, suffers from excessive grain growth above 1100°C and the wire becomes very fragile; the grains become large enough to give a jagged edge to the wire. Because Type B wire does not have a pure platinum leg it gives better behaviour at higher temperatures than the other two types do.

Rare-metal alloys tend to be relatively simple and do not undergo significant metallurgical changes at high temperatures. The rhodium can, however, migrate to the pure platinum wire over a period of time. High-purity insulators must be used and metal sheaths should be avoided unless they are made out of platinum. Platinum thermocouples normally work well in an oxygen atmosphere but not in a reducing atmosphere, especially if hydrogen is present.

Types S and R are very similar and originally were meant to be the same material, but early troubles in obtaining pure platinum and rhodium caused a divergence. Type R has about 10% more voltage output than Type S, but Type S is considered to be slightly more stable and was used as the reference thermocouple for the earlier temperature scales. As a consequence, Type S has a better history of proven performance and is therefore preferred as a reference thermocouple for the calibration of other thermocouple types. With care errors can be kept to a few tenths of a degree up to 1000°C.

Type B thermocouples were designed solely for high-temperature readings. Around room temperature, the Seebeck coefficient is sufficiently low that quite large errors in the reference junction temperature do not cause a significant error in the observed measurement-junction temperature. As a result, Type B thermocouple instrumentation is often supplied with no input from a reference junction temperature and instead a fixed offset is applied to the voltage output to account for a typical room temperature. This kind of room-temperature offset adjustment for the reference-junction compensation of a thermocouple instrument can also be found for other thermocouple types, especially in simple analogue instruments, but it is not a practice to be recommended. The Type B thermocouple was specially designed to allow the practice.

The main disadvantage of the rare metals is their cost. This may be considerable: a reference thermocouple should be continuous from the measurement to reference junction and may require over two metres of wire. Many high-temperature applications need the rare-metal thermocouple for stability but do not need high accuracy. In these cases a cheaper compensation extension cable can be used for the portion of the thermocouple at or near room temperature; such cables are discussed in Section 7.4.3.

7.3.2 Base-metal standard thermocouples

Base-metal thermocouples are the standard types T, J, K, E and N, all of which use nickel in some form (see Table 7.1). Because they all oxidise easily, they are not readily annealed to remove mechanical effects. Also, with higher temperatures the more complex alloys undergo changes which may not be readily reversed. Therefore, overall, base-metal thermocouples do not make as good thermometers as rare-metal thermocouples. However, their lower cost can offset this, especially for harsh environments. A replacement schedule based on the observed rate of degradation can be economically feasible.

As a general rule do not use base-metal thermocouples as all-purpose wide-range thermometers (see Tables 7.3 and 7.4) unless errors of over 10°C are of no concern. Ideally they are best used in fixed locations to measure temperatures over a limited temperature range. This is especially true for higher temperatures. Otherwise great care is needed to ensure that the wires have not been subjected to mechanical forces or to higher temperatures between use. The proper use of a wide-range thermocouple instrument is to connect it to different thermocouples reserved for special purposes and *not* to use a single thermocouple probe as a general-purpose thermometer.

Each thermocouple has its particular niche, but some of these are being taken over by resistance thermometers, either PRTs or thermistors. In general base-metal thermocouples have two advantages: they can be made very small, and they can be made to withstand harsh environments.

Table 7.3. Uses for thermocouple types.

Type	Allowable environment	Comment	Maximum temperature (see Table 7.4) (°C)
B	Oxidising, inert, vacuum for short periods	Avoid metal contact. Most suitable for high temperature. Has low voltage at room temperature	1700
E	Oxidising, inert	Good for subzero temperature. Highest voltage output of common thermocouples	870
J	Oxidising, inert, reducing in partial vacuum	Iron rusts or oxidises quickly	760
K	Oxidising, inert	Subject to 'green-rot' in some atmospheres	1260
N	Oxidising, inert	More stable than Type K at high temperatures.	1300
R&S	Oxidising, inert	Avoid metal contact	1400
T	Oxidising, inert, reducing in partial vacuum	Subzero temperatures. Can tolerate moisture	370

Table 7.4. Upper temperature limits in °C for the various wires with continuous operation.

Type	Wire diameter (mm)				
	3.25	1.53	0.81	0.51	0.33
B				1705	
E	871	649	538	427	427
J	760	593	482	371	371
K	1260	1093	982	871	871
R&S				1482	
T		371	260	204	204

Applies to thermocouples in normal protective ceramic sheathing. Life depends on the type of atmosphere, etc. Operation at higher temperatures for shorter periods may be possible.

Type T, being made out of copper and a simple copper–nickel alloy, can often withstand more handling and is useful for taking temperature surveys for applications such as safety tests on electrical appliances. It is also the preferred thermocouple for low-temperature work, i.e. to −200°C. However, the copper is a good conductor of heat, so thin wire is needed to reduce the heat flow to and from a junction and to ensure that the junction remains isothermal. *Thermal anchoring* of the wire will also help reduce the heat flow, i.e. the wire can be attached to a point at a known temperature near that of the isothermal environment. Copper should not be used above 200°C, not only because of oxidation but also because increasing metal migration can cause contamination. Some manufacturers provide Type T wire made to very tight tolerances.

Type J is the only standard thermocouple suitable for use in a reducing atmosphere at high temperatures and as such finds wide use. In other applications the iron wire can oxidise rapidly if not protected. Unfortunately, Type J is not a unique designation because the German standard was different when the international standards were adopted. As the difference is about 8°C at 200°C care is necessary when obtaining replacement wire. Type J instruments should be checked to find out which standard they require and they then should be marked accordingly. A new letter designation of L appears to be used for the German standard.

Type K was the most successful wire developed for high-temperature use but it also gives useful output down to −200°C. It is therefore tempting to use Type K as a general-purpose thermocouple. However, because of the complex alloys used it is the worst of the thermocouples in preserving its homogeneity. Therefore, other thermocouple types should be used where they are more suitable. There are at least three main problems occurring in Type K, and they are given here as examples of the complex processes that can occur inside a piece of wire and affect its performance.

- Steady drift occurs above 500°C, and more markedly above 1000°C. Oxidation, particularly internal oxidation, changes the wire composition to cause this drift (see Figure 7.5).

- Short-term cyclic changes, as much as 8°C, occur on heating and cooling in the range 250°C to 500°C. This is caused by metallurgical changes in the positive thermoelement, which produce structural inhomogeneities.

- Reversible changes in the voltage make the thermocouple output vary by ±1.5°C over the range 50°C to 250°C, due to a magnetic/non-magnetic transformation in the negative thermoelement.

Note that above 500°C all these effects contribute to the error since at least one part of the thermocouple will be at the lower temperatures.

Type E has the highest Seebeck coefficient of the standard thermocouples and uses the positive thermoelement from Type K with the negative thermoelement of Type T. This gives a better short-term performance than Type K. Type E can be used for high-temperature surveys above the reach of Type T.

Type N was developed to improve on and remove the known problems in Type K. A better performance is obtained for the bare wire; in particular, the high-temperature drift is considerably lower. Figure 7.5 illustrates the typical drift that can be expected for these two types of thermocouples. The drift is that expected under good conditions: in more hostile environments the drift will be much faster. While many problems are reduced

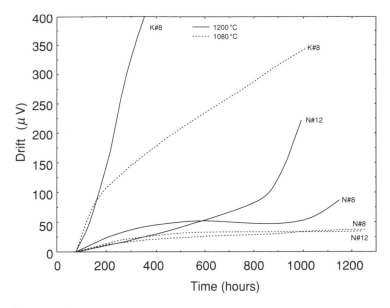

Figure 7.5
High-temperature drift for bare Type K and Type N thermocouple wires.

by using Type N, in some cases the problem has been shifted to higher temperatures. When the wires are fully insulated in metal sheaths, the difference is not so marked and the overall performance depends more on the sheath material than on the thermocouple. Overall, Type N is a significantly better thermocouple than Type K and is becoming widely available. A better neutron immunity is also claimed.

7.3.3 Non-standard thermocouples

Over 200 different thermocouples have been studied and reported on in attempts to find more suitable ways of measuring temperatures in difficult situations. Two areas of industrial importance are the measurement of temperatures up to 3000°C and temperatures of highly reactive gases, especially those rich in hydrogen and carbon monoxide. Hydrogen is a small molecule which can easily pass through hot metals.

Tungsten–rhenium alloys have been found suitable for higher temperatures up to 2300°C and they are not overly affected by hydrogen. Suitable sheaths are needed to prevent oxidation and to protect the wire from mechanical forces as the wires become very embrittled. A fixed position operation is therefore essential and use must be restricted to temperatures over 400°C. Large drifts can occur around 2000°C due to boron in the commonly used boron-nitride sheath. Alumina sheaths cause considerably less drift but restrict the upper temperature to 1800°C. Low drift is found for clean atmospheres up to 1500°C. Various alloy compositions have been tried and Table 7.5 lists the more popular ones.

An alternative high-temperature thermocouple is boron–carbide/graphite (B_4C/C) for temperatures up to 2200°C. The output of these thermocouples is around 290 μV/K. While there is a production variation with these thermocouples, the resulting temperature

Table 7.5. Available tungsten–rhenium alloys for thermocouples.

Alloys	W3% Re/W25% Re	or	W5% Re/W26% Re
Range	0–426°C		425–2315°C
Manufacturing tolerance	± 4.4°C		± 1%

error is only of the same order as that resulting from their use in the hostile environments. Comparison against high-temperature noise thermometers shows drift or variability up to 10% when such a thermocouple is operated around 2000°C. Overall the performance in terms of stability is better than for tungsten–rhenium.

Making a thermocouple out of pure metals rather than alloys should avoid alloy variation as a source of inhomogeneity and this has been proved for platinum–gold thermocouples. Recent investigations of this thermocouple reveal that uncertainties as low as a few hundredths of a degree are possible if care is taken to avoid strain from the differential expansion of the two metals. It is hoped that this thermocouple will be a practical transfer standard which can utilise the increased accuracy of ITS-90 above 630°C. Modern instruments can handle the lower output of this thermocouple with reasonable accuracy.

7.4 CONSTRUCTION

There is no standard way to construct a thermocouple thermometer as they can be adapted to a wide variety of situations. Where possible a thermocouple assembly should be obtained from a known high-quality supplier because specialised materials and techniques can be involved. The main steps involved in construction are covered here to help the user specify the thermocouple when purchasing and also to give you some idea of how to construct a thermocouple yourself if necessary.

7.4.1 Junctions

A prime function of a thermocouple junction is to provide electrical continuity in a reliable manner over the temperature range required, for the period of use.

An important decision to be made when constructing the junction is the appropriate size. The upper limit on the size is determined by the requirement that the junction should always be surrounded by an isothermal environment. Once this limit has been determined, the junction size will depend on the size of the region under investigation. The size will be affected by the method of construction (see Figure 7.6) and the diameter of the wire used. Limits on the wire diameter will depend on the amount of immersion error tolerated (see Section 7.6.1).

Before joining the wires, make sure that they are clean. For the smallest size junction, the two wires can be butt-welded. Otherwise twist the wire pairs together so that thermocouple metal is in contact with thermocouple metal (see Figure 7.6). The wires can be held together more permanently by either soldering for lower temperatures, welding for higher temperatures, spot welding or crimping for speed of operation. Avoid contamination from any acid or flux used and do not subject the wires to undue force. If flames are used, keep them small and avoid contact with any other part of the thermocouple

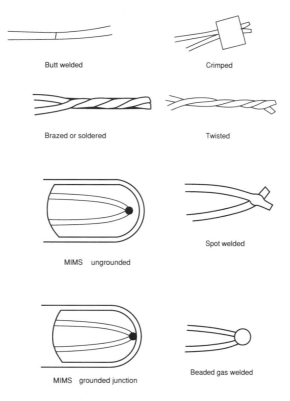

Butt welded   Crimped

Brazed or soldered   Twisted

MIMS   ungrounded

Spot welded

MIMS   grounded junction

Beaded gas welded

Figure 7.6
Construction of thermocouple junctions. A representation
of a variety of methods which have been found satisfactory.
MIMS thermocouples are covered in Section 7.4.5.

wire. All the operations should be done in a neat and tidy manner so that the position of the junction is well-defined, especially if the bare junction is to be exposed.

Reference junctions are sometimes made temporarily for test purposes. In this case other mechanical means can be used to form junctions as outlined in the following section on joins.

7.4.2 Joins

The first rule with joins is: don't use them unless absolutely necessary! They are a major source of difficulty due to the inhomogeneity introduced. That is, the first choice is to have the same single continuous wire for each thermoelement from the measurement junction to the reference junction.

There is only one good reason for a join and that is to connect a specialised thermocouple assembly to the leads which go from the assembly head to the reference junction and instrumentation. Specialised assemblies include high-temperature thermocouples with very heavy wires, totally sealed units protected against corrosive atmospheres, very fine wires to prevent thermal loading and very expensive thermocouple wires such as the rare metals. In any case the specialised assembly should be sufficiently long that the head, when mounted, will be below 50°C, i.e. can be handled with bare skin. Note that in all these cases the lead wire being connected is unlikely to be identical to that in the thermocouple assembly.

A partially acceptable reason for a join is for convenience. For example, in a test rig a large number of changes may need to be made quickly between tests. Procedures need to be in place to check the performance of the thermocouple after any changes have been made.

A poor reason for a join is to repair a break in the wire. Replace the whole wire and if this is not immediately feasible make sure that the repair is well documented so that the wire will be replaced later. This also allows checking if problems arise.

Another poor reason for joins is to switch many thermocouples to one reference junction and instrument. This was common practice in early thermocouple thermometry, but is not recommended here. Each measurement junction should have its own reference junction, i.e. an isothermal region whose temperature is known. Any switching should be made after the reference junction so that it is only copper wires being switched. Even so, remember that this is still part of a thermoelectric circuit and ensure that the following recommendations on joins are followed.

The same principles that apply to junctions apply to joins; namely, they need to be electrically continuous, mechanically strong, clean and in an isothermal environ-

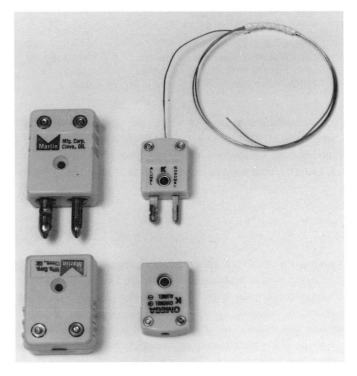

Figure 7.7
Various plugs and sockets for thermocouples.

ment. Unlike a junction, however, a join should never be at high (or low) temperatures and instead should be at a temperature near room temperature, e.g. between 10°C and 50°C.

Joins should be made with thermocouple metal to thermocouple metal and held there in a mechanically stable manner, e.g. by a screw thread. Soldering or welding should not be necessary because there should be very little thermal expansion to loosen the mechanical joint. Ensure that the joins for the pair of wires are as close to each other as possible. Suitable thermocouple connection boxes or heads which are made out of cast metal are available commercially, and can be used to house the joins in order to keep them isothermal.

Keep the location of the join free from draughts and away from hot or cold objects.

Special plugs, sockets and connectors made out of mechanical grades of the thermocouple alloys are readily available commercially and can be used for making quick connections (see Figure 7.7). The plugs and sockets should conform to a standard. While the connection is to a very similar alloy it is best to ensure there are no temperature gradients across the connector.

7.4.3 Extension leads

In spite of the fact that joins are undesirable most practical circuits use joins to connect wire from the assembly head to the reference junctions. Generally this connecting wire will differ from the wire in the thermocouple assembly both in diameter and composition. Thus the thermocouple circuit has a long length of a distributed inhomogeneity. If it was possible to keep the connecting wire isothermal with the joins at either end then there would be no Seebeck voltage from the connecting wire and hence no error. As it is not feasible to ensure isothermal conditions along a long length of wire an alternative approach is needed to reduce the likelihood of error.

A thermocouple *extension lead* is a connecting wire or cable which has been selected not only to keep the error low but also to provide convenience in use, e.g. flexibility. The basis for the selection can be seen by considering equation (7.2). If the temperature gradient along the extension lead can be kept small instead of zero (i.e. the isothermal condition) then the Seebeck voltage produced by the extension lead is small. If in addition the Seebeck coefficient for the extension lead is similar in value to the Seebeck coefficient for the thermocouple wire in the assembly then the additional error contribution due to the use of the extension lead can be kept small and possibly insignificant. The Seebeck coefficents need only match over a limited temperature range if the use of extension leads is restricted. For standard thermocouples the range should be restricted to 10°C to 50°C in line with that for joins.

Two types of extension leads are usually distinguished: *extension leads* and *compensation extension leads*.

Extension leads are made from the same alloys as the thermocouple wires but may have small differences in Seebeck coefficient due to the smaller diameter of the wire. Some manufacturers make the leads to the same tolerance as the single thermocouple wire, but generally they are made to twice the tolerance. They may be supplied as flexible multicore cable.

Compensation extension leads are made from different alloys but match the Seebeck coefficient of the required thermocouple over a limited temperature range. In the case of the platinum-based thermocouples, Types R and S, the leads are made of copper and a copper alloy in a multistranded cable to give flexibility and lower cost. From Figure 7.3 it can be seen that Types T and K match over the room-temperature range and sometimes Type T extension wire is used as a compensation extension lead for Type K thermocouples. However, this is not desirable because the compensation extension lead is a compromise which can introduce more errors than would occur if the proper extension lead was used. Therefore use compensation extension leads only for the expensive rare-metal thermocouples, and then only if long wires are needed.

7.4.4 Sheaths and thermowells

While completely bare wire is sometimes used, especially in the heavier gauges, it is more common to cover the wire to provide at least electrical insulation. A wide variety of insulation materials is available to suit many purposes and the user is advised to consult a catalogue to select an appropriate covering material. See Table 7.6 for a short list of possibilities. For higher-temperature use the thermocouple should be hand assembled from bare wire and ceramic beads. Cleanliness is essential for this operation. Avoid work-hardening the wire during handling. Bare junctions can be used to achieve a low mass or small size. If there is a risk of contamination the wire may need to be replaced frequently or if an increase in size and mass can be tolerated, a sheath can be used to provide protection.

Dimensional constraints on the sheath should be established first in order to select the length and diameter of the sheath. The minimum length will be determined by two thermometric factors: the immersion depth and the temperature gradients. The immersion depth should be at least $5\times$ the diameter of the sheath, and preferably $10\times$ (see Figure 4.4). As the output of the thermocouple is largely determined by the region of maximum temperature gradient it is important that the wire has adequate protection over this region. Therefore the sheath should extend beyond the medium of interest until its temperature is close to room temperature.

The choice of diameter is likely to be a compromise between the time-constant and the thickness to achieve adequate protection. Obviously other physical constraints may be imposed by the size and nature of the physical system.

Table 7.6. Insulating materials for thermocouples.

Material	Range of maximum temperatures*
PVC	65°C to 85°C
Polyurethane	65°C to 85°C
PTFE	190°C to 260°C
Polymer/glass laminate	200°C to 280°C
Glass fibre	400°C to 480°C
Ceramic fibre	800°C to 1200°C
Ceramic beads	1100°C to 1950°C
Magnesia/SS sheath	600°C to 900°C
Magnesia/inconel® sheath	800°C to 1050°C
Magnesia/nicrobel® or Magnesia/nicrosil®	1100°C to 1200°C

*The maximum temperature depends on a number of factors, including the manufacturer, the duration of exposure, the environment and the detailed composition of the material.

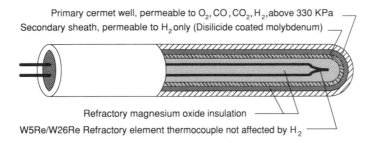

Primary cermet well, permeable to O_2, CO, CO_2, H_2, above 330 KPa
Secondary sheath, permeable to H_2 only (Disilicide coated molybdenum)

Refractory magnesium oxide insulation
W5Re/W26Re Refractory element thermocouple not affected by H_2

Figure 7.8
A multilayer protective sheath required for a high-temperature thermo-couple used in a hostile environment.

For particularly harsh environments it is unlikely that a single sheath will provide all the protection and several layers may be needed. An example is given in Figure 7.8 for tungsten–rhenium wire. Another example is where a platinum-based thermocouple has to go into a metal thermowell, needed for strength, and thus it requires a surrounding ceramic sheath to prevent contamination by the metal. Suppliers of sheathing material provide extensive lists recommending cost-effective materials for particular environments.

Over their lifetime thermocouples need regular checking or calibration and will eventually need replacing. Sheaths and thermowells should be designed to allow this to be done relatively easily and without a major disruption to àn on-going process. One way is to have two thermowells close together so that a calibration can be done in one without disturbing the other, or alternatively provide enough space in one thermowell so that a thin calibrated probe can be inserted alongside.

Complete removal of a thermocouple for calibration is not desirable. Not only will the calibration process assess the output from a different section of wire (see Section 7.8), but it may also heat-treat the wire so that it performs differently in service. The thermocouple will also need tó be reinstalled at exactly the same position.

As an example, consider a uniform-temperature furnace with, say, a 5 cm wall thickness. The 5 cm portion of the thermocouple in the wall generates most of the thermocouple output. The wire in the furnace will develop the larger inhomogeneity with time and moving this part into the wall will affect the thermocouple output. Conversely, moving the thermocouple further in will place relatively fresh wire in the wall and may also change the output closer to specification. In fact, movement of the thermocouple can be used to check on inhomogeneities. Any large change in temperature reading would indicate that wire replacement is needed.

7.4.5 Mineral-insulated metal sheaths

MIMS (mineral-insulated, metal-sheathed) thermocouples are a very convenient form in which to obtain thermocouple cable (see Figure 7.9). They have the protection advantages offered by a metal sheath while still retaining a reasonable amount of flexibility. Various sizes from 0.25 mm to 6 mm diameter are often available from stock with special diameters made to order. Thus it is possible to preserve the size and mass advantage of thermocouples in a protective sheath.

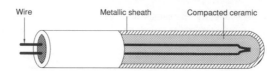

Figure 7.9
Compacted ceramic-insulated 'MIMS' thermocouple
showing its composition.

There are, of course, trade-offs for this convenience:

- The amount of force needed to produce the cable produces strain in the wire and thus can give rise to inhomogeneities. For example, a hard bend in the cable should never be positioned at a temperature gradient.

- Magnesium oxide, the most commonly used insulation material, can absorb moisture and provide an electrical shunt along the length of the cable. It is very difficult to assemble a fully sealed cable; a full seal may not even be desirable, as a pressure build-up in the cable may occur on heating. If the cable is used frequently above room temperature then the cable insulation will remain dry. Prolonged storage of the cable will give time for water to be absorbed and an insulation check should be made before use. The insulation resistance should be over 1 MΩ. If necessary, dry out the cable in an oven until the insulation resistance is restored.

- In the compact structure, migration of metal atoms from the metal sheath to the thermocouple wire can easily occur at higher temperatures and thus contaminate the thermocouple. As a general rule, choose sheath material whose composition is as close as possible to the thermocouple material, providing that there is still adequate chemical protection. Thus, for platinum thermocouple types, platinum can be used as the sheath. For Type K or N thermocouples, Inconel® is better than stainless steel, but the best practice is to use an alloy closely related to the Type N wire, either Nicrosil® or Nicrobell®. As several variants are available, the user will need to exercise due care in selecting the best one for the particular application.

- The MIMS sheath, while convenient, provides only limited protection and also, being flexible, may not provide the best mechanical stability. Use other protection besides that offered by the sheath material, especially where the cable goes through a temperature gradient. Where there is a harsh chemical environment, use additional sleeve material and choose the best MIMS for thermometric stability.

- Electrical effects may interfere with the thermocouple's performance. Long cables will lower the insulation resistance even without moisture. The metal sheath may need to be electrically grounded, depending on the environment. MIMS also allows the junction to be grounded to the sheath. Remember that some furnace materials become semiconductors at high temperatures and could cause stray electric currents to upset the thermocouple instrument.

7.5 INSTRUMENTATION

In order to claim traceability for a thermocouple measurement, two types of measurements are needed: the voltage of the thermocouple and the temperature of the reference junction. The need to know the reference-junction temperature adds a complication to thermocouple instrumentation and therefore many methods have been adopted to make thermocouple instruments convenient to use. Unfortunately, both good and bad methods are available and the user seldom has the information to evaluate this; price and sophistication are no guarantee. Thus the user will not know what confidence to place in the instrumentation. To ensure confidence in a thermocouple measurement the following procedures must be carried out either by the user or by automatic functions of the instrument:

- establish an isothermal reference junction;
- know the temperature of the reference junction;
- use the standard tables or reference functions to determine the Seebeck voltage at the reference junction temperature;
- make an accurate measurement of the Seebeck voltage from the thermocouple;
- add the two voltages together;
- use the standard tables or reference functions to determine the measured temperature.

In principle, a fully integrated instrument approach, if properly implemented in both hardware and software, should give the higher reliability by removing the chance of human error. However, in practice the user may find that the traceability requirement is to have the process directly under human control and properly documented to allow an audit to be made. This is especially true when higher accuracies than normal are called for.

Instrumentation can be assessed on how the reference junction is dealt with. Two broad categories are: instruments with external reference junctions and those with internal reference junctions. Recorders can fall into both categories. Extra instrumentation or equipment is required for the external reference junction. Instrumentation to facilitate the calibration of installed thermocouple instruments is available. Also, an important instrumental consideration is the type of circuit used for the measurement.

7.5.1 Instrument types

Over the years a wide variety of instruments has been used to measure the signal from a thermocouple and convert it to a temperature. Three types will be considered here: voltmeters, thermocouple meters, and chart recorders.

- *Instruments with external reference junctions* Modern digital voltmeters have almost completely supplanted the potentiometer, which used to be the main workhorse for higher-quality thermocouple measurements. Typically a $5\frac{1}{2}$ digit instrument with a resolution to 1 μV is used with an ice-point reference junction.

  Sometimes a lower-precision measurement will be made with an electronic compensating reference junction which may even come with conversion electronics to give an output of, say, 10 mV per degree.

  The voltmeter should have its own calibration as a voltmeter. As most meters have input impedance over 1 MΩ compared with up to 1 kΩ impedance for a thermocouple, there is negligible electrical loading on the thermocouple. The voltmeter should be kept reasonably isothermal, inside its specified temperature range and not subjected to rapid temperature changes. Reverse the leads to the voltmeter to check for thermal stability as the reading should change only its sign and not its numerical value.

Electrical safety needs to be considered where the thermocouple is used near, say, bare electrical heaters. In particular, electrically ground any metal sheaths but avoid any ground loops. Most modern voltmeters work best with differential inputs and hence there is no need to electrically ground any of the thermocouple wires.

- *Instruments with internal reference junctions*   A thermocouple meter combines a reference junction with a voltmeter and has the means to give the reading directly in terms of temperature. Digital equipment for this is available with a wide range of accuracies. In principle the digital equipment should be more capable than most analogue equipment of following good practice for the reference junction and data conversion. Unfortunately, many digital circuits are blind copies of analogue methods, which were designed for convenience and low cost rather than good thermometry, and therefore the method may no longer be appropriate with the new technology. The higher-precision readout on the digital display does not reflect the basic accuracy of the circuits used. In most cases, of course, the errors in the thermocouple used will be greater, typically $\pm 1\%$.

Obtaining good isothermal conditions for the instrument is more important than for the voltmeter because of the included reference junction. In this respect plug-in cards for computers may not always be in suitable environments to provide accurate reference junctions.

A feature of a good thermocouple meter is the ability to monitor the impedance of the thermocouple. A 1 kHz signal can be used for this to avoid thermoelectric voltages interfering with a d.c. resistance measurement. If, say, the impedance is over 1 kΩ, then the display can be blanked out to indicate a likely open circuit. A more useful feature is to have a record of the changes in impedance with time to compare with the temperature record. Aging of the thermocouple can then be followed, as well as any sudden changes which may indicate a fault, e.g. crushing of a thermocouple cable. Otherwise, because of the low voltages involved, it is difficult to determine if the voltage is spurious or not.

Connection to the meter may be through direct wiring or through plugs. Direct wiring to the reference junction will usually be found in cases where the meter can cope with several thermocouple types. Plugs will be used where the meter is intended for one thermocouple type. As thermocouple plugs and sockets will be the correct type for the thermocouple wire it is important not to mix them with other types. Similarly, for a multi-thermocouple meter it is important to switch-select the correct type. Meters will cover a wide range but this does not mean that a single thermocouple probe can be used over that range. In general use one probe for one temperature.

- *Chart recorders*   A chart recorder includes all the features of the thermocouple meter along with a record-keeping function. As such it should be considered as at least two instruments: firstly as a signal processor, and secondly as an instrument to help you record the results.

Calibration of a chart recorder should be in terms of its indicating device and not the paper record. It is the operator's responsibility to ensure that pen, ink and paper make an accurate recording of the output of the chart recorder.

Several factors can influence the accuracy of the record. The pen and ink need regular checking to ensure that the record is actually being written. Paper size varies with humidity and temperature at a different rate than the metal of the chart recorder. Sprocket holes on the chart paper can be distorted or mispunched. The printed scale on the chart may be displaced and usually the chart recorder has a small adjustment to ensure that the pen is indicating the same value on the chart paper as the instrument indicator. Daily procedures may be needed to ensure that the chart records are as accurate as the instrumentation allows. Uncertainties in the recording process are typically half a scale division, i.e. wider than the pen thickness.

Both analogue and digital chart recorders are available. An analogue recorder may use a non-linear scale on the chart paper to convert its output to temperature. Digital recorders often print the reading as well and hence are easier to check. Where colour coding is used for many channels it is essential not to allow confusion by poor lighting or colour blindness.

7.5.2 Reference junctions

The equipment or instrumentation to achieve a satisfactory external reference junction environment will need to be assembled or purchased. Two types of reference junction can be distinguished; those at $0°C$ and those at other temperatures (e.g. room temperature).

In the laboratory the ice-point procedure (Section 3.3.5) makes a reference junction environment of known temperature and uniformity suitable for a small number of thermocouples. If heavy wires or numerous thermocouples are involved, then an ice–water mixture of sufficient capacity can be used. The mixture should be kept well stirred and the temperature monitored with a reference thermometer. Sealed reference junctions can be used to protect the wire or else use an oil-filled thermowell placed in the ice–water mixture. While in the short term pure water may not affect a bare junction, there is the risk of corrosion in the longer term. With some insulation materials there is the risk of contaminating the water and changing its electrical conductivity.

The main difficulty with ice is the need to keep it replenished and stirred. For many industrial surveys an automatic ice point is suitable. Automatic ice points are available commercially that use the expansion of ice on freezing to serve as a control mechanism for thermowells holding reference junctions (Figure 7.10). While the general accuracy may not be as good as that of a well-made ice point, the automatic ice point will work over longer periods and avoid gross errors due to lack of attention.

Use of an ice point implies that there is a temperature gradient of $20°C$ (or thereabouts) over a short length of the thermocouple wire as well as over the leads to the voltmeter. Any risk of inhomogeneities causing errors can be removed if the reference junction is at room temperature. Isothermal conditions can be achieved by thermally anchoring the reference junction to a copper block in a draught-free enclosure. If only low voltages are involved the thermal anchoring can be done through a thin electrical insulator such as Mylar film. The temperature of the block will need to be monitored and used in looking up the standard reference table. Instruments using this approach are available and provide good accuracy. Usually the temperature is monitored with a thermistor or solid-state sensor with the output adjusted to give a temperature reading to a microprocessor which then converts the thermocouple voltage to a temperature reading. Errors in

Figure 7.10
An automatic ice point for thermocouples.

making the adjustment to a reference junction not at 0°C can easily occur, as illustrated
in the following example.

Example 7.1

A Type N thermocouple is used to measure a temperature. A voltage reading of 2050 μV is obtained.
The isothermal reference junction is 18°C at the time of the reading. What is the temperature?

The voltage generated is less than it would be if the reference junction were at 0°C. Equation (7.4) is
relevant and can be expressed in terms of the table values as:

$$E = E_N(t) - E_N(t_R) \qquad (7.6)$$

In this case $E = 2050 \ \mu$V and $t_R = 18$°C.
From the table $E_N(18°C) = 472 \ \mu$V.
Thus $E_N(t) = (2050 + 472) \ \mu$V $= 2522 \ \mu$V.
Again from the table: $t = 91.5$°C.

A common error is to use the table only once to find an apparent temperature difference and add this
to the reference junction temperature. In that case, inferring that 2050 μV was equivalent to a 75.2°C
temperature difference would lead to an incorrect temperature reading of 93.2°C. Unfortunately this simple
practice occurs because a few thermocouple types are sufficiently linear around room temperature for the
two methods to give similar results. Avoid the practice and in all calculation procedures ensure that the
voltages are added.

Exercise 7.1 Using thermocouple tables

(a) Repeat Example 7.1 assuming that a Type B thermocouple was used.

(b) How large is the error if the temperature is calculated by assuming that 2050 μV corresponds to a temperature difference?

Another method of providing a reference junction is to add a compensating voltage electrically to the signal voltage. The compensating voltage must depend on the reference-junction temperature and closely match the thermocouple output around room temperature. These compensating junctions are available commercially as separate devices for use with a voltmeter, or already built-in to the thermocouple meter. Generally such circuits are made simple for low cost, and hence they are not designed to give a high accuracy over a wide temperature range. Separate compensating circuits are needed for each thermocouple type. They represent an analogue solution to a problem which is best solved by digital techniques as outlined above.

Electronic-reference junctions can present traceability problems to the user. A wide variety has been designed but not all by designers with an understanding of thermocouples. The compromises involved are seldom documented, so that the user will have to rely on tests and experience to gain confidence as to whether the electronic junctions meet requirements.

7.5.3 Calibrators

A wide variety of devices called calibrators are available to check thermocouple instruments. A calibrator is essentially a thermocouple meter in reverse. When a temperature is dialled, a voltage is produced at the output which, when connected by thermocouple wire to a thermocouple meter, should cause the meter to give the same temperature reading. A very important point to note is that even though calibrators give temperature readings they cannot provide a temperature calibration. They are designed to check the performance of the instrument. A calibrated instrument is *not* a calibrated thermometer.

Calibrators are normally portable devices used to check thermocouple equipment throughout an industrial plant. Any differences can be used to correct readings made with the meter or the meter may be adjusted to give the same reading as the calibrator. Meter readings made with the calibrator should be taken going both up and down the scale, especially in the case of a recorder which can often have a considerable deadspan (mechanical hysteresis). For a meter resolution of one degree the calibration steps should be every 50°C or so.

Both the meter and the calibrator can be designed for multiple thermocouple types. Therefore it is essential to use the correct thermocouple wire to connect the instruments and to ensure that both are set to the same thermocouple type. Again isothermal connections to both instruments are important. Many calibrators have exposed terminals and these should be protected and allowed to stabilise after connecting the thermocouple wire. If the wrong wire is used then an error will result depending on the temperature difference between the calibrator reference junction and that of the thermocouple meter.

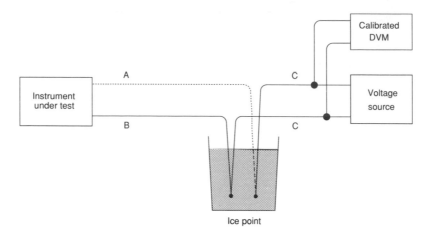

Figure 7.11
Calibration of a thermocouple instrument. The thermocouple instrument is connected by thermocouple wire, of the type undergoing the test, to the reference junction, immersed in an ice point. Buffer wires connect the reference junction to a stable voltage source and a calibrated voltmeter to provide traceability.

The output impedance of the calibrator should be low, that is below 10 Ω. Thermocouple circuits have a low electrical resistance and therefore the design of some thermocouple meters may use a low impedance input so that injecting a voltage from a potentiometer with a 10 kΩ output impedance will cause errors. Modern digital meters should not be prone to this problem.

The general accuracy of calibrators is around ±0.5°C to ±2°C. While this accuracy is sufficient for many thermocouple instruments, a higher precision is sometimes required. Main sources of error are the compensated reference junction and the stability of the voltage source. Some calibrators can have their reference junction bypassed and the voltage injected via an ice point (see Figure 7.11). Further increase in accuracy is possible by using a digital voltmeter to measure the output voltage as shown in Figure 7.11. The voltmeter can also be used to calibrate the calibrator. If the reference junction is bypassed, then copper wires should be used to connect the voltmeter directly and the dialled temperature values should give the voltage shown in the standard tables. With the reference junction in the circuit, the correct thermocouple wire from the calibrator should go to an ice point, with copper wires to the voltmeter. Again the standard table values are used. Note that in all cases the correct thermocouple wire should be used; it must also be in reasonable condition. Any error analysis should consider the effect of variation in the Seebeck voltage from different samples of wire.

7.5.4 Alternative thermocouple circuits

The basic measurement circuit of Figure 7.2 will generally be used with most instruments. However, variations on the thermocouple circuit are commonly found in order to solve instrument problems.

- *Differential circuits*   Because the thermocouple provides a signal related to a temperature difference it can be used as a sensor for a control mechanism which brings two components to the same temperature by ensuring that the differential thermocouple voltage is zero. If the actual temperature difference is needed, then the temperature of one of the junctions is required in order to determine the thermocouple sensitivity. The linearity of the Seebeck coefficient (Figure 7.3) at this temperature will determine how accurately the temperature needs to be known.

  Instead of the preferred circuit of Figure 7.2, the circuit of Figure 7.12 is sometimes used, especially with Type T thermocouples, where the break to the voltmeter is in the copper thermoelement. For other thermocouples this circuit arrangement is not desirable because of the higher thermoelectric effect with the voltmeter terminals. Extremely good isothermal conditions are needed to avoid stray thermoelectric voltages. Swapping the leads on the voltmeter should in principle check this, but in practice the handling of the terminals can upset the thermal balance. Using a switch to change over only moves, to another part of the circuit, the problem of achieving isothermal conditions!

- *Series circuits*   The main practical problem with the differential circuit for small temperature differences is the very low signal voltage, around 1 μV if a temperature differential of less than 0.1°C is to be detected. Higher sensitivity can be obtained by combining circuits in series as in Figure 7.13, to form what is called a *thermopile*. Two disadvantages result: firstly, the thermopile covers a wider area and hence it is more difficult to keep the junctions isothermal, and secondly, error arises from increased heat conduction through the wires unless they are very fine.

- *Parallel circuits*   An apparently simple way to average the results of several thermocouples is to join them in parallel. Two problems result:
  - a voltage average, weighted by the resistance of each circuit, is made, not a temperature average;
  - excessive current may flow through mismatched thermocouples.

The resulting reading will therefore not be at all like a temperature average. Parallel circuits often use a 500 Ω resistor in one leg of each thermocouple to limit the current, but this introduces a major inhomogeneity in the thermocouple wire. The method is suitable only as a monitor of wide temperature fluctuations or variations and hence

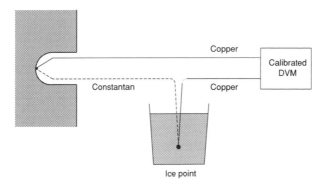

Figure 7.12
Alternative measurement circuit for Type T thermocouples.

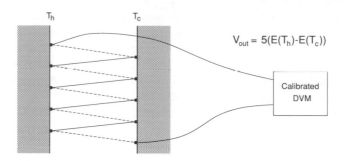

Figure 7.13
Thermocouples combined in series as a thermopile.

only a very low accuracy is expected. Separate thermocouples with a scanner would give a better idea of variations and allow an uncertainty to be calculated as well as an accurate average.

7.6 ERRORS IN THERMOCOUPLES

Reconsider Figure 7.4, which shows a measurement model for a thermocouple thermometer. Unlike other temperature sensors the thermocouple is distributed over a long length and is thus exposed to a wide variation in environmental effects, making an error assessment very difficult. The sensor interfaces to the physical system and to the instrumentation (the measurement and reference junctions respectively) are widely separated but both must be in isothermal environments. Avoidance of major errors from these factors has dominated the construction and instrumentation of thermocouples.

A variety of types of suitable thermocouples have been developed to cope with a range of environmental conditions from mild to harsh. We have already highlighted the main advantages of each thermocouple type in various environments. The main source of error comes from interactions with mechanical and chemical effects. Electrical interference should be small with modern instruments. Interactions with magnetic fields can occur because many wires contain magnetic materials. Thermocouples will also degrade when exposed to nuclear radiation and there is extensive literature on their use in nuclear power plants.

The treatment of errors for thermocouples is further complicated by their wide diversity of uses. They can be a cost-effective solution for many chemically hostile environments or physically difficult situations. To achieve this versatility the accuracy is usually sacrificed. Indeed, if the thermocouple is the only means available for a measurement, then it is almost impossible to assess the accuracy independently; errors of up to 30% have been found in extreme cases when these were analysed later by more reliable methods. The onus will be on the user to make a Type B uncertainty assessment based on experience and knowledge of the system's behaviour.

7.6.1 Thermal errors

The three main thermal effects involved in interfacing the sensor to the temperature of interest are immersion, thermal lag and thermal capacity. These have been covered in

Section 4.4. Because of the wide diversity of probes these parameters are best determined experimentally.

For mechanical reliability these factors are often severely compromised. For example, a thermowell in a pipeline may be short in length and thick in diameter to ensure it does not fatigue and snap off; ideally, the length should be over 5× the diameter. While very little can be done about the thermal lag and capacity it may be possible to improve the effective immersion depth by surrounding the pipe with insulation. Keep the thermocouple connection box outside the insulation so that it is at ambient temperature (i.e. 10°C to 50°C).

7.6.2 Inhomogeneity errors

The discussion on thermocouples has already centred around the inhomogeneity problem which is the major source of error. It is difficult to prove that inhomogeneities have been avoided. While there is on-going research into methods of measuring the inhomogeneity error for a specific thermocouple assembly, the only practical method at present is to calibrate the thermocouple assembly *in situ*. Periodic calibrations will be needed to check on any inhomogeneity arising from use or accidental damage.

Drift and hysteresis are the result of the formation of an inhomogeneity. Usually the change is irreversible, but sometimes reversible effects are found. Details will depend on the thermocouple type, the temperature range and the chemical environment. Figure 7.5 shows the normal drift for Type K and Type N thermocouples. Drift can be very high and frequent replacement is desirable, e.g. for flame temperature measurements a maximum of 8 hours of use is typical. Follow closely any recommendations of a standard or the research results for similar systems.

Where there is no experience, Table 7.2 may be used to make a Type B uncertainty assessment. Use the percentage tolerance for Class 3 as an estimate of the uncertainty at 95% CL. The percentage should be expressed in terms of the maximum temperature difference occurring along the thermocouple wire. For some wire types and some uses this will be quite a pessimistic estimate of the uncertainty. If you wish to claim a better uncertainty then you will need evidence to demonstrate this. A part of the demonstration will be to show that the factors raised in this chapter have been adequately understood and dealt with. For example, the uncertainties could be based on Class 1 or 2 tolerances, providing that there is a test to ensure that a sample of the wire complied with a standard type and class and that the written procedures for the installation were followed.

In assessing errors use the maximum temperature difference to which the wire is exposed.

Example 7.2

A Type K thermocouple instrument is used to monitor a low-temperature bath at −20°C, but the thermocouple wires have to go over a heating pipe at 90°C to reach the bath. What is the likely uncertainty in the measured temperature due to inhomogeneities?

Continued on page 268

— Continued from page 267 —

The Class 3 percentage tolerance for the Type K thermocouple is, from Table 7.2, 1.5%. The maximum temperature difference along the length of wire is 110°C. The uncertainty due to inhomogeneities is estimated as 1.5% of 110°C = ±1.6°C. If the high-temperature excursion is reduced to 20°C then the uncertainty is reduced to ±0.6°C.

7.6.3 Isothermal errors

Several parts of the thermocouple circuit require good isothermal conditions to ensure that any introduced inhomogeneity does not give rise to a significant error, e.g. joins, junctions and instrumentation. Experimental estimation of the likely error may be inferred by the application of a hot-air blower to the suspected parts; less than a 0.2°C change should be observed.

7.6.4 Reference-junction errors

Inadequate knowledge of the reference-junction temperature is probably the next most significant error for thermocouples after the ones caused by inhomogeneity. The use of a well-constructed ice point can remove this error.

Compensating reference junctions will normally have their uncertainty quoted in their specification; ±1°C over the ambient temperature range is typical. The value is likely to be the isothermal value, i.e. the junction needs time to settle down if, say, the temperature changes from 20°C to 15°C. Some instruments can take up to half-an-hour to settle. This can be a problem with hand-held instruments which are used intermittently.

7.6.5 Interference errors

Even though thermocouples are low-impedance devices they will pick up electromagnetic interference. Some of the recommendations of Section 5.4.12 are not possible with thermocouple wires, e.g. twisting the wire pair may create large inhomogeneity errors. Long lengths of thermocouple wire, e.g. 50 m or so, make good radio aerials! They also increase the risk of errors from unknown temperature profiles and accidental damage. Therefore, avoid long lengths of wire by using other signal transmission devices more immune to interference, such as 4 to 20 mA current loops. It is essential that the instrumentation used can cope with any changes in the ambient temperature.

7.6.6 Wire resistance errors

As the signal to be measured is the Seebeck voltage (i.e. open circuit as in Figure 7.2), the wire resistance should have no effect if a potentiometer or good digital voltmeter is used. Analogue meters should not be used as they draw too high a current and the reading becomes dependent on the circuit resistance.

A thermocouple circuit resistance of over 1 kΩ should be considered an open circuit. Monitoring the circuit resistance is useful to check on possible wire damage from chemical or mechanical sources.

7.6.7 Linearisation errors

The defining tables for the thermocouple types (Appendix D) are usually given to a resolution of 0.01°C to allow practical linearisation schemes to be developed to match the tables to better than 0.1°C. Modern digital equipment should easily meet this requirement. More rough forms of linearisation may be found which are accurate to only 1°C.

7.7 CHOICE AND USE OF THERMOCOUPLES

As a rule, thermocouples are not high-precision temperature sensors and hence their main use is for unusual environmental conditions or for high temperatures. Careful selection based on the known properties of the thermocouples is therefore needed. In general a thermocouple should not be considered more accurate than about ±1%, e.g. ±1°C for normal temperatures and ±10°C at 1000°C. Because thermocouples have such a wide range of applications it is difficult to generalise on their use, and the guidelines given here are designed merely as a starting point to find the best thermocouple for your application.

7.7.1 Selection of thermocouple type

First, check your test specification to see if the thermocouple type is specified. Often tests are designed around the properties of a particular thermocouple and substituting another type can give invalid results.

Otherwise the maximum temperature to be measured will be the deciding factor. Tables 7.3 and 7.4 give some of the details for selecting bare wire thermocouples. Table 7.4 assumes a long life, and if a shorter period is acceptable then the range may be extended or thinner wire may be used. With suitable sealed sheathing the range may be extended, but in some cases the sheathing may limit the range.

Above 1700°C none of the standard thermocouple types are suitable and one of the special types will have to be used, e.g. tungsten–rhenium or boron–carbide/graphite. Consult the manufacturer's recommendations for the best use of the special thermocouples. They may also be useful at lower temperatures in chemically hostile environments with appropriate protective sheaths.

From 1500°C to 1700°C a Type B thermocouple would be preferred for a clean environment.

From 1200°C to 1500°C a Type R or Type S thermocouple could also be used in a clean environment, but because of possible grain growth in the platinum leg they may become fragile on long exposure. The advantage of Type R or Type S is that their use can be extended down to 200°C and their accuracy is better. Their main disdavantage is the cost of the wire, the more sensitive instrumentation and the extra care needed in installation. Over time there will be a drift in the output even in a clean environment. If most of the change is due to thermal and mechanical stress, then a clean and anneal as per Section 7.8.3 may help restore the output.

Below 1200°C, where a less expensive solution can be used if high precision is not required, there is a choice between Types K and N. Type N gives a better performance overall but the decision may often rest on availability and cost. In general these thermocouples are sufficiently low priced that a frequent replacement scheme can be considered to keep the thermocouple in good condition in a hostile environment.

Below 700°C, Types J and E can also be considered. Type E will give a better performance than Type K and hence is used in survey work, but it is not so commonly used otherwise. Type J will tolerate a reducing atmosphere and it is the only standard type that will.

Below 200°C and down to −200°C, Type T is a suitable thermocouple wire. The wires are fairly flexible and can be obtained with very fine diameter to reduce the thermal loading on small objects.

Where possible select thermocouples in the most complete state of assembly that will be consistent with your application. In particular, wires should be bought as matching pairs, i.e. as MIMS cable, or as Teflon®-coated or fibreglass-covered duplex cable. If single wires are required, then they should be obtained to specification, so that when paired they will match standard tables. In general, do not mix wires from different manufacturers. For example, the copper wire used in Type T thermocouples is usually less pure than the copper used in modern electrical wire and the manufacturer has the option of varying the constantan alloy or the copper purity in order to match tables.

7.7.2 Acceptance

Ensure that the wire supplied is the wire type ordered and that it is in the continuous lengths required. The length is easily checked, but the wire type is not, because wires are generally not labelled in any consistent fashion.

While colour coding is often used to denote individual wire, wire pairs and extension wire, unfortunately each country of manufacture has a different colour coding system. There is currently no agreed international colour coding for thermocouple wire. Unless you are very sure of the thermocouple's origin then the colour coding may not be helpful. Keep wires, until used, on the original reels if supplied that way, and ensure that any identification labelling will remain attached. Colour coding of plugs and sockets is more consistent but these usually have the type marked on them.

A check can be made on the output voltage of the thermocouple. If the check is made at 50°C, say, then it is not likely to degrade the wire. This should enable most types to be distinguished (see tables in Appendix D), except possibly for Types T and K, or Types R and S. Type T can usually be distinguished by its copper thermoelement and for Type K the negative thermoelement is magnetic, so that a small magnet can be used to check for this. Because of the small difference between Types R and S a full calibration is often the only way to tell them apart. Use only one type, Type S if possible, to avoid confusion throughout your laboratory or plant.

Acceptance tests can be made closer to the temperature of use, if required. In some cases an inhomogeneity test may be called for (see Section 7.7.4).

7.7.3 Assembly

Many of the factors for the successful assembly of a thermocouple and its measurement circuit have been covered in the discussion above. Correct assembly is important for the

traceability of a thermocouple measurement because the errors resulting from mistakes
or failures can give false readings not easily discernible from real readings. Here are the
main points to note:

- Ensure that assembly is done by skilled personnel.
- Ensure that materials used are clean, particularly for high-temperatures.
- Ensure that the materials will withstand the temperature of use.
    - Most materials lose considerable strength well before they collapse at high
      temperatures. The upper rating on a material may be its collapse temperature
      and there may be very little mechanical strength below it.
    - Many of the materials look similar but can have very different temperature
      ratings. Test that they withstand the temperature first.
- Use matched pairs of wires.
- Immerse both junctions in isothermal environments.
- Use sufficient length of wire to enable the connector head to be mounted away from
  the source of heat.
- Do not reverse the polarity of the thermoelements.
    - Test by applying hand heat to see if the temperature reading rises.
- Check that the insulation resistance is adequate.
    - Dry out slowly if moisture causes a low insulation value.
    - Provide an electrical ground if necessary.
- Use proper thermocouple connectors throughout.
- Check the reading at the ice point if at all possible.
- Finally, check the circuit with a hot-air blower.
    - Hold the measurement junction at a fixed temperature; the ice point is ideal.
    - Apply the hot air to all other parts of the circuit and connectors.
    - Any movement in the indicated temperature exposes a problem that needs
      solving. The main causes of problems are reversed connections, wrong wire or
      lack of thermal insulation to keep a join isothermal.

7.7.4 Inhomogeneity tests

Any change in the output of a thermocouple, such as drift, arises from changes in the
wire or in other words an inhomogeneity. Therefore it is useful to have some measure
of what is happening.

There are two methods to check for gross inhomogeneities. One is suitable for new
reels of thermocouple wire and the other for completed circuits. Both methods help
eliminate the worst of the problem but do not ensure good performance.

At least two samples of wire from the reel, usually from the beginning and the end,
can be subjected to a compliance test as in Section 7.8.1. Any significant differences
in the test results for the samples could indicate an inhomogeneity problem, e.g. the
manufacturer did not anneal the wire after it was drawn. The variation in performance
throughout the reel should be smaller than the tolerances of Table 7.2. Further tests may
be needed to determine if the wire is useful or should be discarded.

A completed circuit can be tested by applying local heating to its various parts while the measuring junction is at a constant temperature. The use of a hot-air blower is a convenient way to do this, as indicated in Section 7.7.3. More concentrated heat sources may also be useful, e.g. small flames or a soldering iron, if the insulation material can withstand the heat. Not only will the test indicate unsuspected inhomogeneities, e.g. a badly bent wire, but it will also check that any known inhomogeneity, e.g. a join, is properly installed in an isothermal environment. The test is not good at detecting a distributed inhomogeneity, such as a chemical change along an extensive length of the wire.

There are many applications where a better assessment of homogeneities in a thermocouple wire is desirable if not essential. For instance, calibration of thermocouples *in situ* does require a reference thermocouple of greater accuracy. Also, thermocouples are often the only way to monitor new processes and a good knowledge of the temperature may be essential for the research and development. In order to achieve better accuracy more quantitative measures of the inhomogeneities are essential.

Applying corrections for inhomogeneities is a major undertaking and is beyond the scope of this text. A very careful measurement of the inhomogeneity is needed to ensure that only the inhomogeneity is being assessed and not, say, other electrical effects.

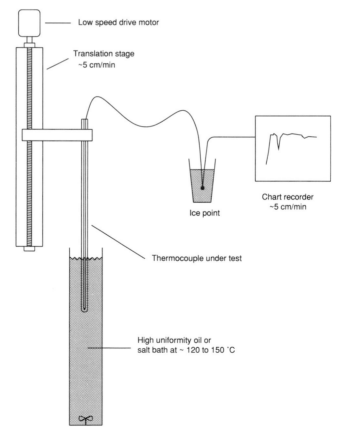

Figure 7.14
A furnace for measuring thermocouple inhomogeneity.

The main limitation of a test is that the test conditions could change the nature of the inhomogeneity.

However, as a quality control over platinum reference thermocouples and a general diagnostic tool, an inhomogeneity test is essential for a calibration laboratory. There are no standard tests for this and there is a variety of approaches to the problem. A straightforward method is to use a two-zone furnace with a sharp transition between the two isothermal zones, as shown in Figure 7.14. The second zone needs to be long enough for the required immersion conditions. For Type K wire the furnace should be at a relatively low-temperature to avoid altering the wire. Type S thermocouples, however, can be tested at 1000°C.

Figure 7.15 shows the output of a thermocouple as it is passed slowly through the vertical furnace. Voltage excursions can be seen as the inhomogeneities pass the transition zone. Figure 7.15 shows the profile for a Type K thermocouple previously exposed to 870°C. As expected, there is a large deviation near the measurement junction. At slightly greater immersion (Zone B) there is a large deviation of around 30% due to green rot forming in the wire. After that there is a Zone (C) which has been altered by the heat and shows significant deviations. Finally, on the far right (Zone D) we encounter essentially the original wire which shows no deviation. Such profiles can give a good indication of when a thermocouple should be replaced. A 'conventional' calibration may not show any significant error (see Section 7.8).

The size of the voltage excursion will depend in part on the thickness of the thermo-couple, the speed of movement, and the sharpness of the transition. However, when these are well-controlled, guidelines can be established as to acceptable levels for reference thermometers.

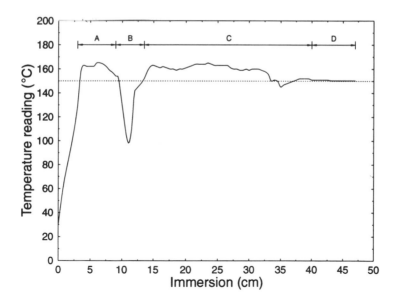

Figure 7.15
The thermocouple reading as it is inserted into the inhomogeneity furnace. The errors due to inhomogeneities vary between +10% and −30%.

7.8 CALIBRATION

When dealing with thermocouple calibration it is important to know which meaning of the word 'calibration' is being used. Frequently 'calibration' is used to refer to the various thermocouple types or the term 'de-calibration' is applied to the growth of inhomogeneities in the wire. Certainly 'de-calibration' means that the calibration certificate is no longer valid. The possibility of undetected inhomogeneities makes the formal calibration of a thermocouple problematical. We will first consider the problem itself before outlining methods for the formal assessment of a thermocouple.

Consider a laboratory thermocouple calibration with reference to Figure 7.16. The figure gives a diagrammatic representation of the temperature along the length of a thermocouple. Curve ABCDEF represents a typical temperature profile during calibration. The measurement junction F is immersed in an isothermal region to E. Then there is a sharp temperature gradient until ambient temperature is reached at D. The lead wires CD to the ice point are approximately isothermal. Then another sharp temperature gradient follows, CB, to reach 0°C before the isothermal region, BA, leading to the reference junction A. All the voltage is generated over the two short lengths ED and CB.

In use, however, a curve similar to GHIJ will usually apply. The thermocouple is immersed in an approximately isothermal region HG; a short temperature drop to ambient occurs along HI and I is joined to a reference junction, J, at ambient temperature by the use of compensating extension wire. All the voltage is generated over the short length HI (hence it can be seen why compensating extension wire can be used from I to J).

The calibration has checked only the two parts of the wire, ED and BC, under unspecified temperature gradients. Even then the length BC may be wire belonging to the calibration laboratory if extension wire is used to connect the instrumentation. To make the calibration at all useful (i.e. traceable) it is necessary to somehow demonstrate that this calibration is also valid for the part HI under unspecified temperature gradients. For example, in Figure 7.15 Section C does not represent the likely behaviour of Section B! That is, we need evidence that the wire is homogeneous. Procedures for handling thermocouple wire become very important to assert traceability. The simple act of the wires twisting and being straightened out again may be sufficient to destroy homogeneity, let alone the danger of reversible and irreversible changes on heating.

Of course in many situations the error or risk of error may be lower than the maximum error tolerated. However, where modern processes require close temperature tolerances,

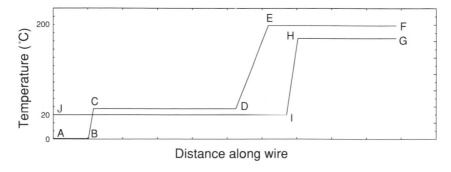

Figure 7.16
Immersion conditions for a thermocouple wire.

thermocouples will have difficulties in achieving the accuracy. It is important that the system design includes consideration of an adequate calibration regime or there may be no way to achieve the desired accuracy.

From the above discussion we can see that the application of the step-by-step calibration procedures of Section 4.3.6 used for the other temperature sensors may not be suitable for thermocouples. The procedure is a good test that the thermocouple wire conforms to the standard table. The difficulty is that after the test we do not know if the wire still conforms or will continue to conform under 'reasonable use'. More information is required in order to claim traceability. Thermocouples cannot be considered good general-purpose thermometers. They are better suited to operating over a single narrow range of temperatures. That is, the use is restricted so that a better estimate of their future behaviour is possible. Only the rare-metal thermocouples come close to being good general-purpose thermometers under reasonable use.

Three possible methods for claiming a calibration are given here. Other methods will be met in practice and they should be evaluated for effectiveness in terms of the concerns raised in this chapter. The first method checks that the thermocouple conforms to type and then uses the known generic history to make a Type B uncertainty assessment. The second method is to calibrate the thermocouples *in situ*. Not all applications will allow this but it is a better method. A detailed example of an oven survey illustrates the method. The third method follows the step-by-step procedure of Section 4.3.6 for rare-metal thermocouples because they have sufficient stability to warrant it.

7.8.1 Conformance or type approval

Conformity testing is an important step to achieve traceability. The generic history will then allow the conversion of the voltage reading into temperature and will also allow a Type B uncertainty assessment to be made. The conversion may be directly from the standard tables or through a deviation function to correct the table values. Procedures must be adopted to ensure the quality of the installed thermocouple.

If thermocouples are made up from wire off a reel then samples along the wire length can be submitted for calibration over the temperature range used. This procedure confirms that the wire as supplied conforms to standard tables. Similarly, sampling of complete assemblies can be carried out.

A replacement regime for the thermocouples should be established based on experience or recommendations, e.g. thermocouples may need to be replaced after every 8 hours of use in high-temperature corrosive atmospheres or may need replacement every 6 months under less harsh conditions.

A variety of methods can be used to check on the thermocouples in use. In a process with a reasonably stable temperature the thermocouple can be replaced with a new one. A significant difference would indicate that all the other thermocouples may need replacing. However, where there are several thermocouples monitoring the same temperature, e.g. separate temperature-controller and recorder probes, half should be of a different age. If one set drifts significantly from the other set of thermocouples, then replacement of one set may be indicated. Another method is to move the thermocouple out by a few centimetres; a drastic change in reading indicates the presence of inhomogeneities and that it is time to replace the thermocouple.

The usefulness of the conformity approach is highly dependent on the application, the type of thermocouple used, and the temperature range covered. It also relies on the ease with which a thermocouple can be replaced. The design of the thermowells should allow for the removal of thermocouples while the plant is operating.

7.8.2 In situ calibration

The only reliable method of reducing errors in thermocouple readings is to calibrate *in situ*. This ensures that the immersion conditions in use are the same as those of the calibration. For base-metal thermocouples an *in situ* calibration may give tenfold improvement in accuracy over a Type B uncertainty assessment, i.e. $\sim$0.1% of temperature. Thus the effect of any inhomogeneity introduced during handling or in use will be removed by the calibration. Again the design of the system needs to allow for this with the provision of thermowells, where a calibration thermometer can be employed without disturbing the thermocouple. This can be achieved by the provision of a second thermowell alongside or by the provision of other means of access to the thermocouple.

The following example shows how this principle can be applied to a temperature uniformity survey within an oven. The user should be able to adapt this to formulate a procedure for an individual situation.

Oven survey procedure

- *Aim*   The survey is to determine the temperature profile throughout the workspace of a hot air oven and to check the calibration of the recording thermometer and temperature-control probe.

- *Preliminary*   Before starting a survey ensure that the controller is working properly. Poor control will result in poor spatial and temporal temperature uniformity. The controller should be set to the temperature of the test and any recording thermometer should read within 1°C of the setting. For a test the controller should be set to temperature and not subsequently adjusted to obtain the 'right' temperature.

    Tuning of the controller may be needed. Ensure that both the controller and recorder instrumentation are functioning correctly and are themselves calibrated.

- *Principle*   The temperature profile for the calibration thermocouples is not to change significantly during the calibration. Therefore the method will apply only to ovens where there is already reasonable temperature uniformity.

- *Equipment*   A multi-channel thermocouple recorder or logger will be needed with good reference junctions. The number of channels will depend on the extent of the survey, either from external specifications or internal requirements. A minimum of 9 survey positions in the oven is needed, one in each corner of the oven plus one in the centre of the oven. More may be needed if details of the uniformity of the central space are required. Two more for the recording and control probes with one to monitor the ambient temperature makes 12 as the minimum number of survey probes.

    Flexible thermocouple wire or cable is needed and of a type to match the range. For best accuracies select from Types T, E or N.

- *Access*   Obtaining access for the survey thermocouple can be a problem if the oven designer has not provided for it. Any hole through to the internal space of the oven may need to be covered subsequently to prevent heat loss through air movement. There should be a means of anchoring the thermocouple cables as they enter the oven so that they do not move during the calibration. Avoid going in through the door, as a slamming door can cause a significant inhomogeneity at the point of maximum temperature gradient! The reference thermometer will also need a means of entry and this thermometer is generally rigid and more fragile. The entry hole for the reference thermometer should be of adequate size and positioned for easy use.

- *Wire Length*   The thermocouple wire length inside the oven should be sufficiently long to reach two sites, namely the centre of the oven and the survey site. Outside of the oven there should be sufficient length so that any joins can be made at ambient temperature.

- *Calibration*   All the survey thermocouples should be brought together in the centre of the oven and thermally mounted to the reference thermometer. The centre is used because it tends to have the best temperature uniformity, but other sites can be used if there are physical restrictions. More care may be needed to ensure an isothermal connection between the thermocouples and the reference thermometer. The reference thermometer may be either a platinum resistance thermometer or platinum thermocouple suitably calibrated.

  Best accuracy is achieved by commencing calibrations at the highest temperatures first. On each step down allow sufficient time at each temperature for thermal stability before making the comparisons.

- *Survey*   When the oven is cool move the thermocouples to their survey sites. Note that in moving the thermocouples the effective immersion depth in the oven will stay the same provided that the bundle of wires is well anchored as it enters the oven space. This ensures that the same length of wire is at the maximum temperature gradient for both the calibration and survey.

  Surveys may be made with or without loading in the oven according to requirements. The work space of the oven needs to be well-specified: usually the space at least 5 cm from any wall is a good guide. Mounting of the survey thermocouples can be facilitated by using a rigid wire frame to hold the thermocouples close to their sites. Wires can shift with heating if they are not secured.

  Survey probes should be mounted alongside the recorder and controller probes. The recorder probe should be inside the designated work space.

  The temperature survey can be done in steps of rising-temperature. Record the temperature rise and its settling to a stable condition for each step. Often a uniformity specification may require that the overshoot should not be excessive and also that the oven comes to an equilibrium condition over a few control cycles.

- *Reporting*   Because the interpretation of the uniformity can become complex if both time and space variations are taken into account, clear and detailed reporting of the results is often called for. To say an oven had a uniformity of $\pm3°C$ at 120°C may not be adequate for a test that requires an object to be heated at $120 \pm 3°C$ for 50 minutes. The uniformity may apply to the oven only after it has been on for 2 hours, whereas the test may require the oven to be loaded when cold and the time period to start when the coldest part of the oven reaches 117°C. At no time after that should any part of the oven be over 123°C.

Depending on the oven survey specifications, there are many variations on the uniformity requirements and the reporting should address these directly.

If required, a calibration certificate may be issued for the recorder and perhaps the controller. A calibration certificate may not be required for the survey thermocouples or the oven because of the short-term nature of the calibration. While a certificate may be issued for the oven it generally will be for the oven complying with a specification. In principle, a certificate of calibration for the oven thermocouples, e.g. those for the recorder and controller, can be issued once a typical history of the oven and its use is established.

7.8.3 Rare-metal thermocouple calibration

The third calibration method is more similar to that for other thermometers but is only relevant to the platinum thermocouples, Types S, R or B. The general principles will also apply to a gold–platinum thermocouple if this proves to be a suitable reference thermometer, not only for the laboratory but also in the industrial environment. While ITS-90 uses high-temperature PRTs, none of the recommended types are suitable for use outside of a calibration laboratory.

The procedures for calibrating rare-metal thermocouples also require a higher level of expertise than for other thermocouples. This is in part because of the need to know how to disassemble and reassemble the thermocouples reliably, and in part because the reproducibility of the thermocouples depends in detail on the annealing procedure. For these reasons few users would have the competence to calibrate the thermocouples in-house, particularly if accuracies of better than ±1°C are required.

The procedure we give here is primarily for calibrating rare-metal thermocouples used as working thermometers. If you are involved in the calibration of reference thermocouples, consult the references at the end of the chapter which explain the procedures in much greater detail.

Step 1 — Initiate record-keeping (as in Section 4.3.6)

Step 2 — General visual inspection (as for Section 4.3.6)

Step 3 — Conditioning and adjustment
There are five main tests to perform at this stage:

- *Disassembly* The thermocouple should be removed from its sheath which usually comprises one or more sections of twin-bore alumina tubing or beads.
- *Cleaning* Reference thermocouples and working thermocouples in good condition (see Step 4, *Detailed visual inspection*, below) should be cleaned with ethanol. Working thermometers that are not visually bright should be cleaned by boiling for 10 minutes in each of distilled water, 20% nitric acid, distilled water, 20% hydrochloric acid and distilled water. Three cycles may be required before the wire is clean.

- *Electric anneal* The thermocouple is then annealed by passing an a.c. electric current through each leg. The pure platinum leg should be annealed at 1100°C and the platinum–rhodium leg at 1450°C. The anneal removes strain due to work-hardening, oxidises residual impurities, and restores the wire to a uniform state of oxidation. For 0.5 mm diameter wires about 12–13 A is required.

- *Reassembly* The thermocouple is then carefully reassembled in its sheath. If the old sheath is not clean it should be replaced by a new sheath which has been baked at 1100°C for a minimum of two hours to drive off potential contaminants.

- *Furnace anneal* After the thermocouple has been reassembled, that part of the thermocouple that will be exposed to temperature gradients is furnace annealed at 1100°C. For reference thermocouples the correct annealing and cooling procedure are critical for best performance (see references).

Step 4 — Generic checks

There are two basic checks that should be carried out with rare-metal thermocouples: a visual inspection and a homogeneity test.

- *Detailed visual inspection* This is carried out at the first stages of the cleaning and annealing phase above. The inspection is primarily to determine the suitability of the thermometer for its purpose. Reference thermocouples are assembled in a single 300–1000 mm length of twin-bore alumina, clean and bright, and do not have any breaks, joins or extension leads attached to them. Thermocouples which are not in this condition are not suitable as reference thermocouples.

  Working thermocouples may be assembled in a variety of sheaths, should be reasonably clean, and not have any breaks or joins. Thermocouples which do not satisfy these requirements have been abused and should not be calibrated.

- *Homogeneity test* This is carried out according to the procedure in Section 7.7.4. The variation in the Seebeck coefficient is assessed to determine the likely uncertainty in use due to inhomogeneities in the wires. For reference thermocouples the maximum observed variation must be less than 0.05% (0.5°C at 1000°C). This test is carried out after the reassembly and furnace anneal.

A homogeneity test may also be carried out prior to the disassembly to assess the user's treatment of the thermocouple.

Step 5 — Intercomparison

In most cases the reference thermometer for the intercomparison will be a rare-metal thermocouple. In some instances where the highest accuracy is required, the reference may be a high-temperature SPRT (Chapters 3 and 5), or a transfer standard radiometer (Chapters 3 and 8), or several of the defined fixed points (Chapter 3).

Whatever the calibration medium, care must be taken to avoid contaminating the thermocouples. In particular there should be no metals, other than platinum and rhodium, in the immediate vicinity of the calibrating furnace.

In all cases the most convenient representation of the determined ITS-90 relationship is as a deviation from the reference function. For reference thermocouples working over a narrow range (e.g. 600°C to 1100°C) a linear or quadratic deviation function will suffice and therefore not as many points are required in the intercomparison. For thermometers working over a wider range or working thermometers, which do not conform so closely to the generic history, a cubic deviation function is more appropriate.

Step 6 — Analysis

The analysis of the intercomparison data proceeds as described in Section 4.3.6 with a least-squares fit used to determine the best values of the coefficients in the deviation function. For reference thermocouples the deviations from the reference function should be less than ±1.5°C for Types R and S, and less than ±2.5°C for Type B.

Step 7 — Uncertainties

The contributing factors to the uncertainty are:

- *Uncertainty in reference thermometer readings* (as in Section 4.3.6).

- *Variations in the stability and uniformity of the calibration medium* (as in Section 4.3.6). This is a difficult assessment when the reference thermometer is a transfer standard radiometer.

- *Departure from the determined ITS-90 relationship* (as in Section 4.3.6).

- *Uncertainty due to hysteresis*   As with all thermocouples rare-metal thermocouples suffer from hysteresis. This is particularly true in the 500°C to 900°C range where the platinum is subject to changes in the state of oxidation. Where the thermal history of the thermocouples is not controlled the uncertainty is about 0.1% (95% CL). Where the annealing and use is controlled the uncertainty is typically 0.02 and 0.05%, depending on the detail of the procedures (see references).

- *Uncertainty due to drift*   The typical drift on a good reference thermocouple used below 1100°C is usually less than 0.2°C, and most of this occurs in the first hours of use. So long as it is used carefully after the initial period the remaining drift is negligible. The use of reference thermocouples is also logged, and once the total exposure exceeds 100 hours the thermocouple is due for recalibration. This ensures that the drift between calibrations is negligible.

  Above 1000°C the drift increases rapidly with temperature, being less for the Type B thermocouple than for Types R and S. Because the drift is highly dependent on use the uncertainty is best set to zero. Instead the user has the responsibility of performing sufficiently frequent verification checks to monitor the drift.

- *Uncertainty due to inhomogeneities*   This uncertainty is assessed as the maximum observed variation in the Seebeck coefficient as determined from the inhomogeneity test at 95% confidence limits. For reference thermocouples this must be less than 0.05%.

- *Total uncertainty*   This is calculated as the quadrature sum of the contributing uncertainties. Since the most significant uncertainties, due to hysteresis and inhomogeneities, are approximately proportional to the output voltage, the uncertainty should be reported as a percentage. For example:

"The uncertainty in the corrected thermocouple readings is estimated as 0.1% of the output voltage at the 95% confidence level."

Step 8 — Complete records (as with Section 4.3.6)

FURTHER READING

Thermocouple Temperature Measurement, P. A. Kinzie, Wiley, New York (1973).
An old but valuable book that describes about 300 different thermocouples and some of their associated problems. It does not include Type N.

Manual on the Use of Thermocouples: ASTM MNL 12, American Society for Testing and Materials, Philadelphia (1993).
A useful handbook that describes the use, assembly and installation of thermocouples. This new edition of an old manual should help correct many of the misconceptions about thermocouples.

Techniques for Approximating the International Temperature Scale of 1990, BIPM (1990).
The Calibration of Thermocouples and Thermocouple Materials, G. W. Burns and M. G. Scroger, NIST special publication 250-35, US Department of Commerce (1989).
Both of these booklets give good guidelines for the calibration of rare-metal thermocouples as well as a useful summary of the properties of base-metal thermocouples.

Temperature-Electromotive Force Reference Functions and Tables for the Letter-designated Thermocouple Types Based on the ITS-90, G. W. Burns *et al.*, NIST Monograph 175, U.S. Department of Commerce (1993).
A key reference on thermocouples. Contains not only tables and reference functions but also detailed information on the properties of each thermocouple type.

Temperature, its Measurement and Control in Science and Industry, American Institute of Physics, New York, Vol. 5 (1982), Vol. 6 (1992).
These proceedings include useful information on the real performance of thermocouples and industrial applications.

8

Radiation Thermometry

8.1 INTRODUCTION

We are all familiar with the dull red glow of coals in a fire and the bright white glow of incandescent lamps. We know that the brighter and whiter a glowing object, the hotter it is. This is the simplest form of radiation thermometry. Although it is simple, temperature discrimination on the basis of colour can be remarkably accurate. Those who work in high temperature processing industries, such as a steel works, can often estimate the temperature to $\pm 50°$C simply on the basis of colour (see Table 8.1).

Because our eyes cannot detect the radiation from bodies cooler than 500°C, most of us associate thermal radiation only with objects that are hot and often dangerous. However, everything around us radiates some electromagnetic energy, and often in quite large quantities. By understanding how this radiation depends on temperature we can make accurate measurements of temperature over a very wide range.

Radiation thermometers have three main features which distinguish them from the other thermometers discussed in this book. Firstly, they are thermodynamic, i.e. they are based on a universal physical law which is known to describe real objects to very high accuracy. Radiation thermometers can and have been built with reference only to the triple point of water. The thermodynamic radiation law is also the basis of the temperature scale above the silver point ($\sim 962°$C).

The second distinguishing feature is that they are non-contact thermometers and can be used to measure the temperatures of remote or moving objects. This makes it possible to measure the temperature inside furnaces, fires and even the Sun and stars, places too hostile or remote for any contact thermometer.

The third feature is that radiation thermometers use the surface of the object of interest as the sensor. This is both a blessing and a bane. While the use of the object as a sensor

Table 8.1. **Temperature versus perceived colour.**

Temperature, °C	Colour
500	red just visible
700	dull red
900	cerise
1000	bright cerise
1100	dull orange red
1250	bright orange yellow
1500	white
1800	dazzling white

overcomes a lot of the difficult questions about immersion and thermal contact that affect other thermometers, it raises even more difficult questions about traceability. How can we make a measurement traceable if it involves a different sensor every time the instrument is used?

8.2 BLACKBODIES AND BLACKBODY RADIATION

Hot objects emit radiation over a wide range of the electromagnetic spectrum. For objects at temperatures of practical interest, most of the radiation is in the infra-red and visible portions of the spectrum. A graphical description of the distribution of thermal radiation is shown in Figure 8.1.

The spectral radiance, plotted on the vertical axis of Figure 8.1 on a logarithmic scale, is a measure of the amount of energy emitted by an object in a given wavelength range. The horizontal scale describes the wavelengths at which the radiation is emitted. The visible portion of the spectrum is also marked, with the violet end of the visible spectrum at 0.4 μm and the red end at 0.7 μm. Radiation at wavelengths shorter than 0.4 μm is described as *ultra-violet* (above violet) or UV, while radiation at wavelengths longer than 0.7 μm is described as *infra-red* (below red) or IR.

We can see from Figure 8.1 that for objects below 500°C (~800 K) all of the radiation is in the invisible infra-red region. As the temperature increases and the radiance curves

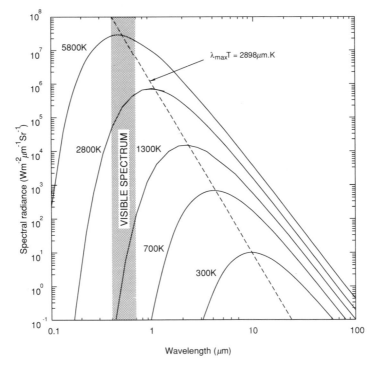

Figure 8.1
Planck's law: the spectral radiance of a blackbody as a function of temperature.

of Figure 8.1 begin to edge into the red end of the visible spectrum, we see objects as red-hot. As the temperature increases further the emission spectrum moves further into the visible and we see objects with the perceived colours shown in Table 8.1. At temperatures above 1500°C to 1800°C, objects become so bright that our eyes have difficuty accurately and comfortably discriminating the colour.

In order to give the mathematical description of the radiation distribution shown in Figure 8.1 we need to introduce the concept of spectral radiance and define the term blackbody.

Spectral radiance, measured as the energy emitted by a surface per unit area, per unit wavelength, per unit solid angle, is the technical term for the optical brightness of a surface. The advantage of using radiance, rather than other optical quantities, is that radiance is independent of the distance to the surface and the size of the surface. Also, in an ideal optical system of lenses and mirrors, the radiance of an object is constant. Instruments that measure radiance allow us to infer temperatures at a distance and, if necessary, to use close-up or telephoto lenses.

A *blackbody* is simply a perfectly black surface: a perfect emitter and absorber of radiation. Those who first encounter the blackbody concept may find it paradoxical; our everyday experience is that bright objects are white, not black. The apparent paradox arises because no visible blackbody radiation is emitted at room temperature. At room temperature the brightest objects are bright because they reflect, not because they emit.

Reflectivity and *emissivity* (the ability to emit) are complementary properties. For opaque objects

$$\text{reflectivity} + \text{emissivity} = 1 \tag{8.1}$$

or using the appropriate symbols:

$$\rho + \varepsilon = 1. \tag{8.2}$$

Figure 8.2 gives a simple pictorial explanation of the relationship between reflectivity and emissivity for two different surfaces.

We shall see later (Section 8.6.3) that good approximations to ideal blackbodies can be made with cavities that are designed to trap and not reflect light.

In 1900, Planck derived the mathematical description of the distribution of blackbody radiation shown in Figure 8.1:

$$L_{\lambda,b} = \frac{c_1}{\lambda^5} \left[\exp\left(\frac{c_2}{\lambda T}\right) - 1 \right]^{-1} \tag{8.3}$$

where L is the radiance, the subscripts λ and b indicate spectral radiance and blackbody respectively, λ is the wavelength of the radiation and T is the temperature of the blackbody in kelvin.

The two constants c_1 and c_2 are known as the *first and second radiation constants*, and their best measured values are currently

$$c_1 = 1.191\,044 \times 10^{-16} \text{ W m}^2$$

and

$$c_2 = 0.014\,387\,69 \text{ m.K.}$$

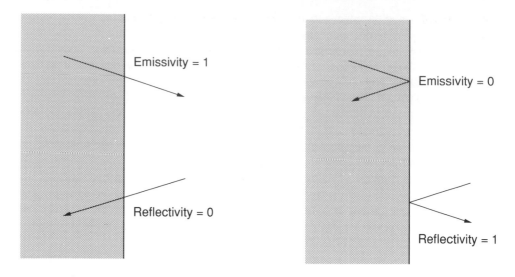

Figure 8.2
The complementary properties of emissivity and reflectivity.

Equation (8.3), *Planck's law*, is used to define the ITS-90 temperature scale above the silver point, 961.78°C. To do this only the second radiation constant is required and it is assigned the value

$$c_2 = 0.014\,388 \text{ m.K.}$$

By assigning a value to c_2 the temperature-scale definition becomes fixed and cannot change simply because research determines a 'better' value. At many points in this chapter we make the approximation $c_2 = 0.0144$ m.K($= (120)^2$ μm.K). This approximation is done in order to establish simple formulae which serve as aids to memory.

In practice, real objects are not blackbodies, but emit less radiation than predicted by Planck's law by the factor ε, the emissivity of the surface. The spectral radiance of a real object is

$$L_\lambda = \varepsilon(\lambda) L_{\lambda,b} \tag{8.4}$$

where $\varepsilon(\lambda)$ indicates that the emissivity may vary with wavelength.

Planck's law (equation (8.3)) is not a simple law, and provides no obvious assistance in acquiring a feel for how radiation thermometers work. We give here some simpler results that are easier to remember and to work with.

All of the curves in Figure 8.1 are characterised by a maximum which occurs at shorter and shorter wavelengths as the temperature increases. The wavelength at which the maximum occurs is

$$\lambda_{\text{max}} = \frac{2989}{T} \ \mu\text{m}. \tag{8.5}$$

At room temperature, for example ($T = 300$ K), the maximum spectral radiance is near 10 μm, and at 3000 K, the temperature of an incandescent lamp filament, the peak occurs at 1 μm. For objects of practical interest most of the radiation is emitted in the infra-red portion of the spectrum. It is interesting to note that the human eye has evolved to match the peak in the solar spectrum near 500 nm.

For several practical reasons, most radiation thermometry is caried out at wavelengths in or near the visible portion of the spectrum. In this portion of the spectrum λ is less than λ_{max}, and Planck's law is approximated to 1% or better by *Wien's law*:

$$L_{\lambda,b} = \frac{c_1}{\lambda^5} \exp\left(\frac{-c_2}{\lambda T}\right). \tag{8.6}$$

Although it is less exact because of the minor simplification, this is a much more 'user-friendly' function than Planck's law for estimating the errors and uncertainties in measurements.

The total radiance of a blackbody, L_b, is found by integrating Planck's law to determine the area under the curves of Figure 8.1:

$$L_b = \frac{\sigma}{\pi} T^4 \tag{8.7}$$

where σ is the Stefan–Boltzmann constant, $\sigma = 5.67051 \times 10^{-8} \mathrm{W\,m^{-2}K^{-4}}$. The total energy emitted by the blackbody in all directions is π times this value, hence energy is emitted by a real surface at the rate of

$$M = \varepsilon \sigma T^4 \mathrm{W\,m^{-2}} \tag{8.8}$$

where ε is the total emissivity. Some examples of the energy emitted by blackbodies are shown in Table 8.2.

Both the table and the fourth-power law in equation (8.7) show that the total radiance increases very rapidly with temperature. At the short-wavelength end of the Planck spectrum the spectral radiance increases spectacularly. For $\lambda < \lambda_{max}$ the spectral radiance follows an approximate power law given by

$$L \propto T^x \tag{8.9}$$

where, using the approximation for c_2,

$$x \approx \frac{12\,1200}{\lambda\ T} \qquad (\lambda \text{ in microns}).$$

Table 8.2. The rate of emission of blackbodies at a range of temperatures.

Temperature, °C	Rate of emission (per square metre)
25 (room temperature)	470 W
230 (melting point of solder)	3.6 kW
500 (a hot stove element)	20 kW
1000 (yellow flame)	150 kW
2500 (lamp filament)	3.4 MW
5800 (sun)	77 MW

At a temperature of 1200 K and a wavelength of 1 μm the spectral radiance is changing as T^{12}. This is typical of the operating regime of most radiation thermometers. One of the earliest radiation thermometers, the disappearing-filament thermometer, operated at 655 nm and at temperatures as low as 600°C: here the power law is T^{25}. This rapid change of radiance with temperature has a good side and a bad side. Over the operating range of a thermometer the measured radiance may vary by a factor of 100 000 times. It is quite difficult to design instruments that operate accurately over such a dynamic range.

The high-power law does, however, make for an extremely sensitive instrument, which is just as well because optical measurements are amongst the most difficult. Even in laboratory conditions it is difficult to measure radiance to better than 1%. A more serious problem is that we are rarely able to determine the emissivity to better than 5%. Were it not for the high-power law, radiation thermometry measurements would have such high uncertainties as to render them useless.

8.3 SPECTRAL-BAND THERMOMETERS

Most radiation thermometers are of the type known as spectral-band thermometers; they measure the radiance over a relatively narrow band of wavelengths within the range 0.5–25 μm. The choice of wavelength depends, amongst other factors, on the temperature range, the environment, and the type of surface under investigation. Discussion of the operating principles, use, errors and calibration of spectral-band thermometers will form the basis of this chapter. The spectral-band grouping includes most industrial radiation thermometers and all primary and transfer standard thermometers.

Figure 8.3 shows a simplified diagram of a spectral-band thermometer. We have deliberately included the radiant target, the surface of interest, as part of the thermometer, since the surface is the sensor. The basic operating principle is to collect radiation from the surface, filter it to select the radiation at the wavelengths of interest, and then measure it with a detector and signal processing system. The two apertures in the system define the target area or field of view, and the acceptance angle of the thermometer (similar

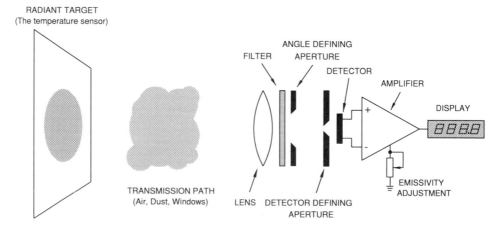

Figure 8.3
Simplified schematic drawing of a spectral-band radiation thermometer showing the basic elements of its construction and operation.

to the f-stop in a camera). The lens is used to focus an image of the target area on to the target-defining aperture. Without the lens, or with the lens not properly focused, the boundary of the target area is not well-defined.

The signal at the output of the detector is a complex function of the dimensions of the apertures, the transmission of the various optical components, and the detector response. For simplicity we assume that the filter has a sufficiently narrow pass-band to ensure that the output of the detector is proportional to Planck's law. We will look more closely at the actual response later when we consider calibration equations.

The spectral radiance measured by the thermometer is

$$L_{\lambda,m} = \varepsilon_\lambda L_{\lambda,b}(T_s) \tag{8.10}$$

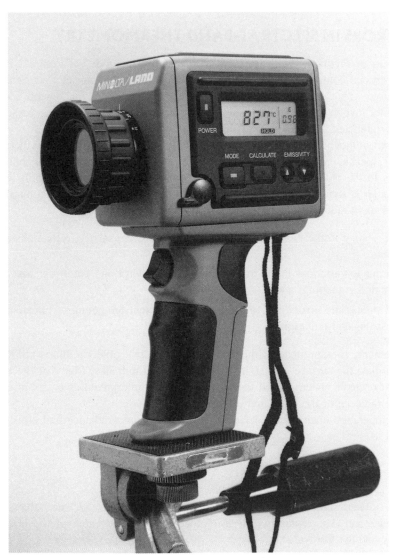

Figure 8.4
An example of a hand-held spectral-band radiation thermometer.

where ε_λ is the spectral emissivity of the surface at the operating wavelength, and T_s is the true temperature of the surface. To determine the temperature we must measure or estimate a value for the surface emissivity; we will call this the *instrumental emissivity*, ε_i. Many radiation thermometers have this instrumental emissivity adjustment built-in. The temperature is calculated by solving

$$L_{\lambda,b}(T_m) = \frac{\varepsilon_\lambda L_{\lambda,b}(T_s)}{\varepsilon_i}. \tag{8.11}$$

Ideally $\varepsilon_i = \varepsilon_\lambda$ so that the measured temperature T_m is the true surface temperature T_s. A typical radiation thermometer of this kind is shown in Figure 8.4.

8.4 ERRORS IN SPECTRAL-BAND THERMOMETRY

Comparison of Figure 8.3 with the general measurement model of Figure 2.10 shows that two crucial components of the thermometer, namely the sensor and transmission path, change with each new measurement. Every measurement made with a radiation thermometer involves the characterisation of these important and often inaccessible parts of the thermometer.

That part of the thermometer which we normally describe as the radiation thermometer is strictly speaking only a *radiometer*. The radiometer, which measures radiance, is analogous to the potentiometer or voltmeter in a thermocouple circuit; only when the potentiometer is attached to the thermocouple do the two, together, form a thermometer.

The errors in radiation thermometry fall into three main groups:

- errors relating to the characterisation of the sensors: emissivity, reflections and fluorescence;
- errors due to variations in the transmission path: absorption, scattering, size-of-object effects and vignetting; and
- signal processing errors due to variations in ambient temperature, linearisation and the instrumental emissivity.

The radiometric measurement of the temperature of a real object requires knowledge of two quantities: the surface emissivity and the spectral radiance. Many of the dominant errors that occur in spectral-band thermometry can be interpreted as errors in either the measured radiance or the estimated emissivity.

The temperature error caused by errors in the measured radiance and emissivity are estimated as

$$\Delta T_m = \frac{\lambda T^2}{c_2} \left(\frac{\Delta L_{\lambda,m}}{L_{\lambda,m}} - \frac{\Delta \varepsilon_\lambda}{\varepsilon_\lambda} \right). \tag{8.12}$$

Here $\Delta L_{\lambda,m}$ represents the difference between the measured and true values of spectral radiance and $\Delta \varepsilon_\lambda$ represents the difference between the value of the instrumental emissivity and the true value, $\varepsilon_i - \varepsilon_\lambda$.

Equation (8.12) is appropriate when the errors are known, but if the errors are unknown, perhaps random, then their relationship to the measurement error is properly expressed in terms of uncertainty:

$$\sigma_{T_m} = \frac{\lambda T^2}{c_2} \left(\frac{\sigma_{L_{\lambda,m}}^2}{L_{\lambda,m}^2} + \frac{\sigma_{\varepsilon_\lambda}^2}{\varepsilon_\lambda^2} \right)^{1/2}. \tag{8.13}$$

This equation is conveniently expressed as

$$\sigma_{T_m} = \lambda \left(\frac{T}{1200} \right)^2 (\rho_{L_{\lambda,m}}^2 + \rho_{\varepsilon_\lambda}^2)^{1/2} \tag{8.14}$$

where λ is in microns and the relative uncertainties, ρ, are in percent. For all three of these equations we can make the following observations:

- The errors and uncertainties increase with operating wavelength, therefore as a general principle choose thermometers with a short operating wavelength.
- The errors and uncertainties increase as the square of the temperature.
- The errors and uncertainties increase as $1/\varepsilon_\lambda$. In general the errors are very large for low-emissivity materials such as metals.

The wavelength dependence of errors can cause confusion when thermometers of differing operating wavelengths are used to measure the same temperature. Indeed, a difference in readings between two thermometers would normally indicate that both are probably in error, since nearly all of the major sources of error are wavelength-dependent and affect all spectral-band thermometers.

In the following sections we treat each source of error in detail. In some cases the errors are beyond the capability of the user to deal with. However, understanding of the sources of error leads to an understanding of the need for the proper care and maintenance of the thermometer. The two most significant errors are those associated with emissivity and reflections, and are dealt with first.

8.4.1 Errors in emissivity

In almost all areas of radiation thermometry the largest source of error is the lack of knowledge about the surface emissivity. Figure 8.5, which shows the spectral emissivity of various samples of inconel®, gives an indication of the problem. Depending on the degree of oxidation, roughness, and wavelength, the emissivity varies between 0.1 and 0.95. And this is a material that is used as an emissivity standard!

For many materials, especially rough and amorphous materials, the practical problems are not as bad as implied by Figure 8.5. It is reasonably easy to identify the material, decide whether it is rough or polished, oxidised or not, and make an estimate of the emissivity. However, to make a good estimate of the emissivity some serious homework is necessary. It is important to know what wavelength the thermometer operates at, and what the material is, and to have access to reliable information on the surface properties of the material. Most manufacturers of radiation thermometers supply a list of materials and their emissivities at the operating wavelength of their thermometer. If all of this information is available, it is usually possible to make an estimate to about ±0.05.

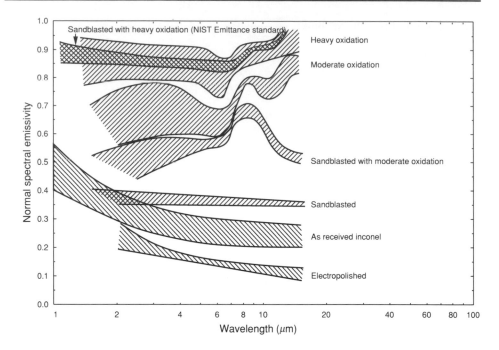

Figure 8.5
The spectral emissivity of inconel®. There is a wide range of emissivities depending on the surface finish.

Example 8.1   Uncertainty in temperature due to uncertainty in emissivity

(a) Estimate the uncertainty in the measurement of the steel temperature in a rolling mill where the temperature is approximately 1000°C and the emissivity of the highly oxidised steel at 1 μm is estimated as 0.80 ± 0.1 (95% CL). Assume the uncertainty in the measured radiance is zero.

Direct substitution into equation (8.13) yields

$$U_{T_m} = \pm 1 \times \left(\frac{1273}{1200}\right)^2 \times \frac{100 \times 0.1}{0.80}$$

$$= \pm 14°C$$

(b) Estimate the uncertainty in the measurement of the temperature of freshly galvanised steel in a galvanising plant where the temperature of the plated steel is about 450°C and the emissivity of molten zinc at 4 μm is estimated to be 0.15 ± 0.05 (95% CL). Assume the uncertainty in the measured radiance is zero.

As above.

$$U_{T_m} = \pm 4 \times \left(\frac{723}{1200}\right)^2 \times \frac{100 \times 0.05}{0.15}$$

$$= \pm 50°C$$

The example above illustrates the importance of knowing the emissivity. At moderately high temperatures, where short-wavelength thermometers operate and the emissivity of materials is usually high, measurements can be made with reasonable accuracy. At lower temperatures, where the longer-wavelength thermometers must be used and materials may have very low emissivities, the uncertainties can be so large as to make the measurement almost useless.

Exercise 8.1

Use equation (8.14) to calculate the uncertainty in a radiation thermometry measurement made near 130°C (400 K) and 10 μm. The uncertainties in the measured radiance and emissivity are 3% and 4% respectively.

Without information on the spectral emissivity of the material, it is almost impossible to make a reasonable estimate of the emissivity from visual assessment alone. Surfaces which are black in the visible portion of the spectrum may well have a low emissivity in the infra-red and vice versa. Two common examples will illustrate the point.

Nowadays, most paints use titanium dioxide as the base pigment. While the pigment is extremely white (i.e. has a low emissivity) in the visible part of the spectrum, it is also very black in the infra-red. Thus the appropriate emissivity setting for a 10 μm thermometer looking at any painted surface is about 0.9. As a general rule most organic materials, e.g. wood, skin and organic fibres, exhibit this type of behaviour.

The opposite effect occurs with metals coated with thin layers of oxide. At short and visible wavelengths the surface can be quite black. At longer wavelengths the oxide layer becomes transparent so that the surface behaves as the pure metal and has a very low emissivity. Figure 8.6 shows some examples. The curves in Figure 8.6 also exhibit some wiggles. These are caused by interference due to thin layers of metal oxide, the same phenomenon seen with oil films on water. These interference phenomena lead to emissivities which are extremely sensitive to the film thickness, viewing angle and operating wavelength. Under these conditions spectral-band radiation thermometers are almost useless.

For most surfaces the emissivity is also dependent on the angle of view. This is shown in Figure 8.7. The drop in emissivity at high angles, i.e. for views near grazing incidence, is a feature in common with all surfaces. As a general rule the emissivities that are published are for normal incidence, i.e. viewing at right angles to the surface, so radiation thermometers should always be used at normal incidence.

To obtain the most accurate information, the spectral emissivity must be measured. The simplest method is to measure the temperature of a sample of the material using an alternative thermometer, such as a calibrated thermocouple, and then to adjust the emissivity setting on the radiation thermometer to give the same reading as the thermocouple. If there are no other errors and the emissivity is the same for all samples of the material, then this technique results in temperature measurements of a similar accuracy to that of the alternative thermometer. Measurements of the emissivity made in this way may be accurate to better than ±0.05 depending on the wavelength.

A second method of measuring emissivity is to make a blackbody to operate at exactly the same temperature as the material. Then adjust the emissivity setting on the thermometer so that the temperature is the same as the reading for the blackbody with

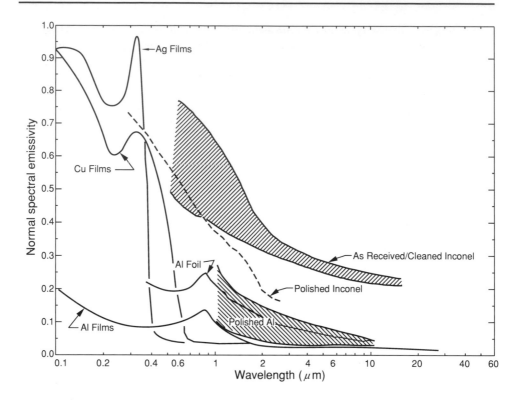

Figure 8.6
The normal spectral emissivity for a number of metals.

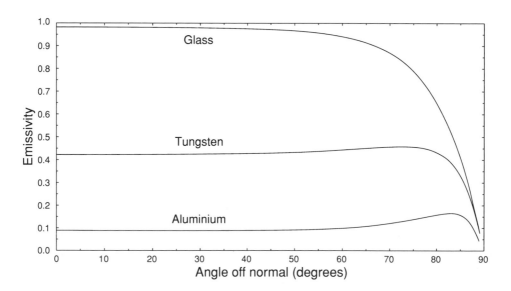

Figure 8.7
The emissivity of glass, aluminium and tungsten as a function of the angle of view.

the instrumental emissivity set to 1.0. The main difficulty with this technique is getting the required uniformity of the blackbody while keeping the other surface free of reflections. The simplest version of the technique is to coat a sample of the surface with black paint or soot, both of which have emissivities in the range 0.9 to 0.95. With this method ±0.05 is about the best accuracy to be expected.

A third, less practical, technique, which is the most accurate, is to measure the spectral reflectance of the surface. This is, however, a job for a specialist with specialised equipment. Accuracies of ±0.02 or better are possible this way. If the measurements are carried out at room temperature they will be in error at high temperatures if the emissivity is temperature dependent.

In recent years a number of radiation thermometers have emerged on the market that use an infra-red laser to measure the emissivity of the surface *in situ*. Their operation is based on the relationship between the emissivity and hemispherical reflectance (equation (8.2)). In practice these thermometers measure the retro-reflectance of the surface. Estimating the hemispherical reflectance from the retro-reflectance is analogous to estimating the volume of earth in a hill from one measurement of the height. The measurement is quite good for highly diffusing surfaces, such as those with a fine powdery texture, or for very rough surfaces. But for other surfaces the measurements can be very poor, certainly worse than a well-informed estimate.

In general the emissivity of surfaces can rarely be determined with the accuracy desired for accurate radiation thermometry. The exception is for objects which behave as blackbodies. As a general guide the thermometer manufacturer's list of emissivity versus material will enable estimates to within ±0.05 for rough or diffuse surfaces, and to within ±0.1 for surfaces with any gloss, polish or film associated with them.

8.4.2 Reflection errors

Since radiation thermometers infer temperature from measured radiance, anything that adds to the surface radiance will cause the thermometer to be in error. The most important source of additional radiance is radiation reflected from other sources in the vicinity. Unfortunately, radiation thermometers are most useful in high-temperature processing industries, where there are invariably reflections from flames, electric heaters, and furnace walls, etc. At low-temperatures the problem may be worse because the whole environment behaves as a very large blackbody at 300 K. In fact it is quite hard to find an application where reflections are not a problem.

The most difficult aspect of the reflection problem is that we do not naturally associate other hot objects, that may be some distance from the surface of interest, with the surface itself. It is a matter of discipline to be aware of all objects in the space above a surface and methodically assess the likelihood of a reflection error caused by that object.

The most effective way of eliminating reflection errors is to eliminate the source of extraneous radiation. One of the most important sources of radiation in measurements made outdoors is the Sun. If a radiation thermometer is being used to detect hot spots, such as thermal leaks in buildings, then the measurements should be made when the surface is shaded, or at night. In many cases, it is possible to shade the surface artificially. This technique can also be used inside furnaces to shade heaters or flames that are in close proximity to a surface.

More often than not the interfering source is too large or too hot to shade. This occurs in many high-temperature processing industries where a product is preheating in a large firebox and radiation thermometers are used to determine when the product has reached the required temperature. Fortunately, in these cases it is relatively easy to estimate the magnitude of the errors.

If we assume that the firebox walls are at a uniform temperature, T_w, then the firebox behaves as a blackbody cavity with an emissivity of 1.0 (see Exercise 8.2). The radiance of a small object within the firebox then comprises two parts:

$$L_{\lambda,m} = \varepsilon_\lambda L_{\lambda,b}(T_s) + (1 - \varepsilon_\lambda)L_{\lambda,b}(T_w). \tag{8.15}$$

The first part of the equation represents the thermal emission from the surface (equation (8.10)), the second is the radiance due to reflections originating from the firebox walls. Now, depending on the wall temperature T_w, there are three strategies for handling the reflection.

Strategy 1: Assume negligible error and set $\varepsilon_i = \varepsilon_\lambda$

This is the strategy employed during normal use of a radiation thermometer. It is the appropriate strategy for situations where the firebox temperature is much less than that of the object, or the source of extraneous radiation is small. Figure 8.8 shows the errors that occur for this situation with the object at 900°C and various firebox wall temperatures. A simple approximation of equation (8.15), based on Wien's law, shows that the error in the measured temperature when T_w is near T_s is

$$\Delta T_m = T_m - T_s = \left(\frac{1 - \varepsilon_\lambda}{\varepsilon_\lambda}\right)\left[\frac{\lambda T_s^2}{c_2} + T_w - T_s\right]. \tag{8.16}$$

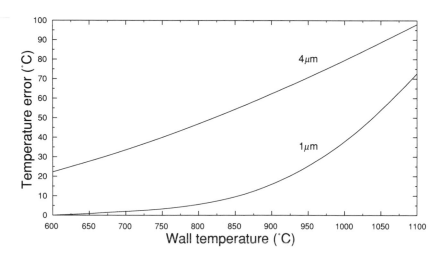

Figure 8.8

Temperature errors due to reflections versus firebox wall temperature for thermometers using the $\varepsilon_i = \varepsilon_\lambda$ strategy. The surface of interest has an emissivity of 0.85 and a temperature of 900°C. For wall temperatures less than the surface temperature short-wavelength thermometers have less error.

Both Figure 8.8 and the approximation show that the errors increase with wavelength and the square of the temperature. The error also decreases with increasing emissivity.

Strategy 2: Assume blackbody conditions and set $\varepsilon_i = 1.0$

In many applications the object and the firebox have very similar temperatures. Under these conditions the object/firebox system behaves as a blackbody. Indeed, if we substitute $T_w = T_s$ in equation (8.15), then

$$L_{\lambda,m} = L_{\lambda,b}(T_s) \tag{8.17}$$

so that the effective emissivity of the object is 1.0. Setting $\varepsilon_i = 1.0$ is therefore a good strategy when the firebox and the object are at similar temperatures. Figure 8.9 shows the same situation as in Figure 8.8, except that $\varepsilon_i = 1.0$. The central region of the graph near $T_w = 900°C$ shows that the strategy is quite effective and that the errors are almost independent of wavelength. A good approximation for the error in this region is

$$\Delta T_m = (1 - \varepsilon_\lambda)(T_w - T_s). \tag{8.18}$$

This is the equation of the dotted line in Figure 8.9. This measurement strategy also has the advantages of eliminating the uncertainties in the emissivity, and operating the thermometer under the same conditions as those under which it is calibrated (namely $\varepsilon_i = 1$).

It might be thought that the situations where this strategy can be employed with confidence are rare. However, nearly all measurements made indoors near room temperature fall into this category. Measurements in any temperature-controlled environment, such as coolstores, furnaces and kilns, are also candidates for this strategy.

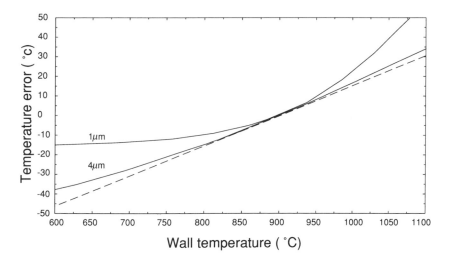

Figure 8.9
Temperature errors for radiation thermometers employing the $\varepsilon_i = 1.0$ strategy. The errors are almost independent of wavelength when $T_w = T_s$. At very long wavelengths the error approaches the dotted line (equation (8.18)).

Strategy 3: Apply corrections for the reflections

There are now a number of thermometers on the market, both hand-held and fixed installation types, that measure the firebox wall radiance and the radiance of the surface, and apply corrections for the reflected radiance. However, for $T_w > T_s$ this strategy is very sensitive to the operating wavelength and the uncertainties in both the emissivity of the surface and the wall radiance, particularly as the wall is rarely uniform. For these systems there is an optimum operating wavelength that minimises the uncertainty in the corrected result. For $T_w > T_s$ this wavelength is near

$$\lambda_{\text{opt}} \approx \frac{c_2(T_w - T_s)}{T_w T_s}$$

$$= \left(\frac{1200}{T_s} \right) \left(\frac{1200}{T_w} \right) \left(\frac{T_w - T_s}{100} \right) \ \mu\text{m}. \tag{8.19}$$

For $T_w \leq T_s$ the shortest practical wavelength should be chosen. This is also a good guide to the best operating wavelength for the $\varepsilon_i = 1.0$ strategy.

For applications where the temperature difference between the wall and object is less than 200 K the optimum wavelength is well within the normal range of operating wavelengths for that temperature. As the temperature difference $T_w - T_s$ increases the best operating wavelength also increases. Usually the best operating wavelength is within the range of commercially available instruments.

Exercise 8.2

By considering the total radiance of a closed isothermal (i.e. with a uniform temperature) cavity as the sum of emitted and reflected parts (equation (8.15)) show that the effective emissivity is 1.0.

Exercise 8.3

Compare the accuracy of measurements of the temperature of a hotplate using the $\varepsilon_i = \varepsilon_\lambda$ and $\varepsilon_i = 1.0$ strategies (Section 8.4.2). The temperature of the hotplate is expected to be 37°C and has an emissivity of 0.3. The operating wavelength of the radiation thermometer is 12 μm. Room temperature is 300 K.

8.4.3 Non-thermal emission

Another, less frequent, source of error is *fluorescence*, which arises because thermal energy excites impurities in the object, which then emit radiation in a very narrow band of wavelengths. This can happen with glass, for example, and Figure 8.10 shows the phenomenon for diamond.

If this type of non-blackbody emission occurs within the pass-band of the radiation thermometer, then the measured radiance will be high. The problem is most likely in relatively pure materials that are partially transparent in the operating pass-band of the thermometer.

Avoiding errors due to fluorescence is difficult unless equipment is available to measure the whole spectrum. The best strategy is to use well-established procedures and

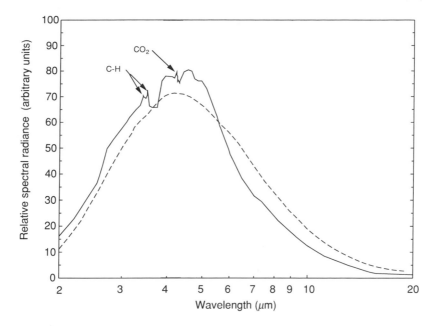

Figure 8.10
The emission spectrum of a sample of diamond. The dotted curve shows the emission spectrum for an object with constant spectral emissivity and a temperature of 400°C. As well as the blackbody radiation there is additional radiation due to fluorescence. The C-H emission is at the same wavelength as the C-H absorption band in plastics (see Section 8.7.4).

operating wavelengths. This relies heavily on the fact that others have found such procedures reliable.

8.4.4 Absorption errors

One of the great advantages of radiation thermometers is that they can measure temperature at a distance. However, this involves using the intervening space between the object and the thermometer as the transmission path for the radiation, and unfortunately most gases, including air, are not completely transparent. Most of the absorption in air (Figure 8.11) is due to water and carbon dioxide. In some situations, particularly near flame exhausts, the concentrations of both water and carbon dioxide can be much higher and the absorption greater. This leads to a low value for the measured temperature. In some cases, particularly near flames, the gas may emit radiation at these wavelengths and cause the temperature reading to be high.

Nearly all spectral-band thermometers are designed to avoid the worst of the absorption bands. However, some broadband thermometers, for example operating over 0.8 to 1.2 μm, will not be completely immune to these effects. Some environments, such as inside oil- or gas-fired furnaces, have very high concentrations of carbon dioxide and water which are very strong absorbers. Again the choice of thermometer and operating procedures should be based on well-established industry practices.

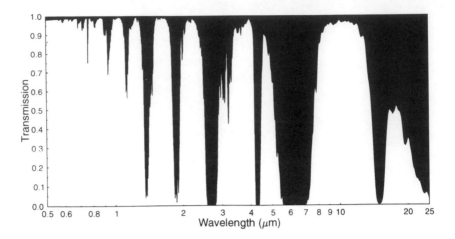

Figure 8.11

The transmittance of 300 m of air at sea level. The area above the transmittance curve is shaded to emphasise wavelengths for which the atmosphere is opaque. Those parts of the spectrum for which the atmosphere is transparent (unshaded) are known as *windows*. Some of the most useful windows for spectral-band radiation thermometry are near 0.65 μm, 0.9 μm, 1.05 μm, 1.35 μm, 1.6 μm, 2.2 μm, 4 μm and 10 μm.

8.4.5 Scattering errors

Dust in the transmission path of a radiation thermometer has three detrimental effects. Firstly, it scatters radiation out of the transmission path. This causes a decrease in the measured radiance from the object of interest, and therefore a decrease in the temperature reading. Secondly, it may scatter radiation from other sources into the transmission path and increase the temperature reading. And thirdly, the dust may itself emit blackbody radiation, so that the thermometer reading will be affected by the dust temperature. Examples of dust include smoke, luminous flames, water fog, carbon, metal ore and silica. The 2–3% loss in transmission in the windows in Figure 8.11 is due to atmospheric scattering.

The general problem of the scattering of radiation from small particles is extremely complicated and depends on the size of the particles, on whether they transmit or absorb, and on the wavelength of the radiation. The only useful general principle is that the problem can often, but not always, be reduced by using thermometers that operate at longer wavelengths.

8.4.6 Size-of-object effects

All radiation thermometers collect the radiation from a well-defined conical zone in front of the thermometer (see Figure 8.12). The size of the zone is defined by the two defining apertures and is known as the field of view. Ideally the zone has a sharp boundary so that radiation from outside the cone has no effect on the reading. Also, the field of view must be completely filled to give an accurate reading. In practice, there are three effects,

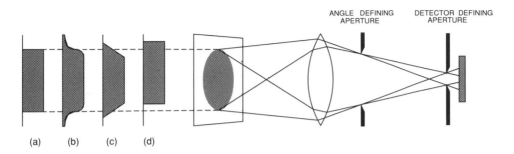

Figure 8.12
Size-of-object effects. (a) an ideal target profile, (b) a target profile broadened by flare, (c) a target profile due to poor focus, and (d) misalignment.

as shown in Figure 8.12, which contribute to the blurring of the field-of-view boundary. These three effects are discussed below.

Flare

Flare is the most serious of the size-of-object effects. It is caused by the scattering of radiation within the radiation thermometer, in particular, by dust, scratches and density imperfections on, or in, the front lens of the thermometer. An analogous problem is the glare caused by dust on the windscreen of a car when driving towards the Sun. When driving away from the Sun, so that the windscreen is shaded from the sun, there is usually no problem. As the analogy suggests, flare is a serious problem when there are other bright radiation sources near the field of view of the thermometer. In many cases flare is difficult to distinguish from a reflection error.

Flare is usually prevalent in long-wavelength thermometers operating at 4 μm or more. In these thermometers the increased effect is due to density variations in the lenses which scatter the radiation, and the increased sensitivity due to the longer wavelength. It is also more prevalent in systems with narrow fields of view or high magnification. As a guide the errors for a short-wavelength ($\sim$ 1 μm) industrial thermometer operating at 1000°C are typically 2–5°C, and for long-wavelength thermometers ($\sim$ 5 μm) about 10–25°C.

Usually the only way to minimise flare is to use *sight tubes*. These are tubes which are black on the inside and mounted on the front of the thermometer, to restrict the radiation falling on the lens to that within the field of view. The lens hoods on camera lenses and car windscreen visors perform the same function. When employing sight tubes it is important that the tube does not impinge on the field of view as this will cause vignetting (see below).

Because scratches and dust also cause flare, it is essential that radiation thermometers are maintained with due care. In particular, the front lens should be cleaned regularly with an air-brush or high-quality lens tissue. On no account should abrasive materials be used to clean a lens. Permanently installed radiation thermometers may require an air purge facility which supplies cool filtered air over the front lens to both cool the lens and prevent dust from settling.

Poor focus

Lenses and mirrors in radiation thermometers are used to focus an image of the object of interest on to the target-defining aperture. Radiation from the portion of the image over the aperture then passes on to the detector. If the thermometer is not well focused, then the boundaries of the target area are not well-defined. This is shown in the poor-focus target profile in Figure 8.12(c). For systems with a fixed focus or no lenses, the field of view must be well overfilled in order to get an accurate reading.

Optical aberrations and alignment

A radiation thermometer with the ideal target profile shown in Figure 8.12(a) is not realisable. In practice, imperfections in the optical components, inter-element reflections, and slight misalignment of the optical components all lead to very slight blurring of the target image.

Usually these effects are negligible for practical purposes. However, there are two examples of misalignment that lead to large size-of-object problems. The first occurs if, for example, the thermometer is dropped or knocked, so that some of the components become seriously misaligned. Of course, serious misalignments of this kind usually invalidate the calibration as well.

The second problem occurs in long-wavelength thermometers where a separate visual telescope or sighting laser is provided to sight the target. If the two optical paths are not exactly aligned, or not in focus at the same time, then the field of view may have to be overfilled considerably to get an accurate reading.

To minimise size-of-object effects, always overfill the field of view as much as possible with neighbouring objects at a temperature the same as or similar to that of the object of interest. In particular, avoid having objects at a much higher temperature than the object of interest near the field of view.

8.4.7 Ambient temperature dependence

All radiation thermometers suffer some sensitivity to the ambient temperature due to any one of three causes. The one cause that affects all radiation thermometers is the change in the detector sensitivity with temperature. In most radiation thermometers there is an electronic means of compensating the change in sensitivity. However, if the ambient temperature changes quickly the compensation is unlikely to track the change in detector temperature exactly. For this reason thermometers should be allowed to settle in a new environment for up to an hour to ensure that the whole instrument has come to equilibrium. This is a more serious problem for low-temperature instruments: in part because of the longer wavelength used and in part because of the types of detectors used.

In very narrow-band instruments the wavelength response is determined by the pass-bands of interference filters which are extremely sensitive to temperature. Examples of instruments that include interference filters are laboratory instruments, which often have their own temperature control, ratio thermometers (see Section 8.7.2) and special thermometers for the glass and plastics industries.

Low-temperature and long-wavelength radiation thermometers are probably the most susceptible to ambient temperature changes. This is because the signal from a detector in a radiation thermometer actually depends on two radiances:

$$L_{\lambda,m} = L_{\lambda,b}(T_s) - L_{\lambda,b}(T_d) \tag{8.20}$$

where $L_{\lambda,b}(T_d)$ is the spectral radiance of the detector. For high-temperature applications the detector radiance is negligible and is ignored. However, for thermometers working below 200°C, and especially those working near 20°C or lower, the detector radiance is significant and may be greater than that of the object of interest. In these instruments it is necessary to measure or compensate for the detector radiance in order to achieve an accurate measurement. Again these instruments are susceptible to rapid changes in the ambient temperature.

8.4.8 Vignetting

In all radiation thermometers the acceptance angle and the target area are defined by the two apertures (see Figure 8.12). Anything that further restricts the cone of radiation accepted by the thermometer will cause the thermometer to read low, since there will be less radiation falling on the detector. In particular, all parts of the front lens of the thermometer must have an unobstructed view of all of the target. Obstruction of the field of view, known as *vignetting*, occurs often in high-temperature applications where the thermometer is sighted through small peepholes in furnace walls. Vignetting also occurs when sight tubes are misaligned.

8.4.9 Linearisation

All direct-reading radiation thermometers include some form of linearisation in their electronic systems. This is necessary to convert the signal from the detector, which is an extremely non-linear function of temperature, into a signal that is proportional to temperature. As with other direct-reading thermometers (Section 4.6), this linearisation is at best an approximation. For most industrial thermometers the residual errors of 1°C to 5°C are negligible in comparison to the errors introduced through reflections, flare, and uncertainty in the emissivity.

8.4.10 Instrumental emissivity

Most spectral-band thermometers include an emissivity adjustment to compensate for the emissivity of the surface of interest. In its simplest form the adjustment is a dial on the side of the thermometer with a scale marked typically from 0.2 to 1.0. Except when the setting is at 1.0, the accuracy of the dial and scale limits the precision in the setting to about ±0.02. This uncertainty is additional to the uncertainty in the knowledge of the emissivity (Section 8.4.1).

In higher-accuracy applications, thermometers with a digital emissivity setting are preferred since the uncertainty in the setting is reduced to ±0.005. This applies particularly to long-wavelength thermometers, which are more susceptible to the error.

Quite a number of fixed-installation thermometers have the instrumental emissivity set at the time of manufacture. For these instruments the uncertainty is probably negligible. There will, however, be an error if the factory setting is different from what is required.

8.5 USE AND CARE OF RADIATION THERMOMETERS

8.5.1 Choosing a radiation thermometer

Probably the first question that should be asked is, 'Is the radiation thermometer the best option?'. Generally, if a good contact thermometer can be used for the application it is almost certainly capable of higher accuracy than a radiation thermometer in the same situation. Situations where radiation thermometers are appropriate include:

- where the target object is moving;
- where, because of vibration or corrosion, the environment is too hostile for a contact thermometer;
- where the temperature is very high, especially above 1500°C, or 1100°C if the installation is long term;
- where a fast response is required;
- where remote measurement is required;
- where a contact thermometer would disturb the heat balance around the object.

Where several of these factors are involved, the radiation thermometer may be the only choice.

Once it has been decided that a radiation thermometer is required, a suitable thermometer must be chosen from the literally hundreds available. This is a bewildering problem for those unfamiliar with the variety of operating wavelengths, applications and options. Some suggestions are made here as a guide.

In the first instance determine the specifications required of the instrument:

- *Temperature range*   The temperature range should be chosen conservatively. The accuracy of wide-range instruments is generally less than that of narrower-range instruments.
- *Accuracy*   When determining the accuracy required, some thought should be given to the likelihood of errors due to reflections, uncertainty in the surface emissivity and flare. If the likelihood of error is high, then short-wavelength, high-quality instruments are to be preferred. Instruments with digital rather than analogue adjustments for the emissivity compensation are also preferred, especially for long-wavelength thermometers.
- *Operating wavelength*   In choosing the operating wavelength, the shortest wavelength is usually best. However, there are three situations where longer wavelengths may prove advantageous. Firstly, if there are reflections from large distributed sources and the thermometer is to be used in the $\varepsilon_i = 1.0$ mode (Section 8.4.2), then the operating wavelength should be chosen according to equation (8.19). Long-wavelength thermometers also reduce the reflection errors caused by very hot sources such as

the Sun. Secondly, if there is fine dust, smoke or visible flames in the vicinity (indicating that carbon dust is present), then a slightly longer-wavelength thermometer may be less susceptible to scattering errors. Thirdly, if the thermometer is to be used in the plastics or glass industry, then the radiation thermometer should operate at wavelengths where the glass or plastic is opaque (see Section 8.7.4).

- *Field of view* The field of view is determined by the size of the target and the distance to it from the most convenient observation or mounting point. Some attention may need to be given to the choice of observation point if there are other radiation sources and flames about.

- *Response time* The response time of radiation thermometers varies from about 0.001 to 10 seconds, with most industrial thermometers in the range 0.1 to 10 seconds. Some manufacturers are prepared to set the response time according to requirements.

- *Mode of readout* Some manufacturers provide voltage outputs that simulate thermocouples as well as the usual analogue, digital, current, and voltage output modes.

- *Special environmental considerations* One of the most important factors is the environment of the thermometer. If it is exposed to dust then air purge systems may be required; if it is exposed to high ambient temperatures then water cooling and a high ambient temperature rating will be required. Other possible options include explosion-proofing and radiation shields.

- *Application/manufacturer* Often the best advice on the choice of thermometer can be obtained from a manufacturer. Many of the larger manufacturers have links with or cater for particular industries, e.g. glass, petrochemicals, plastics and steel industries. If you are working in an industry which has always used radiation thermometers, it is likely that one of the larger manufacturers has specialised in supplying your industry. In this case you should first determine which is the manufacturer with the strongest association with your industry. They will be able to advise on environmental concerns and on choice of wavelength if, for example, scattering due to dust is a problem.

- *Calibration* Thought should also be given to the calibration of the instrument. If this is the only radiation thermometer you have, then the calibration overheads may be quite high. The thermometer should be calibrated when new, when one year old, then when necessary (judged by the observed drift in regular one-point checks) up to a maximum of five years.

  Unlike other thermometers, radiation thermometers are not easily checked at the ice point, so additional equipment may be required for traceability. There are now on the market a number of relatively low-cost blackbodies which are suitable for both regular verification checks and calibrations. Also, a number of manufacturers offer an annual calibration service.

8.5.2 Care and maintenance

The physical care and maintenance of radiation thermometers is straightforward: they should be treated as you would treat a very expensive camera. Usually the manufacturer makes quite strong recommendations in the manual for cleaning and general care. The lens should be cleaned periodically with high-quality lens tissue or an air brush, to remove dust. If absolutely necessary, ethanol and lens tissue can be used to remove grease.

8.5.3 Using the thermometer

When using a radiation thermometer it is useful to go through a simple checklist to make sure that none of the possible sources of error are overlooked, and that everything practical is done to minimise errors.

- *Emissivity* Know the emissivity of the surface. Spend some time in advance of the measurement looking at the thermometer manufacturer's guide and samples of the material so that you have a good estimate of the emissivity. The choice of emissivity may well be controlled by a QA system so that there is uniform practice within the company.

- *Reflections* Systematically look about the space in the hemisphere above the surface for bright objects that may be sources of additional radiation, e.g. sun, flames, heaters, furnace walls and incandescent lamps. If possible, shield the sources. If the object is in near blackbody conditions the $\varepsilon_i = 1$ strategy will minimise the errors (see Section 8.4.2).

- *Environment* Avoid taking a thermometer into areas where there is a lot of dust or the ambient temperature is high. Remember that lenses are prone to fracture if they are exposed to too high a temperature too quickly. If you cannot keep your hand on the thermometer in use, it is too hot. This is true particularly when a manufacturer has supplied a radiation shield to help keep the thermometer body (but not the lens) cool.

- *Absorption* Make sure there are no windows, smoke, dust, or haze in the field of view of the thermometer. If there are visible flames nearby there may be some carbon dust in the field of view.

- *Flare* Check that there are no objects brighter than the object of interest near the field of view of the thermometer. If the flare risk is high a sight tube may be necessary. It may also be possible to use a cool object in the foreground to shield the thermometer from hot objects outside the field of view.

- *Field of View* Ensure that the field of view is completely filled, preferably well overfilled, and as uniform as possible.

- *Safety and exposure to bright sources* Never ever sight a radiation thermometer on the Sun. Quite apart from the potential damage to the instrument, it is also likely that your eye could be permanently damaged.

    With most radiation thermometers exposure to very bright sources, particularly sources like the Sun that contain a lot of ultraviolet radiation, may cause permanent damage to the thermometer. The thermal stress resulting from exposure may be sufficient to break any of the optical components. Also, ultraviolet radiation has sufficient energy to photodegrade many detector materials, and change the detector characteristics.

- *Record-keeping* More than most thermometers, radiation thermometers are affected by the manner in which they are used. For a measurement to be traceable it must be repeatable, i.e. documented in sufficient detail for a similarly competent person to be able to verify and, if necessary, modify the results of the measurement. The documentation may be casual, as in a laboratory book, or more formal, as required by a QA system. The record should identify the instrument used, include the choice of emissivity setting and the rationale for that choice, state who took the measurement, the position from where the measurement was taken, and any significant features in the environment that are of concern — dust, bright objects, etc.

8.6 CALIBRATION OF RADIATION THERMOMETERS

Historically radiation thermometers have been used mainly for monitoring industrial processes rather than for making accurate quantitative measurements. With the trend towards more accurate and traceable measurements, the calibration procedures for radiation thermometers are having to. evolve rapidly. In particular, calibrations of direct-reading thermometers must establish the accuracy of the emissivity compensation adjustment as well as determining the radiance-temperature relationship.

Traceability of radiation thermometers to the ITS-90 may be obtained through three chains as shown in Figure 8.13.

Via a tungsten strip lamp

This is the traditional means of disseminating the radiation thermometry scale. The lamp is calibrated by the national standards laboratory against a fixed point blackbody with

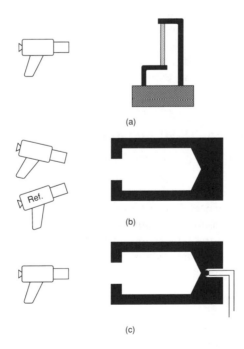

(a)

(b)

(c)

Figure 8.13
The three basic traceability chains for the calibration of radiation thermometers: (a) via a tungsten strip lamp; (b) via a transfer standard radiometer; (c) via a thermocouple or resistance thermometer.

the aid of a transfer standard thermometer. The lamp can then be used as the radiance source for calibrating other radiation thermometers, particularly disappearing-filament thermometers, which operate near 655 nm and use tungsten filaments.

The lamp may also be used in conjunction with a transfer standard radiometer (operating at 655 nm) to establish the temperature of a blackbody. This is the best method for calibrating thermometers operating at wavelengths much longer than 655 nm or with fields of view much greater than 2 mm.

Via a transfer standard radiometer

With the improvement in the stability of transfer standard radiometers it is no longer necessary to use strip lamps to maintain the temperature scale. Once calibrated against a fixed point blackbody and fully characterised, the thermometer can be used to establish the radiance temperature of a blackbody which is used to calibrate thermometers of all wavelengths.

Via a thermocouple or resistance thermometer

At temperatures below 960°C, where the ITS-90 is defined in terms of platinum resistance thermometers, the temperature of a blackbody can be determined by using a calibrated thermocouple or resistance thermometer. This is a very convenient and cost-effective way of measuring blackbody temperature but it is subject to significant errors and uncertainty due to temperature gradients in the cavity.

This method may also be used above 960°C and up to 1700°C with rare-metal thermocouples. However, this involves an extra step in the calibration chain which may increase the uncertainty further.

8.6.1 Realisation of the ITS-90 radiation scale

Above the silver point ($\sim$962°C) the ITS-90 defines temperature in terms of Planck's radiation law (equation (8.3)). The primary thermometer, which compares unknown radiance with the radiance of a fixed point blackbody at silver, gold ($\sim$1064°C) or copper points ($\sim$1085°C), is known as a *transfer standard radiometer*. The thermometer may be used to transfer the scale to a tungsten strip lamp or, if sufficiently stable, it may be used directly to determine the temperature of a blackbody. A fuller description of the realisation of the ITS-90 radiation scale is given in Chapter 3.

8.6.2 Calibration equations

So far we have assumed that the measured radiance is proportional to Planck's law for all spectral-band radiometers. In practice the finite bandwidth of the filter leads to significant departures from Planck's law. For transfer standard radiometers the departure is usually characterised by an effective operating wavelength that is temperature dependent. For narrow-band thermometers the effective wavelength is quite closely described by the simple relationship

$$\lambda_e = A + B/T. \tag{8.21}$$

A further simplification is to use this directly in Wien's law; this leads to a simple calibration equation for the thermometer response:

$$V(T) = C \exp\left(\frac{-c_2}{AT + B}\right), \tag{8.22}$$

where $V(T)$ is the measured output (voltage) of the thermometer. This equation is used for many radiation thermometers including some transfer standard thermometers. For thermometers with bandwidths of 50 nm or less equation (8.22) will fit the thermometer response to within a few tenths of a degree. For wideband thermometers a similar equation is used which includes higher-order terms to account for the greater departures from Wien's law:

$$\log(V(T)) = A + \frac{B}{T} + \frac{C}{T^2} + \frac{D}{T^3}. \tag{8.23}$$

This equation is amenable to least squares fitting, and will fit most spectral-band responses to a degree or better.

These two equations are used either directly or as the basis for the electronic linearisation schemes in most direct-reading radiation thermometers. For direct-reading radiation thermometers the equation

$$\Delta T_{\text{cor}} = a + bT + cT^2 + dT^3 \tag{8.24}$$

which was developed in Chapter 4 for direct reading thermometers, should be used. Note that the T^2 term will usually be significant since most errors have a quadratic dependence (Section 8.4).

8.6.3 Practical blackbodies

The principles underlying the design of practical blackbodies are useful for several reasons. The most important application is, of course, the manufacture of blackbodies used for calibrating radiation thermometers. Additionally reflection errors can be more easily understood or eliminated by invoking blackbody concepts. As we indicated in Section 8.1, there are two main factors affecting the design of blackbodies, namely the emissivity of the blackbody and its temperature uniformity.

Practical blackbodies are not surfaces, but cavities. Because cavities trap and absorb rather than reflect light, they have a much higher emissivity than any real surface. Figure 8.14 shows a simple example. An upper limit of the cavity emissivity can be very easily determined by considering its reflectance. Consider a ray of light which enters the cavity and strikes the back surface. The amount of light reflected back out of the aperture depends on two factors: firstly, the reflectance of the surface; and secondly, the size of the aperture relative to the hemisphere above the back surface. By considering only the rays which undergo a single reflection, the effective reflectance of the cavity is estimated as

$$\rho_{\text{eff}} = \rho_s \frac{r^2}{R^2} \tag{8.25}$$

where ρ_s is the reflectance of the cavity material, r is the radius of the aperture and R is the distance between the back of the cavity and the aperture. Since the emissivity and

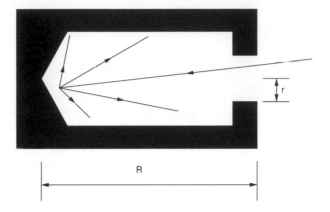

Figure 8.14
A simple representation of a blackbody cavity.

reflectance of the cavity are related (by equation (8.2)) we estimate the emissivity as

$$\varepsilon_{\mathrm{eff}} = 1 - (1 - \varepsilon_s)\frac{r^2}{R^2}. \tag{8.26}$$

This formula is an upper limit for the emissivity since the light escaping via two or more reflections is assumed not to have escaped. We can see from the equation that three of the factors affecting the quality of a blackbody are

- the emissivity of the surface (ε_s);
- the size of the aperture (r);
- the size of the cavity (R).

Example 8.2   Estimating the emissivity of a blackbody cavity

Estimate the emissivity of a cavity manufactured from inconel[®]. The cavity has a 1 cm diameter aperture and is 10 cm long. The emissivity of rough and heavily oxidised inconel is about 0.9.

From the information supplied $\varepsilon_s = 0.9$, $r = 0.5$ cm, and $R = 10$ cm. Hence

$$\varepsilon_{\mathrm{eff}} = 1 - 0.1 \times (0.5)^2/(10)^2$$

$$= 0.999\,75.$$

   Remember that this is an optimistic estimate. In practice the extra reflections might double the reflectance of the cavity.

Exercise 8.4

Calculate the effective emissivity of the cavity formed by a glass jar 200 mm deep and 100 mm wide at the mouth. The emissivity of glass at 8 μm is 0.85.

Exercise 8.5

Consider the blackbody cavity of Figure 8.14.

(a) By assuming that radiation falling on the rear surface is scattered equally in all directions, show that the reflectance of the cavity is $\rho_s r^2 / 2R^2$.

(b) In practice the radiation is not scattered equally in all directions: it is scattered according to the projected area of the rear surface, i.e. the scattered flux is proportional to $\cos\alpha$, where α is the angle from the normal. (This is *Lambert's law*.) Hence show that the reflectance of the cavity is $\rho_s r^2 / R^2$ (equation (8.26)), which is twice that calculated in (a). (*Note.* Part (b) is difficult.)

To be a good blackbody, the cavity must also be uniform in temperature. This can be seen by considering the cavity radiance due to the surface emission and the first reflection. Using equation (8.15) for reflection errors we estimate the total radiance of the rear of the cavity as

$$L_{\text{tot}} = \varepsilon_s L_{\lambda,b}(T_c) + (1 - \varepsilon_s)L_{\lambda,b}(T_w), \tag{8.27}$$

where T_c is the nominal cavity temperature, and T_w is the cavity wall temperature. We can relate this radiance (see Exercise 8.6) to the *radiance temperature*, T_λ, of the cavity:

$$T_\lambda = T_c + (1 - \varepsilon_s)(T_w - T_c). \tag{8.28}$$

This tells us what a radiation thermometer with an emissivity setting of 1.0 will read when viewing the blackbody. We can also determine the uncertainty in the radiance temperature in terms of the uncertainty in the cavity-wall temperature.

Example 8.3   *The uncertainty in radiance temperature due to non-uniformity of the cavity-wall temperature*

A blackbody is monitored by a thermocouple mounted in the rear wall of the cavity. Experiments with a fine rare-metal thermocouple show that the temperature gradient is such that the front of the cavity is 6°C cooler than the rear of the cavity. Calculate the uncertainty in the cavity temperature if the emissivity of the cavity material is 0.9.

The easiest way of characterising the radiance temperature is to estimate the two extremes of the likely range of values.

Maximum temperature   Since those portions of the cavity closest to the rear wall are at a temperature very near to that of the rear wall, the maximum radiance temperature is

$$T_{\lambda,\text{max}} = T_c.$$

Minimum temperature   The average wall temperature is 3.0°C lower than the rear wall temperature so we could expect, from equation (8.28), that the minimum radiance temperature is

$$T_{\lambda,\text{min}} = T_c - 0.1 \times 3.0.$$

Treating these two values as the limits of a rectangular distribution, we estimate that the radiance temperature is

$$T_\lambda = T_c - 0.15 \pm 0.15°C \quad (95\% \text{ CL}).$$

Exercise 8.6

(a)   By defining the radiance temperature T_λ by $L_{\lambda,b}(T_\lambda) = L_{\text{tot}}$, derive equation (8.28).

$$\left(\text{Hint. expand } L_{\lambda,b}(T) = L_{\lambda,b}(T_0) + \frac{\partial L}{\partial T}(T - T_0).\right)$$

Non-uniformity is a serious problem in most blackbodies because of the convection currents in the air near the aperture of the cavity. The currents cause a cool stream of air to enter the cavity, which disturbs the heat balance, generating a gradient in the cavity walls.

The last factor to consider in the evaluation of the performance of a blackbody is the effect of the loss of energy that is radiated by the aperture. Since the energy must be continuously replaced by heaters around the cavity, there must be a temperature gradient across the walls of the cavity. If the cavity uses a reference thermocouple mounted in the cavity wall to determine the radiance temperature, then it will be in error due to the gradient. If it is assumed that the energy is lost uniformly by all parts of the cavity, then the wall temperature gradient can be estimated as

$$\frac{\partial T}{\partial x} = \sigma(T_c^4 - T_a^4)\frac{a}{A} \times \rho_{\text{th}} \tag{8.29}$$

where σ is the Stefan–Boltzmann constant, T_a is the ambient temperature around the cavity, a is the aperture area, A is the internal area of the cavity and ρ_{th} is the thermal resistivity of the blackbody material.

In high-temperature blackbodies, this effect is the most significant source of error. Not only does it contribute to the non-uniformity of the cavity, it also makes accurate measurement of the radiance temperature difficult by any means other than a transfer standard radiometer. For the highest-precision work the aperture must be as small as possible to reduce the radiation loss to a minimum.

Example 8.4   Blackbody wall gradient due to aperture loss

Estimate the temperature gradient in the wall of a spherical blackbody 20 cm in diameter with a 2 cm diameter aperture. The cavity is made from inconel® (thermal resistivity: 4°C cm/W) and must operate at 1100°C.

At 1100°C the radiation received by the cavity is negligible compared to that radiated, therefore the T_a term can be ignored: substitution of the values for the other variables leads to

$$\frac{\partial T}{\partial x} = 5.7 \times 10^{-12} \times (1100 + 273)^4 \frac{\pi(0.01)^2 \times 4}{\pi(0.1)^2};$$

hence

$$\frac{\partial T}{\partial x} = 3.2°\text{C/cm}.$$

The factor of 10^{-4} has been inserted in the Stefan–Boltzmann constant in order to change the units from square metres to square centimetres.

Exercise 8.7

A small fixed point blackbody at the silver point uses a graphite cavity 10 mm in diameter and 80 mm long, with a 5 mm diameter aperture. Estimate the temperature gradient across the 5 mm thick graphite wall. The thermal resistivity of graphite is approximately 1°C cm/W.

Overall we have identified three main factors contributing to errors in blackbody cavities. In order of significance 'they are: temperature gradients due to energy loss, errors in radiance temperature or emissivity due to temperature non-uniformity, and the reduction in effective emissivity due to the cavity reflectance. Collectively this limits the accuracy of blackbodies' radiance temperatures as measured by a contact thermometer to about ±0.5%.

8.6.4 Tungsten strip lamps

Tungsten strip or ribbon lamps have been used for many years to maintain the radiation-thermometry portion of the temperature scale. They are also a convenient and often lower-cost source of spectral radiance than the equivalent blackbody.

Strip lamps consist of a tungsten ribbon up to 5 mm wide and 50 mm long supported in a pyrex or silica envelope (see Figure 8.15). They are usually mounted on a substantial base which is cooled to minimise the influence of the ambient temperature on the lamp. The envelope may be filled with an inert gas to prevent oxidation and contamination of the filament. Vacuum lamps are suitable for operation between 700°C and 1700°C while gas-filled lamps are suitable from 1500°C to 2300°C.

The lamps are calibrated in terms of the filament current required to achieve a specified radiance temperature at a specific wavelength (often 655 nm). The radiance temperature is the temperature of a blackbody that would have the same spectral radiance as the lamp. The radiance temperature of a lamp is strongly wavelength-dependent and corrections must be applied if the lamp is used at other wavelengths. Lamps may also be calibrated at several wavelengths. Since the emissivity of the tungsten filament is about 0.4, the radiance temperature of the lamp is typically 40 to 300°C less than the true temperature of the filament.

The difference between the true temperature of the tungsten filament and the radiance temperature is approximately (equation (8.13))

$$T_s - T_\lambda = \frac{\lambda T^2}{c_2}(1 - \varepsilon(\lambda)). \qquad (8.30)$$

Hence the difference in radiance temperature for thermometers operating at different wavelengths is

$$T_{\lambda 2} - T_{\lambda 1} = \frac{T^2}{c_2}\left[\lambda_1(1 - \varepsilon(\lambda_1)) - \lambda_2(1 - \varepsilon(\lambda_2))\right]. \qquad (8.31)$$

This is the correction that must be applied when a tungsten lamp that has been calibrated at one wavelength is used at another wavelength. Figure 8.16 shows the variation of the emissivity of tungsten with temperature and wavelength.

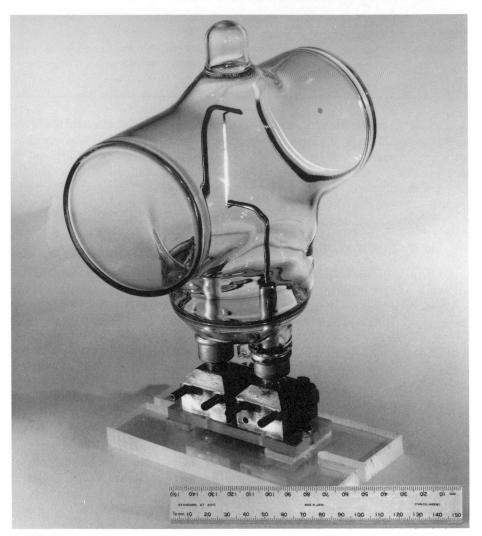

Figure 8.15
A tungsten strip lamp.

Example 8.5   Radiance temperature correction

A tungsten strip lamp calibrated at 655 nm is used to calibrate a radiation thermometer which operates at 900 nm. Calculate the radiance temperature correction for the lamp at 1200 K.

By applying equation (8.31) and the approximation for c_2 directly we obtain

$$\Delta T_\lambda = 100 \left(\frac{T}{1200} \right)^2 [0.655(1 - \varepsilon(0.655)) - 0.9(1 - \varepsilon(0.9))]°C.$$

Continued on page 315

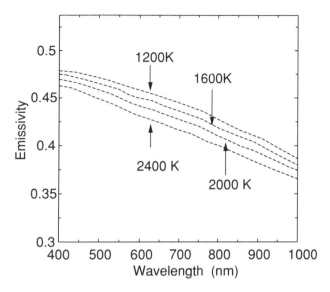

Figure 8.16
The emissivity of tungsten filament versus temperature and wavelength.

Continued from page 314

Now substituting values of 0.46 and 0.41 for the spectral emissivity of tungsten at 655 nm and 900 nm respectively, the correction is calculated as

$$\Delta T_\lambda = 100 \left(\frac{1200}{1200} \right)^2 (0.655(1 - 0.46) - 0.900(1 - 0.41))\,^\circ\text{C};$$

hence

$$\Delta T_\lambda = -17.7^\circ\text{C}.$$

That is, the radiance temperature at 900 nm (the longer wavelength) is 18°C less than that at 655 nm (the shorter wavelength).

Exercise 8.8

(a) By following Example 8.5, calculate the radiance temperature correction for the lamp at 1600 K and 2000 K.

(b) Estimate the uncertainty in each of the corrections at 1200 K, 1600 K and 2000 K.

The typical uncertainty in radiance-temperature corrections is 2% to 3% due to the uncertainty in the emissivity. The uncertainty in the correction is much larger at longer wavelengths and where the operating wavelength of the thermometer is not known to high accuracy.

The spectral-radiance calibration for a lamp applies to a small area of the filament marked by a notch in one edge midway along its length. Since the radiance of the tungsten filament depends slightly on the angle of view (Figure 8.7), there is a second mark on the envelope behind the filament to aid the alignment of the thermometer. Current for the lamp should be provided by a high-stability current source capable of supplying several tens of amperes (depending on the lamp design). Note that the lamps are sensitive to the polarity of the d.c. current, and the required polarity is usually marked on the base. Connecting the lamp with the wrong polarity will change the emissivity of the lamp and invalidate the calibration. Equilibrium is reached within 30 minutes of turning the lamp on, with shorter settling times of several minutes for small changes in filament current.

Good quality lamps, properly annealed, are capable of reproducing spectral radiance to better than 0.1% for several hundred hours. This corresponds to reproducibility in radiance temperatures of better than 0.1°C around 1000°C.

8.6.5 Calibrating a radiation thermometer

The calibration of radiation thermometers follows the basic guide in Chapter 4. Here we give a simple procedure which highlights the additional elements relevant to radiation thermometers.

Step 1 — Initiate record-keeping (as in Section 4.3.6)

Step 2 — General visual inspection (as in Section 4.3.6)

Step 3 — Conditioning and adjustment
Only if necessary and according to the manufacturer's manual.

Step 4 — Generic checks
Before the intercomparison proper, there are a number of checks that should be carried out on the thermometer. The results of the checks can be expected to be similar for all thermometers of the same make and model number. Departure from the typical behaviour is a departure from the generic history and may be indicative of faults or damage.

- *Detailed visual inspection*   Remember that a radiation thermometer is an instrument that should be treated like an expensive camera. Check the lens for dust, grease and scratches: if necessary, clean it. If the thermometer is battery-powered, check that the battery is charged. Check that the thermometer radiance measurement and emissivity compensation both work. This can be done by viewing a desk-lamp with a frosted bulb.
- *Stability and settling*   Over the first hour of settling after exposure to a blackbody, monitor the reading of the thermometer to ensure that the thermometer is stable. If possible, also change the ambient temperature to determine the sensitivity to ambient temperature. These checks are particularly important for long-wavelength and for very narrow-band thermometers.

    If the thermometer's sensitivity to ambient temperature changes is large, then it may be necessary to assess the resulting uncertainty in use for inclusion in the total uncertainty of calibration. Extreme sensitivity may be indicative of faulty ambient temperature compensation.

- *Size-of-object effects* With the blackbody set to the highest-temperature of the calibration range, adjust a variable aperture between the thermometer and the blackbody to a diameter a little greater than the specified field of view for the thermometer. Ensure that the thermometer is properly focused. Record the reading. Now open the aperture to at least two times the specified field of view and record this reading. The difference between the two readings is a measure of the uncertainty due to size-of-object effects.

  This test is impossible to carry out effectively unless the environment around the blackbody is cool relative to the blackbody, hence the test is carried out at the highest calibration temperature. A laboratory (a closed cavity) behaves as a blackbody at room temperature so the variable aperture and surrounds will have a similar radiance to a room-temperature blackbody. To be sure that the size of the source is well-defined, the reference blackbody should be about 200°C hotter than ambient temperature. This test may be most conveniently carried out at the end of the intercomparison.

  With most radiation thermometers the field of view must be overfilled by 2 to 3 times before the reading is independent of the size of the blackbody aperture. This may be the most significant contributor to the total uncertainty for long-wavelength thermometers. Large flare effects may be indicative of lens damage or misalignment. For example, damage to the protective film on the lens of a 10 μm thermometer has been observed to cause errors of more than 10°C at 140°C.

- *Emissivity calibration (if appropriate)* Traditionally, radiation thermometers are calibrated only against blackbodies, yet in almost all applications they are used with emissivity compensation on surfaces that are not blackbodies. Clearly those measurements cannot be traceable unless the accuracy of the emissivity compensation is confirmed. The accuracy of the compensation mechanism is checked as follows.

  Set the blackbody to the highest calibration temperature and record the reading of the thermometer with the emissivity compensation set to 1.0. Now place a calibrated neutral density filter between the thermometer and the blackbody so that the filter overfills the field of view by at least a factor of 2. Now adjust the emissivity compensation to obtain the same reading as without the filter. Record the emissivity setting. Ideally the emissivity setting is the same as the transmittance of the filter. Use a range of neutral density filters with transmittances in the range 0.2 to 1.0.

  Suitable filters include absorbing glass for short-wavelength thermometers and rotating sectored discs for longer wavelength thermometers. Calibrated wire mesh may also be used. Care should be taken to ensure that the filters are cool relative to the blackbody and that there are no reflections (e.g. from room lighting) from the filter surface.

  Since this test is carried out at the high-temperature end of the calibration range it may be more conveniently carried out after the radiance calibration.

Step 5 — Intercomparison

This portion of the calibration compares the temperatures measured by the thermometer with those from a calibrated radiance source. As discussed in Section 8.6, there are three basic ways to carry out the intercomparison: tungsten strip lamp (with or without transfer standard radiometer); blackbody plus transfer standard radiometer; and blackbody plus calibrated contact thermometer. Intercomparisons with a transfer standard radiometer are more accurate but also more time-consuming, since the positions of the two thermometers must be swapped repeatedly to view the blackbody.

The radiation thermometer(s) must be mounted to view the blackbody with the specified field of view overfilled by at least a factor of 2. This ensures that size-of-object effects have the least effect on the readings. Throughout the calibration care should be taken to ensure that the front lens of the thermometer is not exposed to direct radiation from sources other than the blackbody. Radiation from other lamps or room lighting may cause additional error in thermometers that are prone to flare. As with all optical measurements, the calibration should be carried out in a darkened laboratory. With tungsten strip lamps a darkened laboratory is essential.

Two of the calibration equations for radiation thermometers (equations (8.23) and (8.24)) require four constants to be fitted to calibration data, and therefore a minimum of 12 data points are required in the intercomparison. Unlike other thermometers, radiation thermometers do not normally exhibit hysteresis, so it does not matter whether the temperature range is covered in an ascending or descending sequence. Blackbody furnaces often settle faster through an ascending sequence, making this the preferred option.

The achievable accuracy in the intercomparison depends strongly on the traceability chain chosen. When a blackbody and contact thermometer are used to provide the reference radiance the uncertainties are quite large because of the temperature gradients in the blackbody cavity and radiation losses (see Section 8.6.3).

When a transfer standard radiometer or a tungsten strip lamp is used for the calibration the uncertainty can be either very small or large, depending on the operating wavelength of the thermometer under test. Transfer standard thermometers and strip lamps are calibrated in terms of radiance temperature, that is, an emissivity of 1.0 is assumed. If the emissivity of the radiance source is not 1.0 then the radiance temperature is wavelength dependent. This is a particular problem with strip lamps which have an emissivity of about 0.4. When strip lamps are used a correction must be applied to correct for the wavelength dependence (Section 8.6.4).

Step 6 — Analysis

The first part of the analysis is the least-squares fit which provides the following information:

- By showing that a thermometer fits the calibration equation well, a successful least-squares fit confirms that the thermometer is well behaved and conforms to the generic history for that type of thermometer. There should not be unexplained jumps in the errors, or large consistent patterns in the residual errors in the fit. There should also be sensible values for the calculated corrections.
- The variance of the residual errors in the fit measures both the random error in the intercomparison and the unpredictable departures of the thermometer from the calibration equation. This effectively measures the repeatability of the thermometer.
- By using a relatively large number of calibration points compared to the number of parameters in the fit and demonstrating that all the points fit the calibration function, we show that the fitted function is suitable for interpolation between calibration points.

Step 7 — Uncertainties

The contributing factors to the calibration uncertainty are:

- *Uncertainty in the reference thermometer readings* This uncertainty is easily assessed since it is reported on the reference-thermometer or tungsten strip lamp certificate. The value may need to be adjusted to the required confidence limits.

- *Variations in the stability and uniformity of the calibration medium* This depends on which of the three traceability chains is employed.

    (1) *Transfer standard radiometer and tungsten strip lamp* In this case the major source of uncertainty is in the radiance temperature correction as shown in Example 8.5.

    (2) *Transfer standard radiometer and blackbody* In this case the transfer standard radiometer measures the radiance as seen by the thermometer under test, so there is minimal error in the intercomparison. If the emissivity of the blackbody is uncertain, then there will be an uncertainty in the radiance temperature of the cavity. This leads to an additional uncertainty in the calibration of

    $$\sigma_{T_\lambda} = \frac{|\lambda_1 - \lambda_2|T^2}{c_2}\sigma_\varepsilon, \tag{8.32}$$

    where λ_1 and λ_2 are the effective wavelengths of the transfer standard radiometer and the thermometer under test. This uncertainty is usually significant only for very long-wavelength thermometers and low-precision blackbodies.

    (3) *Contact thermometer and blackbody* The uncertainty in the radiance temperature of a blackbody monitored by a contact thermometer depends on two factors: the degree of temperature uniformity within the cavity (equation (8.28) and Example 8.3), and the combination of the proximity of the contact thermometer to the cavity and the energy radiated by the cavity (equation (8.29) and Example 8.4). As a guide, it is usually difficult to reduce the uncertainty below 0.5% (e.g. $\pm 5°C$ at 1000°C).

- *Departure from the determined ITS-90 relationship* This is the uncertainty in the fit to the calibration equation, namely the standard deviation of the residual errors.

- *Uncertainty due to drift* For thermometers employing interference filters the drift may be as large as 1% (in radiance) or more per year. Otherwise the drift is negligible for most broad-band thermometers so long as the thermometer is well maintained. Therefore, assume that uncertainty due to drift is zero.

- *Uncertainty due to hysteresis* There should be no hysteresis effects in radiation thermometers other than those caused by the response time of the thermometer, which is usually less than a few seconds. Set this uncertainty to zero. If hysteresis is observed then the instrument is faulty.

- *Uncertainty due to flare* This is usually the largest source of calibration uncertainty in working thermometers. In use a thermometer will be used to measure the temperatures of a variety of objects of different sizes, and with surrounds both hotter and colder than the object of interest. The best approach is to calibrate the thermometer with the specified field of view overfilled by at least a factor of two (in diameter). Assess the uncertainty as the difference in reading between the situations when the field of view is filled exactly and when overfilled. Thus the fitted ITS-90 relationship corresponds to near-ideal use, and the uncertainty covers use in environments where the surrounds of the object are hotter or colder.

For reference and transfer standard radiometers this uncertainty is set to zero since the thermometer is always used in ideal conditions to measure the temperature of a blackbody. There may be a small size-of-object effect, so the blackbody aperture size should be reported on the certificate as a calibration condition. The user of the thermometer can then assess any uncertainty due to different usage.

- *Total uncertainty*   The total uncertainty in the calibration is assessed as the quadrature sum of all the contributing uncertainties with confidence intervals of 95%. For all spectral-band thermometers the uncertainty will have a predominantly T^2 dependence. For wide-range thermometers the uncertainty should be reported for one temperature within the calibration range with an indication of how to determine the uncertainty at other temperatures. For example, if all the uncertainties are evaluated at 200°C and total uncertainty is ±7°C (95% CL) then the uncertainty may be reported as

'the uncertainty in the corrected readings of the thermometer at T kelvin is estimated as $\pm7(T/473)^2$ K at a 95% confidence level'.

Step 8 — Complete records (As with Section 4.3.6)

8.7 OTHER RADIATION THERMOMETERS

8.7.1 The disappearing-filament thermometer

The disappearing-filament thermometer is one of the earliest examples of a spectral-band radiation thermometer. It uses the observer's eye to compare the surface radiance against a known radiance — a hot tungsten filament. The temperature of the filament is adjusted until it has the same radiance as the surface in the background and disappears. The current through the filament is the indicator of the surface temperature.

The disappearing-filament thermometer uses very short wavelengths, about 650 nm, so that instrumental uncertainties, including the emissivity dependence, are minimised. The main difficulties lie with the observer. Firstly, it takes a good deal of practice before the measurements made by one observer are highly repeatable. Secondly, variations in the response of the eye from different observers and at different states of dark adaption also affect the accuracy. When used to view uniform objects, in a darkened room so that the eye is properly dark adapted, the thermometer is capable of accuracies of better than ±5°C.

The third and most important factor affecting the accuracy is the uniformity of the field of view. Any non-uniformity in the surface radiance will betray the presence of the filament and give the eye sufficient information to reconstruct the filament outline so that it never quite disappears. This image processing done by the eye is unconscious and cannot be completely overcome by training. Errors of several hundred degrees can occur for small non-uniform objects.

The disappearing-filament thermometer is calibrated in terms of radiance temperature and has no emissivity adjustment. The error introduced when measuring the temperatures of non-blackbody surfaces is serious only when the emissivity is low. For example, for $\varepsilon = 0.6$ the error is about −25°C at 900°C and −100°C at 2000°C.

The temperature range of the disappearing-filament thermometer is determined by the sensitivity of the eye. The lowest operating temperature is about $600°C$. The upper temperature range can be extended from $1400°C$ to as far as $4000°C$ by using filters to reduce the radiance to a level where the eye is both comfortable and most sensitive.

8.7.2 The ratio thermometer

In some applications the uncertainty in the emissivity seriously limits the utility of spectral-band thermometers. This is particularly true in some parts of the steel and aluminium industries where the emissivities are not only low but extremely variable. One of the worst examples is in the manufacture of galvanised steel, where the emissivity varies from a little over 0.1 to 0.7 in a single process. Under these conditions ratio thermometers are a useful alternative to spectral-band thermometers. They are also useful in some applications where smoke, dust or windows affect spectral-band measurements.

Ratio thermometers, also known as dual-wavelength and two-colour thermometers, measure radiance at two wavelengths and determine the ratio

$$R = \frac{\alpha(\lambda_1)}{\alpha(\lambda_2)} \frac{\varepsilon(\lambda_1)}{\varepsilon(\lambda_2)} \frac{L_{\lambda,b}(\lambda_1, T_s)}{L_{\lambda,b}(\lambda_2, T_s)}. \tag{8.33}$$

If it is assumed that the absorption $\alpha(\lambda)$ and the emissivity $\varepsilon(\lambda)$ are constant over the wavelength range including λ_1, and λ_2, then the ratio R depends only on the temperature.

The independence from emissivity and absorption is obtained at the expense of sensitivity. This can be seen firstly from the Wien law approximation for R:

$$R = \left(\frac{\lambda_2}{\lambda_1}\right)^5 \exp\left(\frac{c_2}{T_2}\left(\frac{1}{\lambda_2} - \frac{1}{\lambda_1}\right)\right) \tag{8.34}$$

and secondly from the propagation-of-uncertainty formula which gives the uncertainty in temperature versus the uncertainty in R:

$$\sigma_{T_m} = \frac{\lambda_1 \lambda_2}{\lambda_1 - \lambda_2} \frac{T^2}{c_2} \left(\frac{\sigma_R}{R}\right). \tag{8.35}$$

The similarity of these equations to those for the single wavelength spectral-band thermometers suggests that the performance of the ratio thermometer would be similar to that of a spectral band thermometer with an operating wavelength of

$$\lambda_e = \frac{\lambda_1 \lambda_2}{\lambda_1 - \lambda_2}. \tag{8.36}$$

However, this is misleading. While the sensitivity is ten to twenty times worse than a good spectral-band thermometer, some of the most significant errors are also much less. Firstly, the most significant error in spectral band thermometry, the uncertainty in the emissivity, has been eliminated. And secondly, many of the instrumental errors that affect the radiance measurement in spectral band thermometers are common to both channels of the ratio thermometer and so do not affect the ratio (see Exercise 8.9).

Overall the performance of ratio thermometers on surfaces that have a high emissivity and are grey (i.e. constant emissivity with wavelength) is perhaps two to three times worse than good spectral-band measurements. On the other hand, if the surface is grey and has a low or highly variable (with time or temperature) emissivity, then the ratio thermometer is clearly better. Ratio thermometers also find application where the object is small and a spectral-band thermometer would be susceptible to size-of-object effects.

Example 8.6   Comparison of spectral-band and ratio thermometers

A steel galvanising plant is monitoring temperatures near 450°C. The emissivity of the freshly plated steel varies from about 0.15 to 0.7 as molten zinc forms the protective alloy surface. Compare the performance of a spectral-band thermometer operating at 2.2 μm and a ratio thermometer operating at wavelengths of 2.2 and 2.4 μm.

Spectral-band thermometer   Based on equation (8.13) and a nominal value of emissivity of 0.4, the variation of the reading error is estimated to be

$$U_T = \pm 2.2 \times \frac{(273 + 450)^2}{1200^2} \times 100 \times \frac{0.3}{0.4} °C$$

$$U_T = \pm 60°C.$$

Ratio thermometer   Based on equation (8.35) and a variation of spectral emissivity of 1% between 2.2 and 2.4 μm, the variation in the temperature error is estimated to be

$$U_T = \pm \frac{2.2 \times 2.4}{2.4 - 2.2} \times \frac{(273 + 450)^2 \times 1}{1200^2} °C$$

$$U_T = \pm 10°C.$$

Thus the ratio thermometer can accommodate quite large changes in emissivity so long as the spectral emissivities at the two wavelengths are the same. In practice the emissivity variations are often larger than 1%.

Exercise 8.9

(a)   Consider a temperature measured with a ratio thermometer. Estimate the uncertainty in temperature due to the uncertainty in the measured radiances $L_{\lambda,b}(\lambda_1, T_s)$, $L_{\lambda,b}(\lambda_2, T_s)$. Assume that the relative uncertainty σ_L/L is the same in both measurements and that the correlation coefficient of the errors in the measured radiances is r.

$$\left[\text{Ans. } \sigma_{T_m} = \frac{\lambda_1 \lambda_2}{\lambda_1 - \lambda_2} \frac{T^2}{c_2} (2(1-r))^{\frac{1}{2}} \frac{\sigma_L}{L} \right].$$

(b)   Now by including the dependence on the emissivities show that the total uncertainty is

$$\sigma_{T_m} = \frac{\lambda_1 \lambda_2}{\lambda_1 - \lambda_2} \frac{T^2}{c_2} \left[2(1-r) \frac{\sigma_L^2}{L^2} + \frac{\sigma_\varepsilon^2}{\varepsilon^2} \right]^{1/2}$$

where σ_ε characterises the likely difference between $\varepsilon(\lambda_1)$ and $\varepsilon(\lambda_2)$.

8.7.3 Total radiation thermometers

Total radiation thermometers measure the total radiance of a surface. Because of the problems with atmospheric absorption they are capable of accurate operation only when very close to the surface of interest. One of the best-known examples is the *gold cup thermometer* shown schematically in Figure 8.17.

In use the gold-plated hemisphere is placed against the surface to form a blackbody cavity. This eliminates the need to know the surface emissivity. A small aperture in the hemisphere allows radiation to be exchanged between the cavity and the detector. The net response of the detector is

$$V(T) = g(T_s^4 - T_d^4), \tag{8.37}$$

where T_s and T_d are the temperatures of the surface and detector respectively and g is a constant. The dependence of the response on the detector temperature is common to all radiation thermometers (equation (8.20)). In practice the thermometer is rarely used below 200°C so that the uncertainty in the detector temperature is unimportant. When used below 200°C considerable care is required to obtain accurate and repeatable results.

Errors may also arise because the detector responsivity (included in the constant g) is temperature dependent. This dependence is usually compensated by a simple thermistor circuit. However, the direct dependence on T_d cannot be compensated so easily.

In use the thermometer can suffer from large errors due to the surface heating. The surface of a hot object loses energy by radiation to cool surroundings. When the surface is covered by the thermometer the heat loss is reduced to almost zero, so that the local

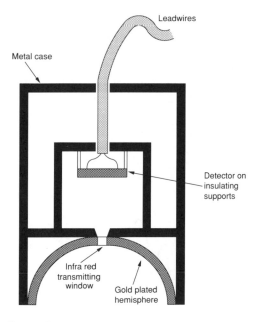

Figure 8.17
A simple schematic diagram of a gold cup thermometer.

temperature of the surface rises. The errors may be as large as 20°C to 40°C depending on the size of the thermometer and the properties of the surface.

There is a practical upper limit for the head temperature of the gold cup thermometer. Rubber and plastic components in the head and the temperature-compensation circuit limit the head temperature to the range −20°C to 50°C. Both of these factors, and the need to minimise the heating errors, limit the measurement period to 2 to 6 seconds. Thus the thermometer is only for intermittent use, and restricted to surfaces within arm's reach. Although strictly a contact thermometer it can be held a few millimetres off moving surfaces with minimal loss in accuracy.

One useful application of the gold cup thermometer is the measurement of total emissivity. By replacing the gold hemisphere with a very black hemisphere the detector response becomes

$$V(T) = \varepsilon_s g(T_s^4 - T_d^4). \tag{8.38}$$

Thus the ratio of the two measurements is the emissivity of the surface. Knowledge of the total emissivity may be useful when estimating the spectral emissivity of surfaces.

Overall, the gold cup thermometer is capable of accuracies similar to that of a base-metal thermocouple, ±3/4%, and covers a range from 200°C up to 1300°C.

8.7.4 Special purpose thermometers for plastic and glass

The plastics industry, and to a lesser extent the glass industry, present some interesting temperature-measurement problems. How, for example, can we measure the temperature of a fast-moving plastic film less than 0.05 mm thick? Radiation thermometers would seem to be the obvious choice except that many plastics are transparent and most radiation thermometers would see straight through the film. Fortunately organic materials such as plastics exhibit absorption lines in their spectra (see Figure 8.18). In these narrow regions of the spectra the plastics are opaque and have extremely high emissivities, typically 0.97. Therefore a spectral-band thermometer with the filter passband centred on one of the absorption lines will make an accurate temperature measurement, indeed more accurate than most radiation thermometry measurements because of the high emissivity of the plastic.

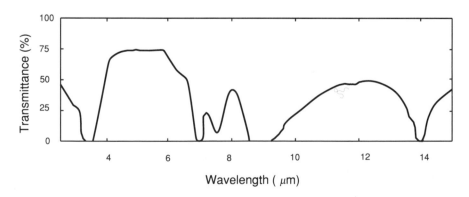

Figure 8.18
The spectral absorption of a sample of polyethylene film.

Table 8.3. Infra-red absorption for polymers.

	C-H Band 3.43 μm	Ester Band 7.95 μm
Acrylic	X	X
Cellulose acetate	X*	X
Fluoroplastic (FEP)		X
Polyester (PET)	X*	X
Polyimide		X
Polyurethane	X	X
Polyvinyl chloride	X	X
Polycarbonate	X	X
Polyamide (nylon)	X	X
Polypropylene	X	
Polyethylene	X	
Polystyrene	X	
Ionomer	X	
Polybutylene	X	
Glassine	X	
Cellophane	X	
Paints	X	
Epoxy resins	X	

*For films $\geq$ 0.5 mm

Table 8.3 shows the two most commonly used absorption lines and the plastics which absorb there. Thermometers which operate at the 3.43 μm band require filter bandwidths of 50 nm or less, while those operating at the 7.95 μm band should have bandwidths of 100 nm or less. The narrow bandwidths are particularly important for very thin and visually transparent (non-pigmented) films. A simple check to make sure that the thermometer cannot see through the film is to move a highly polished metal sheet behind the film in the field of view of the thermometer. The metal mirror effectively doubles the thickness of the film seen by the thermometer. If the film is sufficiently thick the thermometer reading should not change.

Similar temperature measurement problems also occur in the glass industry except that the absorption bands in glass are very much broader. The most useful band for radiation thermometry is from about 5 μm to 8 μm. For relatively thick or pigmented glasses, 5 μm thermometers are suitable; for thin transparent glasses the thermometer should operate nearer the 8 μm end of the band where the absorption is much stronger.

Radiation thermometers operating in narrow bandwidths must use interference filters to select the bandwidth. This makes them more susceptible than wider-band thermometers to rapid drifts in calibration and changes in the ambient temperature. Drifts of 1% to 2% per year are not unusual. Narrow-band thermometers therefore have slightly higher maintenance and calibration demands.

8.7.5 Fibre-optic thermometers

In principle, a radiation thermometer can be made by exploiting any temperature-dependent optical property: e.g. transmittance, reflectance, scattering, and fluorescence, as well as radiance. The wide variety of fibre-optic thermometers is such that few of these techniques remain untried. However, the main attraction of fibre-optic thermometers lies not in the physical principles used, but in their ability to measure temperature in situations inaccessible to other thermometers. They are, for example, used increasingly in medical applications where the small sensor size and chemical immunity are important, and in the heavy electrical industries where their immunity to electromagnetic interference is important.

Other advantages over conventional radiation thermometers include confinement of the optical path, which eliminates scattering errors, and well-defined optical properties of the sensor. The main disadvantage of fibre-optic thermometers is the cost. Despite intensive efforts to reduce the cost of manufacture, the good performance of the thermometers invariably relies on a number of critical and often expensive components.

There are two types of fibre-optic thermometer which seem to be the most practical. The first uses fluorescence decay times to measure the temperature of phosphor located at the end of the fibre. It is extremely stable with time and immune to many environmental conditions, especially ambient radiation. The typical temperature range is from $-200°$C to $250°$C with accuracies of $\pm 1°$C.

The second type is essentially a radiance meter like conventional spectral-band radiation thermometers except that the end of the fibre is covered to form a small blackbody cavity. With the use of sapphire fibres for the hot portion of the fibre these thermometers are useful from $250°$C 2o $2000°$C. Typical accuracies are about 1% to 2%.

FURTHER READING

Theory and Practice of Radiation Thermometry, D. P. DeWitt and G. D. Nutter, Wiley Interscience, New York (1988).
This book is very strong on the theory and operating principles of radiation thermometers and has specific applications chapters for the steel, aluminium, glass, plastics and energy management industries.

Radimetric Calibration: Theory and Methods, C. L. Wyatt, Accademic Press, New York (1978).
An older book with a solid technical description of the evaluation of the performance characteristics of near infra-red systems.

Supplementary Information for the International Temperature Scale of 1990, BIPM (1990).
Techniques for Approximating the International Temperature Scale of 1990, BIPM (1990).
These two booklets give a simple overview of the procedures and equipment required to establish ITS-90. Especially useful for the bibliography.

Temperature, its Measurement and Control in Science and Industry, American Institute of Physics, New York, Vol. 5 (1982), Vol. 6 (1992).
These proceedings from the 10-yearly temperature symposia are a very useful snapshot of thermometry practice. Strong on examples of difficult industrial measurements.

Appendix A
Further Information for Least-Squares Fitting

A.1 NORMAL EQUATIONS FOR CALIBRATION EQUATIONS

A.1.1 Deviation function for direct-reading thermometers

$$\Delta t = A + Bt + Ct^2 + Dt^3 \tag{4.10}$$

$$\begin{pmatrix} A \\ B \\ C \\ D \end{pmatrix} = \begin{pmatrix} N & \Sigma t_i & \Sigma t_i^2 & \Sigma t_i^3 \\ \Sigma t_i & \Sigma t_i^2 & \Sigma t_i^3 & \Sigma t_i^4 \\ \Sigma t_i^2 & \Sigma t_i^3 & \Sigma t_i^4 & \Sigma t_i^5 \\ \Sigma t_i^3 & \Sigma t_i^4 & \Sigma t_i^5 & \Sigma t_i^6 \end{pmatrix}^{-1} \begin{pmatrix} \Sigma \Delta t_i \\ \Sigma t_i \Delta t_i \\ \Sigma t_i^2 \Delta t_i \\ \Sigma t_i^3 \Delta t_i \end{pmatrix}.$$

A.1.2 Extended Callendar equation for platinum resistance thermometers

$$W(t) = \frac{R(t)}{R(0°C)} = 1 + At + Bt^2 + Dt^3 \tag{5.43}$$

$$\begin{pmatrix} A \\ B \\ D \end{pmatrix} = \begin{pmatrix} \Sigma t_i^2 \Sigma t_i^3 \Sigma t_i^4 \\ \Sigma t_i^3 \Sigma t_i^4 \Sigma t_i^5 \\ \Sigma t_i^4 \Sigma t_i^5 \Sigma t_i^6 \end{pmatrix}^{-1} \begin{pmatrix} \Sigma t_i [W(t_i) - 1] \\ \Sigma t_i^2 [W(t_i) - 1] \\ \Sigma t_i^3 [W(t_i) - 1] \end{pmatrix}$$

A.1.3 Callendar–Van Dusen equation for platinum resistance thermometers

$$W(t) = 1 + At + Bt^2 + Ct^3(t - 100) \tag{5.42}$$

$C = 0$ above $0°C$.

$$\begin{pmatrix} A \\ B \\ C \end{pmatrix} = \begin{pmatrix} \Sigma t_i^2 & \Sigma t_i^3 & \sum_{t_i<0} t_i^4(t_i - 100) \\ \Sigma t_i^3 & \Sigma t_i^4 & \sum_{t_i<0} t_i^5(t_i - 100) \\ \sum_{t_i<0} t_i^4(t_i - 100) & \sum_{t_i<0} t_i^5(t_i - 100) & \sum_{t_i<0} t_i^6(t_i - 100)^2 \end{pmatrix}^{-1} \times$$

$$\begin{pmatrix} \Sigma t_i [W(t_i) - 1] \\ \Sigma t_i^2 [W(t_i) - 1] \\ \sum_{t_i < 0} t_i^3 (t_i - 100)[W(t_i) - 1] \end{pmatrix}.$$

A.1.4 The thermistor equation

$$\frac{1}{T} = a_0 + a_1 \log(R) + a_2 \log^2(R) + a_3 \log^3(R) \tag{5.51}$$

$$\begin{pmatrix} a_0 \\ a_1 \\ a_2 \\ a_3 \end{pmatrix} = \begin{pmatrix} N & \Sigma \log R_i & \Sigma \log^2 R_i & \Sigma \log^3 R_i \\ \Sigma \log R_i & \Sigma \log^2 R_i & \Sigma \log^3 R_i & \Sigma \log^4 R_i \\ \Sigma \log^2 R_i & \Sigma \log^3 R_i & \Sigma \log^4 R_i & \Sigma \log^5 R_i \\ \Sigma \log^3 R_i & \Sigma \log^4 R_i & \Sigma \log^5 R_i & \Sigma \log^6 R_i \end{pmatrix}^{-1} \begin{pmatrix} \Sigma \dfrac{1}{T_i} \\ \Sigma \dfrac{1}{T_i} \log R_i \\ \Sigma \dfrac{1}{T_i} \log^2 R_i \\ \Sigma \dfrac{1}{T_i} \log^3 R_i \end{pmatrix}$$

A.2 A PASCAL MATRIX INVERSION PROCEDURE ("MATINV.P")

procedure invert_matrix(**var** A:arraytype; **var** det: real; rho : integer);
{
This procedure uses the Gauss–Jordan algorithm with full pivot to perform matrix inversion. The inverted matrix is assembled in the space vacated by the input matrix. The routine may be used for both symmetric and non-symmetric matrices. If the value of the determinant returned by the procedure is zero then the matrix is singular and the inversion is not completed.
Usage: invert_matrix(A,det,rho);
   where: A is a rho × rho real array of arraytype,
           rho is the number of undetermined parameters in the fit,
           det is the determinant of the matrix.
Declarations: (these must be global)
const
   rhomax = 10;   (the maximum number of undetermined parameters)
type
   arraytype = **array**[1 . . rhomax, 1 . . rhomax] **of** real;
}
var
   diag,row,col : integer;
   Amax,save   : real;
   introw,intcol : **array**[1 . . rhomax] **of** integer;   {pivot interchange vectors}
begin;
     det   := 1.0;

```
for diag := 1 to rho do
begin                                    {find largest element in rest of matrix}
    Amax   := 0.0;
    for row := diag to rho do
    for col := diag to rho do
    if abs(Amax) < abs(A[row,col]) then
    begin
        Amax        := A[row,col];
        introw[diag] := row;
        intcol[diag] := col;
    end;

    det := det * Amax;
    if (det = 0.0) then return           {check that matrix is not singular}

    else begin
        row        := introw[diag];      {interchange rows and colums}
        if row > diag then               {to preserve precision}
        for col   := 1 to rho do
        begin
            save         := A[diag,col];
            A[diag,col] := A[row,col];
            A[row,col]  := - save;
        end;
        col        := intcol[diag];
        if col >  diag then
        for row  := 1 to rho do
        begin
            save         := A[row, diag];
            A[row,diag] := A[row,col];
            A[row,dol]  := - save;
        end;                             {interchange complete}

    A[diag,diag] := 1.0;                 {now carry out Gauss-Jordan operations}
    for col        := 1 to rho do
        A[diag,col] := A[diag,col]/Amax;
    for row        := 1 to rho do
        if not (row = diag) then
        begin
            save         := A[row,diag];
            A[row,diag] := 0.0;
            for col      := 1 to rho do
                A[row,col] := A[row,col] - save * A[diag,col];
    end;
end;
```

```
        end;

    for diag     := rho downto 1 do        {now restore the order of the matrix}
    begin
        col      := introw[diag];
        if col > diag then
        for row := 1 to rho do
        begin
            save         := A[row,diag];
            A[row,diag] := - A[row,col];
            A[row,col]  := save;
        end;
        row      := intcol[diag];
        if row > diag then
        for col  := 1 to rho do
        begin
            save         := A[diag,col];
            A[diag,col] := -A[row,col];
            A[row,col]  := save;
        end;
    end;
end;            {invert_matrix}
```

A.3 LEAST-SQUARES FIT PROGRAM IN PASCAL FOR EXAMPLE 2.12

```
program rescal (lst,output,input);
{
program to demonstrate least squares
}
const
    Nmax = 20;                              {maximum of 20 data points}
    rhomax = 10;                            {maximum number of parameters}
    rho = 3;                                {3 unknown parameters for PRTs}
type
    arraytype = array[1 . . rhomax, 1 . . rhomax] of real;
var
    T : array[1 . . Nmax] of real;         {arrays for data}
    R : array[1 . . Nmax] of real;
    Tpred : array[1 . . Nmax] of real;     {arrays for results}
    Rpred : array[1 . . Nmax] of real;
    error : array[1 . . Nmax] of real;
    aa : array[1 . . rhomax] of real;      {array of parameters}
    b : array[1 . . rhomax] of real;       {array for summation}
```

```pascal
  A : arraytype;                                    {covariance matrix}
  det : real;                                       {determinant of matrix}
  N : integer;                                      {number of data points}
  i,j, :integer;                                    {array index}
  ssfit :real;                                      {variance of resistance error}
  sst : real;                                       {variance of temperature error}
  lst : text;                                       {output file for printing}
#include "matinv.p"
begin                                               {first, input the data}
    writeln('enter the number of data points (', Nmax, ' maximum)');
    readln(N);
    for i := 1 to N do
    begin
        writeln('enter temperature and resistance for the ', i, 'th measurement');
        readln(T[i],R[i]);
    end;

    for i := 1 to rho do                            {set matrix and array to zero}
    begin
        b[i]  := 0.0;
        aa[i] := 0.0;
        for j := 1 to rho do
            A[i,j] := 0.0;
    end;

    for i := 1 to N do                              {now assemble matrix and array}
    begin
        b[1] := b[1] + R[i];
        b[2] := b[2] + R[i] * T[i];
        b[3] := b[3] + R[i] * T[i] * T[i];
        A[1,2] := A[1,2] + T[i];
        A[1,3] := A[1,3] + T[i] * T[i];
        A[2,3] := A[2,3] + T[i] * T[i] * T[i];
        A[3,3] := A[3,3] + T[i] * T[i] * T[i] * T[i];
    end;
    A[1,1] := N;
    A[2,1] := A[1,2];
    A[2,2] := A[1,3];
    A[3,1] := A[1,3];
    A[3,2] := A[2,3];

    invert_matrix(A,det,rho);                       {invert matrix}

    if (det = 0.0) then                             {check for zero determinant}
    begin
        writeln('matrix singular, execution halted');
```

```
        halt;
   end;

   for i := 1 to rho do                        {calculate parameters}
   for j := 1 to rho do
       aa[i] := aa[i] + A[i,j] * b[j];

   ssfit  := 0.0;                              {calculate variance of errors}
   sst    := 0.0;
   for i := 1 to N do
   begin
       Rpred[i] := aa[i] + T[i]*(aa[2] + T[i]*aa[3]);
       ssfit      := ssfit + sqr(R[i] − Rpred[i]);
       Tpred[i] := 0.0;                        {solve T(R) relationship}
       for j := 1 to 5 do
             Tpred[i] := (R[i] − aa[1])/(aa[2] + aa[3]*Tpred[i]);
       sst    := sst + sqr(T[i] − Tpred[i]);
   end;
   ssfit := ssfit/(N - rho);
   sst   := sst/(N - rho);

   rewrite(lst);                              {presentation of results}
   writeln(lst,'SUMMARY FOR PLATINUM RESISTANCE THERMOMETER');
   writeln(lst);
   writeln(lst,'Reading':10,'Measured':15,'Measured':15,'Predicted':15,'Residual':15);
   writeln(lst,'number':10,'resistance':15,'temperature':15,'temperature':15,'error':15);
   writeln(lst);
   for i := 1 to N do
       writeln(lst,i:8, ' ', R[i]:15:4,T[i]:15:4,Tpred[i]:15:4,T[i] - Tpred[i]:15:4);
   writeln(lst);
   writeln(lst,'standard error-of-fit =',sqrt(ssfit):10:4,' ohms');
   writeln(lst);
   writeln(lst,'standard deviation of residual errors =',sqrt(sst):14:4,' C');
   writeln(lst);
   writeln(lst, 'fitted paramters:');
   writeln(lst);
   for i := 1 to rho do
       writeln(lst,'a',i-1:1,' = ',aa[i]:15,' +- ',sqrt(A[i,i]*ssfit):8);
   writeln(lst);
   writeln(lst,'R(0 C) = ',aa[1]:7:4);
   writeln(lst,'A = ',aa[2]/aa[1]:12);
   writeln(lst,'B = ',aa[3]/aa[1]:10);
end.
```

Appendix B
The Differences Between ITS-90 and IPTS-68

The numerical differences between the ITS-90 and the IPTS-68 scales up to 1064.18°C have been fitted in three ranges by the following power-series polynomials.

(1) From 13.8033 K to 83.8058 K
   (accuracy approximately ±0.001 K)

$$(T_{90} - T_{68})/\text{K} = \sum_{i=0}^{12} a_i[(T - 40)/40]^i.$$

(2) From −200°C to 630.6°C
   (accuracy approximately ±0.0015°C up to 0°C and ±0.001°C above 0°C)

$$(t_{90} - t_{68})/^\circ\text{C} = \sum_{i=1}^{8} b_i[t/630]^i.$$

(3) From 630.6°C to 1064.18°C
   (accuracy approximately ±0.01°C)

$$(t_{90} - t_{68})/^\circ\text{C} = \sum_{i=0}^{7} c_i[(t - 900)/300]^i.$$

The coefficients a_i, b_i and c_i, are

i	a_i	b_i	c_i
0	−0.005903	—	−0.00317
1	0.008174	−0.148759	−0.97737
2	−0.061924	−0.267408	1.25590
3	−0.193388	1.080760	2.03295
4	1.490793	1.269056	−5.91887
5	1.252347	−4.089591	−3.23561
6	−9.835868	−1.871251	7.23364
7	1.411912	7.438081	5.04151
8	25.277595	−3.536296	—
9	−19.183815	—	—
10	−18.437089	—	—
11	27.000895	—	—
12	−8.716324	—	—

At temperatures above 1064.18°C the differences are represented by

$$(t_{90} - t_{68})/°C = -0.25[(t + 273.15)/1337.33]^2.$$

Wherever possible it is recommended that IPTS-68 calibrations should be converted to the ITS-90 scale directly, using the resistance ratios at the fixed points and the equations in the text of the scale.

Table of numerical differences, $T_{90} - T_{68}$, as published by the BIPM (1990)

$(T_{90} - T_{68})/K$

T_{90}/K	0	1	2	3	4	5	6	7	8	9
10					−0.006	−0.003	−0.004	−0.006	−0.008	−0.009
20	−0.009	−0.008	−0.007	−0.007	−0.006	−0.005	−0.004	−0.004	−0.005	−0.006
30	−0.006	−0.007	−0.008	−0.008	−0.008	−0.007	−0.007	−0.007	−0.006	−0.006
40	−0.006	−0.006	−0.006	−0.006	−0.006	−0.007	−0.007	−0.007	−0.006	−0.006
50	−0.006	−0.005	−0.005	−0.004	−0.003	−0.002	−0.001	0.000	0.001	0.002
60	0.003	0.003	0.004	0.004	0.005	0.005	0.006	0.006	0.007	0.007
70	0.007	0.007	0.007	0.007	0.007	0.008	0.008	0.008	0.008	0.008
80	0.008	0.008	0.008	0.008	0.008	0.008	0.008	0.008	0.008	0.008
90	0.008	0.008	0.008	0.008	0.008	0.008	0.008	0.009	0.009	0.009

T_{90}/K	0	10	20	30	40	50	60	70	80	90
100	0.009	0.011	0.013	0.014	0.014	0.014	0.014	0.013	0.012	0.012
200	0.011	0.010	0.009	0.008	0.007	0.005 ·	0.003	0.001		

$(t_{90} - t_{68})/°C$

$t_{90}/°C$	0	−10	−20	−30	−40	−50	−60	−70	−80	−90
−100	0.013	0.013	0.014	0.014	0.014	0.013	0.012	0.010	0.008	0.008
0	0.000	0.002	0.004	0.006	0.008	0.009	0.010	0.011	0.012	0.012

$t_{90}/°C$	0	10	20	30	40	50	60	70	80	90
0	0.000	−0.002	−0.005	−0.007	−0.010	−0.013	−0.016	−0.018	−0.021	−0.024
100	−0.026	−0.028	−0.030	−0.032	−0.034	−0.036	−0.037	−0.038	−0.039	−0.039
200	−0.040	−0.040	−0.040	−0.040	−0.040	−0.040	−0.040	−0.039	−0.039	−0.039
300	−0.039	−0.039	−0.039	−0.040	−0.040	−0.041	−0.042	−0.043	−0.045	−0.046
400	−0.048	−0.051	−0.053	−0.056	−0.059	−0.062	−0.065	−0.068	−0.072	−0.075
500	−0.079	−0.083	−0.087	−0.090	−0.094	−0.098	−0.101	−0.105	−0.108	−0.112
600	−0.115	−0.118	−0.122	−0.125	−0.08	−0.03	0.02	0.06	0.11	0.16
700	0.20	0.24	0.28	0.31	0.33	0.35	0.36	0.36	0.36	0.35
800	0.34	0.32	0.29	0.25	0.22	0.18	0.14	0.10	0.06	0.03
900	−0.01	−0.03	−0.06	−0.08	−0.10	−0.12	−0.14	−0.16	−0.17	−0.18
1000	−0.19	−0.20	−0.21	−0.22	−0.23	−0.24	−0.25	−0.25	−0.26	−0.26

$t_{90}/°C$	0	100	200	300	400	500	600	700	800	900
1000		−0.26	−0.30	−0.35	−0.39	−0.44	−0.49	−0.54	−0.60	−0.66
2000	−0.72	−0.79	−0.85	−0.93	−1.00	−1.07	−1.15	−1.24	−1.32	−1.41
3000	−1.50	−1.59	−1.69	−1.78	−1.89	−1.99	−2.10	−2.21	−2.32	−2.43

Appendix C
Resistance Thermometer Reference Tables (IEC Proposed 1993)

$$R(t) = R(0°C)(1 + At + Bt^2 + Ct^3(t - 100))$$
$$R(0°C) = 100\ \Omega$$
$$A = 3.9083 \times 10^{-3}/°C$$
$$B = -5.775 \times 10^{-7}/°C^2$$
$$C = -4.183 \times 10^{-12}/°C^4$$

$T_{90}[°C]$	−0.0	−5.0	−10.0	−15.0	−20.0
−200.0	18.52				
−175.0	29.22	27.10	24.97	22.83	20.68
−150.0	39.72	37.64	35.54	33.44	31.34
−125.0	50.06	48.00	45.94	43.88	41.80
−100.0	60.26	58.23	56.19	54.15	52.11
−75.0	70.33	68.33	66.31	64.30	62.28
−50.0	80.31	78.32	76.33	74.33	72.33
−25.0	90.19	88.22	86.25	84.27	82.29
0.0	100.00	98.04	96.09	94.12	92.16

$T_{90}[°C]$	0.0	5.0	10.0	15.0	20.0
0.0	100.00	101.95	103.90	105.85	107.79
25.0	109.73	111.67	113.61	115.54	117.47
50.0	119.40	121.32	123.24	125.16	127.08
75.0	128.99	130.90	132.80	134.71	136.61
100.0	138.51	140.40	142.29	144.18	146.07
125.0	147.95	149.83	151.71	153.58	155.46
150.0	157.33	159.19	161.05	162.91	164.77
175.0	166.63	168.48	170.33	172.17	174.02
200.0	175.86	177.69	179.53	181.36	183.19
225.0	185.01	186.84	188.66	190.47	192.29
250.0	194.10	195.91	197.71	199.51	201.31
275.0	203.11	204.90	206.70	208.48	210.27

continued overleaf

continued

$T_{90}[°C]$	0.0	5.0	10.0	15.0	20.0
300.0	212.05	213.83	215.61	217.38	219.15
325.0	220.92	222.68	224.45	226.21	227.96
350.0	229.72	231.47	233.21	234.96	236.70
375.0	238.44	240.18	241.91	243.64	245.37
400.0	247.09	248.81	250.53	252.25	253.96
425.0	255.67	257.38	259.08	260.78	262.48
450.0	264.18	265.87	267.56	269.25	270.93
475.0	272.61	274.29	275.97	277.64	279.31
500.0	280.98	282.64	284.30	285.96	287.62
525.0	289.27	290.92	292.56	294.21	295.85
550.0	297.49	299.12	300.75	302.38	304.01
575.0	305.63	307.25	308.87	310.49	312.10
600.0	313.71				

Tolerances (in °C)

Class A	0.15 + 0.2%
Class B	0.3 + 0.5%

Appendix D
Thermocouple Reference Tables

D.1 REFERENCE FUNCTIONS

The coeficients for the reference functions for each of the IEC letter-designated thermocouple types is given on the following pages. The tables given have been formulated using these equations.

Except for the Type K thermocouple in the range 0°C to 1372°C, the reference functions are of the form:

$$E = \sum_{i=0}^{n} a_i t_{90}^i,$$

where t_{90} is in degrees Celsius and E is in the thermocouple output in microvolts. For Type K in the above range the reference function is of the form

$$E = \sum_{i=0}^{n} b_i t_{90}^i + c_1 \exp\left[-0.5\left(\frac{t_{90} - 126.9686}{65}\right)^2\right],$$

where t_{90} is in degrees Celsius and E is in microvolts.

D.2 INVERSE FUNCTIONS

The coefficients of inverse functions for each of the thermocouple types is also given. The inverse functions are of the form

$$t_{90} = \sum_{i=0}^{n} d_i E^i.$$

These inverse functions are approximate. The errors in temperatures calculated with these functions, relative to the reference functions, are less than 0.06°C. The functions should not be extrapolated beyond the specified ranges.

D.3 TYPE B

Output in μV.

$T_{90}[°C]$	0	10	20	30	40
0	0	−2	−3	−2	−0
50	2	6	11	17	25
100	33	43	53	65	78
150	92	107	123	141	159
200	178	199	220	243	267
250	291	317	344	372	401
300	431	462	494	527	561
350	596	632	669	707	746
400	787	828	870	913	957
450	1 002	1 048	1 095	1 143	1 192
500	1 242	1 293	1 344	1 397	1 451
550	1 505	1 561	1 617	1 675	1 733
600	1 792	1 852	1 913	1 975	2 037
650	2 101	2 165	2 230	2 296	2 363
700	2 431	2 499	2 569	2 639	2 710
750	2 782	2 854	2 928	3 002	3 078
800	3 154	3 230	3 308	3 386	3 466
850	3 546	3 626	3 708	3 790	3 873
900	3 957	4 041	4 127	4 213	4 299
950	4 387	4 475	4 564	4 653	4 743
1 000	4 834	4 926	5 018	5 111	5 205
1 050	5 299	5 394	5 489	5 585	5 682
1 100	5 780	5 878	5 976	6 075	6 175
1 150	6 276	6 377	6 478	6 580	6 683
1 200	6 786	6 890	6 995	7 100	7 205
1 250	7 311	7 417	7 524	7 632	7 740
1 300	7 848	7 957	8 066	8 176	8 286
1 350	8 397	8 508	8 620	8 731	8 844
1 400	8 956	9 069	9 182	9 296	9 410
1 450	9 524	9 639	9 753	9 868	9 984
1 500	10 099	10 215	10 331	10 447	10 563
1 550	10 679	10 796	10 913	11 029	11 146
1 600	11 263	11 380	11 497	11 614	11 731
1 650	11 848	11 965	12 082	12 199	12 316
1 700	12 433	12 549	12 666	12 782	12 898
1 750	13 014	13 130	13 246	13 361	13 476
1 800	13 591	13 706	13 820		

Type B Reference Function Coefficients.

	0°C to 630.615° C		630.615°C to 1820° C
a_0	0	a_0	$-3.8938168621E + 03$
a_1	$-2.4650818346E - 01$	a_1	$2.8571747470E + 01$
a_2	$5.9040421171E - 03$	a_2	$-8.4885104785E - 02$
a_3	$-1.3257931636E - 06$	a_3	$1.5785280164E - 04$
a_4	$1.5668291901E - 09$	a_4	$-1.6835344864E - 07$
a_5	$-1.6944529240E - 12$	a_5	$1.1109794013E - 10$
a_6	$6.2990347094E - 16$	a_6	$-4.4515431033E - 14$
		a_7	$9.8975640821E - 18$
		a_8	$-9.3791330289E - 22$

Type B Inverse Function Coefficients.

	250°C to 700°C 291 μV to 2431 μV		700°C to 1820°C 2431 μV to 13820 μV
d_0	$9.8423321E + 01$	d_0	$2.1315071E + 02$
d_1	$6.9971500E - 01$	d_1	$2.8510504E - 01$
d_2	$-8.4765304E - 04$	d_2	$-5.2742887E - 05$
d_3	$1.0052644E - 06$	d_3	$9.9160804E - 09$
d_4	$-8.3345952E - 10$	d_4	$-1.2965303E - 12$
d_5	$4.5508542E - 13$	d_5	$1.1195870E - 16$
d_6	$-1.5523037E - 16$	d_6	$-6.0625199E - 21$
d_7	$2.9886750E - 20$	d_7	$1.8661696E - 25$
d_8	$-2.4742860E - 24$	d_8	$-2.4878585E - 30$

Tolerances (whichever is greater)

Class 2:   1.5°C or 0.25%   for 600°C to 1700°C
Class 3:   4°C or 0.5%   for 600°C to 1700°C

Properties

Nominal composition: Platinum–30% Rhodium vs. Platinum–6% Rhodium

Type B is well suited for use in oxidising or inert atmospheres at high temperatures. It suffers less grain growth than either of Types R or S. Although it is slightly more immune to contamination than Types R or S it is still susceptible to contamination. In particular, Type B should not be exposed to metallic vapours or reducing environments.

Type B has very low output at low temperatures (< 200°C). For low-accuracy applications no cold junction compensation is necessary if the cold junction can be kept between 0°C and 50°C.

D.4 TYPE E

Output in μV.

T_{90} [°C]	−0	−10	−20	−30	−40
−250	−9 718	−9 797	−9 835		
−200	−8 825	−9 063	−9 274	−9 455	−9 604
−150	−7 279	−7 632	−7 963	−8 273	−8 561
−100	−5 237	−5 681	−6 107	−6 516	−6 907
−50	−2 787	−3 306	−3 811	−4 302	−4 777
0	0	−582	−1 152	−1 709	−2 255

T_{90} [°C]	0	10	20	30	40
0	0	591	1 192	1 801	2 420
50	3 048	3 685	4 330	4 985	5 648
100	6 319	6 998	7 685	8 379	9 081
150	9 789	10 503	11 224	11 951	12 684
200	13 421	14 164	14 912	15 664	16 420
250	17 181	17 945	18 713	19 484	20 259
300	21 036	21 817	22 600	23 386	24 174
350	24 964	25 757	26 552	27 348	28 146
400	28 946	29 747	30 550	31 354	32 159
450	32 965	33 772	34 579	35 387	36 196
500	37 005	37 815	38 624	39 434	40 243
550	41 053	41 862	42 671	43 479	44 286
600	45 093	45 900	46 705	47 509	48 313
650	49 116	49 917	50 718	51 517	52 315
700	53 112	53 908	54 703	55 497	56 289
750	57 080	57 870	58 659	59 446	60 232
800	61 017	61 801	62 583	63 364	64 144
850	64 922	65 698	66 473	67 246	68 017
900	68 787	69 554	70 319	71 082	71 844
950	72 603	73 360	74 115	74 869	75 621
1 000	76 373				

Tolerances (whichever is greater)

Class 1:	1.5 °C or 0.4%	for −40 °C to 800 °C
Class 2:	2.5 °C or 0.75%	for −40 °C to 900 °C
Class 3:	2.5 °C or 1.5%	for −200 °C to 40 °C

Type E Reference Function Coefficients.

	−270°C to 0°C		0°C to 1000°C
a_0	0	a_0	0
a_1	5.8665508708E+01	a_1	5.8665508710E+01
a_2	4.5410977124E−02	a_2	4.5032275582E−02
a_3	−7.7998048686E−04	a_3	2.8908407212E−05
a_4	−2.5800160843E−05	a_4	−3.3056896652E−07
a_5	−5.9452583057E−07	a_5	6.5024403270E−10
a_6	−9.3214058667E−09	a_6	−1.9197495504E−13
a_7	−1.0287605534E−10	a_7	−1.2536600497E−15
a_8	−8.0370123621E−13	a_8	2.1489217569E−18
a_9	−4.3979497391E−15	a_9	−1.4388041782E−21
a_{10}	−1.6414776355E−17	a_{10}	3.5960899481E−25
a_{11}	−3.9673619516E−20		
a_{12}	−5.5827328721E−23		
a_{13}	−3.4657842013E−26		

Type E Inverse Function Coefficients.

	−200°C to 0°C −8825μV to 0μV		0°C to 1000°C 0μV to 76373μV
d_0	0	d_0	0
d_1	1.6977288E−02	d_1	1.7057035E−02
d_2	−4.3514970E−07	d_2	−2.3301759E−07
d_3	−1.5859697E−10	d_3	6.5435585E−12
d_4	−9.2502871E−14	d_4	−7.3562749E−17
d_5	−2.6084314E−17	d_5	−1.7896001E−21
d_6	−4.1360199E−21	d_6	8.4036165E−26
d_7	−3.4034030E−25	d_7	−1.3735879E−30
d_8	−1.1564890E−29	d_8	1.0629823E−35
		d_9	−3.2447087E−41

Properties

Nominal composition: Chromel−constantan   90% Nickel−10% Chromium vs. 55% Copper−45% Nickel

Type E has the highest output of all common thermocouples and is suited for use at low-temperatures ($< 0°C$). It is best used in strongly oxidising or inert atmospheres and will stand limited use in vacuum and reducing environments. It will not withstand prolonged use in marginally oxidising environments. In the medium temperature range ($< 500°C$) it has a higher reproducibility than Type K.

D.5 TYPE J

Output in μV.

T_{90}[°C]	−0	−10	−20	−30	−40
−200	−7 890	−8 095			
−150	−6 500	−6 821	−7 123	−7 403	−7 659
−100	−4 633	−5 037	−5 426	−5 801	−6 159
−50	−2 431	−2 893	−3 344	−3 786	−4 215
0	0	−501	−995	−1 482	−1 961

T_{90}[°C]	0	10	20	30	40
0	0	507	1 019	1 537	2 059
50	2 585	3 116	3 650	4 187	4 726
100	5 269	5 814	6 360	6 909	7 459
150	8 010	8 562	9 115	9 669	10 224
200	10 779	11 334	11 889	12 445	13 000
250	13 555	14 110	14 665	15 219	15 773
300	16 327	16 881	17 434	17 986	18 538
350	19 090	19 642	20 194	20 745	21 297
400	21 848	22 400	22 952	23 504	24 057
450	24 610	25 164	25 720	26 276	26 834
500	27 393	27 953	28 516	29 080	29 647
550	30 216	30 788	31 362	31 939	32 519
600	33 102	33 689	34 279	34 873	35 470
650	36 071	36 675	37 284	37 896	38 512
700	39 132	39 755	40 382	41 012	41 645
750	42 281	42 919	43 559	44 203	44 848
800	45 494	46 141	46 786	47 431	48 074
850	48 715	49 353	49 989	50 622	51 251
900	51 877	52 500	53 119	53 735	54 347
950	54 956	55 561	56 164	56 763	57 360
1 000	57 953	58 545	59 134	59 721	60 307
1 050	60 890	61 473	62 054	62 634	63 214
1 100	63 792	64 370	64 948	65 525	66 102
1 150	66 679	67 255	67 831	68 406	68 980
1 200	69 553				

Type J Reference Function Coefficients.

	−210°C to 760°C		−210°C to 760°C	760°C to 1200°C
a_0	0		a_0	2.9645625681E + 05
a_1	5.0381187815E + 01		a_1	−1.4976127786E + 03
a_2	3.0475836930E − 02		a_2	3.1787103924E + 00
a_3	−8.5681065720E − 05		a_3	−3.1847686701E − 03
a_4	1.3228195295E − 07		a_4	1.5720819004E − 06
a_5	−1.7052958337E − 10		a_5	−3.0691369056E − 10
a_6	2.0948090697E − 13			
a_7	−1.2538395336E − 16			
a_8	1.5631725697E − 20			

Type J Inverse Function Coefficients.

	−210°C to 0°C −8095 μV to 0 μV		0°C to 760°C 0 μV to 42919 μV		760°C to 1200°C 42919 μV to 69553 μV
d_0	0	d_0	0	d_0	−3.11358187E + 03
d_1	1.9528268E − 02	d_1	1.978425E − 02	d_1	3.00543684E − 01
d_2	−1.2286185E − 06	d_2	−2.001204E − 07	d_2	−9.94773230E − 06
d_3	−1.0752178E − 09	d_3	1.036969E − 11	d_3	1.70276630E − 10
d_4	−5.9086933E − 13	d_4	−2.549687E − 16	d_4	−1.43033468E − 15
d_5	−1.7256713E − 16	d_5	3.585153E − 21	d_5	4.73886084E − 21
d_6	−2.8131513E − 20	d_6	−5.344285E − 26		
d_7	−2.3963370E − 24	d_7	5.099890E − 31		
d_8	−8.3823321E − 29				

Tolerances (whichever is greater)

Class 1:	1.5°C or 0.4%	for −40°C to 750°C
Class 2:	2.5°C or 0.75%	for −40°C to 750°C

Properties

Nominal composition: Iron–constantan   Iron vs. 55% Copper–45% Nickel

Type J is one of the few common thermocouples that is suited to use in reducing environments. It is also suited to use in oxidising and inert atmospheres. In oxidising and sulphurous atmospheres above 500°C the iron leg is prone to rapid corrosion. Type J is not recommended for use at low temperatures.

D.6 TYPE K

Output in μV.

T_{90}[°C]	−0	−10	−20	−30	−40
−250	−6 404	−6 441	−6 458		
−200	−5 891	−6 035	−6 158	−6 262	−6 344
−150	−4 913	−5 141	−5 354	−5 550	−5 730
−100	−3 554	−3 852	−4 138	−4 411	−4 669
−50	−1 889	−2 243	−2 587	−2 920	−3 243
0	0	−392	−778	−1 156	−1 527

T_{90}[°C]	0	10	20	30	40
0	0	397	798	1 203	1 612
50	2 023	2 436	2 851	3 267	3 682
100	4 096	4 509	4 920	5 328	5 735
150	6 138	6 540	6 941	7 340	7 739
200	8 138	8 539	8 940	9 343	9 747
250	10 153	10 561	10 971	11 382	11 795
300	12 209	12 624	13 040	13 457	13 874
350	14 293	14 713	15 133	15 554	15 975
400	16 397	16 820	17 243	17 667	18 091
450	18 516	18 941	19 366	19 792	20 218
500	20 644	21 071	21 497	21 924	22 350
550	22 776	23 203	2 3629	2 4055	24 480
600	24 905	25 330	25 755	26 179	26 602
650	27 025	27 447	27 869	28 289	28 710
700	29 129	29 548	29 965	30 382	30 798
750	31 213	31 628	32 041	32 453	32 865
800	33 275	33 685	34 093	34 501	34 908
850	35 313	35 718	36 121	36 524	36 925
900	37 326	37 725	38 124	38 522	38 918
950	39 314	39 708	40 101	40 494	40 885
1 000	41 276	41 665	42 053	42 440	42 826
1 050	43 211	43 595	43 978	44 359	44 740
1 100	45 119	45 497	45 873	46 249	46 623
1 150	46 995	47 367	47 737	48 105	48 473
1 200	48 838	49 202	49 565	49 926	50 286
1 250	50 644	51 000	51 355	51 708	52 060
1 300	52 410	52 759	53 106	53 451	53 795
1 350	54 138	54 479	54 819		

Tolerances (whichever is greater)

Class 1:	1.5°C or 0.4%	for −40°C to 1000°C
Class 2:	2.5°C or 0.75%	for −40°C to 1200°C
Class 3:	2.5°C or 1.5%	for −200°C to 40°C

Type K Reference Function Coefficients.

−270°C to 0°C		0°C to 1372°C	
a_0	0	b_0	$-1.7600413686E + 01$
a_1	$3.9450128025E + 01$	b_1	$3.8921204975E + 01$
a_2	$2.3622373598E - 02$	b_2	$1.8558770032E - 02$
a_3	$-3.2858906784E - 04$	b_3	$-9.9457592874E - 05$
a_4	$-4.9904828777E - 06$	b_4	$3.1840945719E - 07$
a_5	$-6.7509059173E - 08$	b_5	$-5.6072844889E - 10$
a_6	$-5.7410327428E - 10$	b_6	$5.6075059059E - 13$
a_7	$-3.1088872894E - 12$	b_7	$-3.2020720003E - 16$
a_8	$-1.0451609365E - 14$	b_8	$9.7151147152E - 20$
a_9	$-1.9889266878E - 17$	b_9	$-1.2104721275E - 23$
a_{10}	$-1.6322697486E - 20$		
		c_1	$1.185976E + 02$

Type K Inverse Function Coefficients.

−200°C to 0°C −5891 μV to 0 μV		0°C to 500°C 0 μV to 20644 μV		500°C to 1372°C 20644 μV to 54886 μV	
d_0	0	d_0	0	d_0	$-1.318058E + 02$
d_1	$2.5173462E - 02$	d_1	$2.508355E - 02$	d_1	$4.830222E - 02$
d_2	$-1.1662878E - 06$	d_2	$7.860106E - 08$	d_2	$-1.646031E - 06$
d_3	$-1.0833638E - 09$	d_3	$-2.503131E - 10$	d_3	$5.464731E - 11$
d_4	$-8.9773540E - 13$	d_4	$8.315270E - 14$	d_4	$-9.650715E - 16$
d_5	$-3.7342377E - 16$	d_5	$-1.228034E - 17$	d_5	$8.802193E - 21$
d_6	$-8.6632643E - 20$	d_6	$9.804036E - 22$	d_6	$-3.110810E - 26$
d_7	$-1.0450598E - 23$	d_7	$-4.413030E - 26$		
d_8	$-5.1920577E - 28$	d_8	$1.057734E - 30$		
		d_9	$-1.052755E - 35$		

Properties

Nominal composition: Chromel–Alumel 90% Nickel–10% Chromium vs. 95% Nickel–2% Aluminium–2% Manganese–1% Silicon.

Type K is the most common thermocouple type and the most irreproducible, showing spurious errors of up to 8°C in the 300°C to 500°C range and steady drift above 700°C. It is suited to oxidising and inert atmospheres but suffers from 'green rot' and embrittlement in marginally oxidising atmospheres. The main advantages of Type K are the wide range, the low cost and ready availability of instrumentation.

D.7 TYPE N

Output in μV

$T_{90}[°C]$	−0	−10	−20	−30	−40
−250	−4313	−4336	−4345		
−200	−3990	−4083	−4162	−4226	−4277
−150	−3336	−3491	−3634	−3766	−3884
−100	−2407	−2612	−2808	−2994	−3171
−50	−1269	−1509	−1744	−1972	−2193
0	0	−260	−518	−772	−1023

$T_{90}[°C]$	0	10	20	30	40
0	0	261	525	793	1065
50	1340	1619	1902	2189	2480
100	2774	3072	3374	3680	3989
150	4302	4618	4937	5259	5585
200	5913	6245	6579	6916	7255
250	7597	7941	8288	8637	8988
300	9341	9696	10054	10413	10774
350	11136	11501	11867	12234	12603
400	12974	13346	13719	14094	14469
450	14846	15225	15604	15984	16366
500	16748	17131	17515	17900	18286
550	18672	19059	19447	19835	20224
600	20613	21003	21393	21784	22175
650	22566	22958	23350	23742	24134
700	24527	24919	25312	25705	26098
750	26491	26883	27276	27669	28062
800	28455	28847	29239	29632	30024
850	30416	30807	31199	31590	31981
900	32371	32761	33151	33541	33930
950	34319	34707	35095	35482	35869
1000	36256	36641	37027	37411	37795
1050	38179	38562	38944	39326	39706
1100	40087	40466	40845	41223	41600
1150	41976	42352	42727	43101	43474
1200	43846	44218	44588	44958	45326
1250	45694	46060	46425	46789	47152
1300	47513				

Type N Reference Function Coefficients

	−270°C to 0°C		0°C to 1300°C
a_0	0	a_0	0
a_1	$2.6159105962E + 01$	a_1	$2.5929394601E + 01$
a_2	$1.0957484228E - 02$	a_2	$1.5710141880E - 02$
a_3	$-9.3841111554E - 05$	a_3	$4.3825627237E - 05$
a_4	$-4.6412039759E - 08$	a_4	$-2.5261169794E - 07$
a_5	$-2.6303357716E - 09$	a_5	$6.4311819339E - 10$
a_6	$-2.2653438003E - 11$	a_6	$-1.0063471519E - 12$
a_7	$-7.6089300791E - 14$	a_7	$9.9745338992E - 16$
a_8	$-9.3419667835E - 17$	a_8	$-6.0863245607E - 19$
		a_9	$2.0849229339E - 22$
		a_{10}	$-3.0682196151E - 26$

Type N Inverse Function Coefficients

	−200°C to 0°C −3990 μV to 0 μV		0°C to 600°C 0 μV to 20613 μV		600°C to 1300°C 20613 μV to 47513 μV
d_0	0	d_0	0	d_0	$1.972485E + 01$
d_1	$3.8436847E - 02$	d_1	$3.86896E - 02$	d_1	$3.300943E - 02$
d_2	$1.1010485E - 06$	d_2	$-1.08267E - 06$	d_2	$-3.915159E - 07$
d_3	$5.2229312E - 09$	d_3	$4.70205E - 11$	d_3	$9.855391E - 12$
d_4	$7.2060525E - 12$	d_4	$-2.12169E - 18$	d_4	$-1.274371E - 16$
d_5	$5.8488586E - 15$	d_5	$-1.17272E - 19$	d_5	$7.767022E - 22$
d_6	$2.7754916E - 18$	d_6	$5.39280E - 24$		
d_7	$7.7075166E - 22$	d_7	$-7.98156E - 29$		
d_8	$1.1582665E - 25$				
d_9	$7.3138868E - 30$				

Tolerances (whichever is greater)

Class 1:	1.5°C or 0.4%	for −40°C to 1000°C
Class 2:	2.5°C or 0.75%	for −40°C to 1200°C
Class 3:	2.5°C or 1.5%	for −200°C to 40°C

Properties

Nominal composition: Nicrosil–Nisil   14.2% Nickel–84.4% Chromium–1.4% Silicon vs. 95.5% Nickel–4.4% Silicon–0.1% Magnesium.

Type N is a nominal replacement for Type K with a very similar temperature range but much higher reproducibility. It is suited to oxidising and inert environments and limited exposure in vacuum and reducing environments. In MIMS form with nicrosil® or nicrobel® or nicrobel sheathing it is the most stable of the base-metal thermocouples for the 300°C to 1200°C range. The wire and instrumentation are becoming more available.

D.8 TYPE R

Output in μV.

$T_{90}[^\circ C]$	0	10	20	30	40
0	0	54	111	171	232
50	296	363	431	501	573
100	647	723	800	879	959
150	1 041	1 124	1 208	1 294	1 381
200	1 469	1 558	1 648	1 739	1 831
250	1 923	2 017	2 112	2 207	2 304
300	2 401	2 498	2 597	2 696	2 796
350	2 896	2 997	3 099	3 201	3 304
400	3 408	3 512	3 616	3 721	3 827
450	3 933	4 040	4 147	4 255	4 363
500	4 471	4 580	4 690	4 800	4 910
550	5 021	5 133	5 245	5 357	5 470
600	5 583	5 697	5 812	5 926	6 041
650	6 157	6 273	6 390	6 507	6 625
700	6 743	6 861	6 980	7 100	7 220
750	7 340	7 461	7 583	7 705	7 827
800	7 950	8 073	8 197	8 321	8 446
850	8 571	8 697	8 823	8 950	9 077
900	9 205	9 333	9 461	9 590	9 720
950	9 850	9 980	10 111	10 242	10 374
1 000	10 506	10 638	10 771	10 905	11 039
1 050	11 173	11 307	11 442	11 578	11 714
1 100	11 850	11 986	12 123	12 260	12 397
1 150	12 535	12 673	12 812	12 950	13 089
1 200	13 228	13 367	13 507	13 646	13 786
1 250	13 926	14 066	14 207	14 347	14 488
1 300	14 629	14 770	14 911	15 052	15 193
1 350	15 334	15 475	15 616	15 758	15 899
1 400	16 040	16 181	16 323	16 464	16 605
1 450	16 746	16 887	17 028	17 169	17 310
1 500	17 451	17 591	17 732	17 872	18 012
1 550	18 152	18 292	18 431	18 571	18 710
1 600	18 849	18 988	19 126	19 264	19 402
1 650	19 540	19 677	19 814	19 951	20 087
1 700	20 222	20 356	20 488	20 620	20 749
1 750	20 877	21 003			

Type R Reference Function Coefficients.

	−50°C to 1064.18°C		1064.18°C to 1664.5°C		1664.5°C to 1768.1°C
a_1	$5.28961729765E + 00$	a_0	$2.95157925316E + 03$	a_0	$1.52232118209E + 05$
a_2	$1.39166589782E - 02$	a_1	$-2.52061251332E + 00$	a_1	$-2.68819888545E + 02$
a_3	$-2.38855693017E - 05$	a_2	$1.59564501865E - 02$	a_2	$1.71280280471E - 01$
a_4	$3.56916001063E - 08$	a_3	$-7.64085947576E - 06$	a_3	$-3.45895706453E - 05$
a_5	$-4.62347666298E - 11$	a_4	$2.05305291024E - 09$	a_4	$-9.34633971046E - 12$
a_6	$5.00777441034E - 14$	a_5	$-2.93359668173E - 13$		
a_7	$-3.73105886191E - 17$				
a_8	$1.57716482367E - 20$				
a_9	$-2.81033625251E - 24$				

Type R Inverse Function Coefficients.

	−50°C to 250°C −226 μV to 1923 μV		250°C to 1064°C 1923 μV to 11361 μV		1064°C to 1664.5°C 11361 μV to 19739 μV		1664.5°C to 1768.1°C 19739 μV to 21103 μV
d_0	0	d_0	$1.334584505E + 01$	d_0	$-8.199599416E + 01$	d_0	$3.406177836E + 04$
d_1	$1.8891380E - 01$	d_1	$1.472644573E - 01$	d_1	$1.553962042E - 01$	d_1	$-7.023729171E + 00$
d_2	$-9.3835290E - 05$	d_2	$-1.844024844E - 05$	d_2	$-8.342197663E - 06$	d_2	$5.582903813E - 04$
d_3	$1.3068619E - 07$	d_3	$4.031129726E - 09$	d_3	$4.279433549E - 10$	d_3	$-1.952394635E - 08$
d_4	$-2.2703580E - 10$	d_4	$-6.249428360E - 13$	d_4	$-1.191577910E - 14$	d_4	$2.560740231E - 13$
d_5	$3.5145659E - 13$	d_5	$6.468412046E - 17$	d_5	$1.492290091E - 19$		
d_6	$-3.8953900E - 16$	d_6	$-4.458750426E - 21$				
d_7	$2.8239471E - 19$	d_7	$1.994710149E - 25$				
d_8	$-1.2607281E - 22$	d_8	$-5.313401790E - 30$				
d_9	$3.1353611E - 26$	d_9	$6.481976217E - 35$				
d_{10}	$-3.3187769E - 30$						

Tolerances (whichever is greater)

Class 1:	1.0°C or $[1 + 0.3\%(t - 1100)]$	for 0°C to 1600°C
Class 2:	1.0°C or 0.25%	for 0°C to 1600°C

Properties

Nominal composition: Platinum–13% Rhodium vs. Platinum

Type R is suited for use at high temperature in oxidising and inert atmospheres. It may also be used intermittently in vacuum. At temperatures above 1100°C prolonged use results in grain growth in the platinum leg, making the thermocouple fragile. Type R is very prone to contamination, especially from metal vapours. Types R and S are the two most accurate of the designated thermocouples for high temperatures (200°C to 1400°C).

D.9 TYPE S

Output in μV.

$T_{90}[^\circ C]$	0	10	20	30	40
0	0	55	113	173	235
50	299	365	433	502	573
100	646	720	795	872	950
150	1 029	1 110	1 191	1 273	1 357
200	1 441	1 526	1 612	1 698	1 786
250	1 874	1 962	2 052	2 141	2 232
300	2 323	2 415	2 507	2 599	2 692
350	2 786	2 880	2 974	3 069	3 164
400	3 259	3 355	3 451	3 548	3 645
450	3 742	3 840	3 938	4 036	4 134
500	4 233	4 332	4 432	4 532	4 632
550	4 732	4 833	4 934	5 035	5 137
600	5 239	5 341	5 443	5 546	5 649
650	5 753	5 857	5 961	6 065	6 170
700	6 275	6 381	6 486	6 593	6 699
750	6 806	6 913	7 020	7 128	7 236
800	7 345	7 454	7 563	7 673	7 783
850	7 893	8 003	8 114	8 226	8 337
900	8 449	8 562	8 674	8 787	8 900
950	9 014	9 128	9 242	9 357	9 472
1 000	9 587	9 703	9 819	9 935	10 051
1 050	10 168	10 285	10 403	10 520	10 638
1 100	10 757	10 875	10 994	11 113	11 232
1 150	11 351	11 471	11 590	11 710	11 830
1 200	11 951	12 071	12 191	12 312	12 433
1 250	12 554	12 675	12 796	12 917	13 038
1 300	13 159	13 280	13 402	13 523	13 644
1 350	13 766	13 887	14 009	14 130	14 251
1 400	14 373	14 494	14 615	14 736	14 857
1 450	14 978	15 099	15 220	15 341	15 461
1 500	15 582	15 702	15 822	15 942	16 062
1 550	16 182	16 301	16 420	16 539	16 658
1 600	16 777	16 895	17 013	17 131	17 249
1 650	17 366	17 483	17 600	17 717	17 832
1 700	17 947	18 061	18 174	18 285	18 395
1 750	18 503	18 609			

Type S Reference Function Coefficients.

	−50°C to 1064.18°C		1064.18°C to 1664.5°C		1664.5°C to 1768.1°C
a_1	$5.40313308631E+00$	a_0	$1.32900444085E+03$	a_0	$1.46628232636E+05$
a_2	$1.25934289740E-02$	a_1	$3.34509311344E+00$	a_1	$-2.58430516752E+02$
a_3	$-2.32477968689E-05$	a_2	$6.54805192818E-03$	a_2	$1.63693574641E-01$
a_4	$3.22028823036E-08$	a_3	$-1.64856259209E-06$	a_3	$-3.30439046987E-05$
a_5	$-3.31465196389E-11$	a_4	$1.29989605174E-11$	a_4	$-9.43223690612E-12$
a_6	$2.55744251786E-14$				
a_7	$-1.25068871393E-17$				
a_8	$2.71443176145E-21$				

Type S Inverse Function Coefficients.

	−50°C to 250°C −236 µV to 1874 µV		250°C to 1064°C 1874 µV to 10332 µV		1064°C to 1664.5°C 10332 µV to 17536 µV		1664.5°C to 1768.1°C 17536 µV to 18694 µV
d_0	0	d_0	$1.291507177E+01$	d_0	$-8.087801117E+01$	d_0	$5.333875126E+04$
d_1	$1.84949460E-01$	d_1	$1.466298863E-01$	d_1	$1.621573104E-01$	d_1	$-1.235892298E+01$
d_2	$-8.00504062E-05$	d_2	$-1.534713402E-05$	d_2	$-8.536869453E-06$	d_2	$1.092657613E-03$
d_3	$1.02237430E-07$	d_3	$3.145945973E-09$	d_3	$4.719686976E-10$	d_3	$-4.265693686E-08$
d_4	$-1.52248592E-10$	d_4	$-4.163257839E-13$	d_4	$-1.441693666E-14$	d_4	$6.247205420E-13$
d_5	$1.88821343E-13$	d_5	$3.187963771E-17$	d_5	$2.081618890E-19$		
d_6	$-1.59085941E-16$	d_6	$-1.291637500E-21$				
d_7	$8.23027880E-20$	d_7	$2.183475087E-26$				
d_8	$-2.34181944E-23$	d_8	$-1.447379511E-31$				
d_9	$2.79786260E-27$	d_9	$8.211272125E-36$				

Tolerances (whichever is greater)

Class 1:	$1.0°C$ or $[1 + 0.3\%(t - 1100)]$	for 0°C to 1600°C
Class 2:	$1.0°C$ or 0.25%	for 0°C to 1600°C

Properties

Nominal composition: Platinum–13% Rhodium vs. Platinum

Type S is suited for use at high temperature in oxidising and inert atmospheres. It may also be used intermittently in vacuum. At temperatures above 1100°C prolonged use results in grain growth in the platinum leg, making the thermocouple fragile. Type S is very prone to contamination, especially from metal vapours. Types R and S are the two most accurate of the designated thermocouples for high temperatures (200°C to 1400°C).

D.10 TYPE T

Output in μV.

$T_{90}[°C]$	-0	-5	-10	-15	-20
-250	-6180	-6209	-6232	-6248	-6258
-225	-5950	-6007	-6059	-6105	-6146
-200	-5603	-5680	-5753	-5823	-5888
-175	-5167	-5261	-5351	-5439	-5523
-150	-4648	-4759	-4865	-4969	-5070
-125	-4052	-4177	-4300	-4419	-4535
-100	-3379	-3519	-3657	-3791	-3923
-75	-2633	-2788	-2940	-3089	-3235
-50	-1819	-1987	-2153	-2316	-2476
-25	-940	-1121	-1299	-1475	-1648
0	0	-193	-383	-571	-757

$T_{90}[°C]$	0	5	10	15	20
0	0	195	391	589	790
25	992	1196	1403	1612	1823
50	2036	2251	2468	2687	2909
75	3132	3358	3585	3814	4046
100	4279	4513	4750	4988	5228
125	5470	5714	5959	6206	6454
150	6704	6956	7209	7463	7720
175	7977	8237	8497	8759	9023
200	9288	9555	9822	10092	10362
225	10634	10907	11182	11458	11735
250	12013	12293	12574	12856	13139
275	13423	13709	13995	14283	14572
300	14862	15153	15445	15738	16032
325	16327	16624	16921	17219	17518
350	17819	18120	18422	18.725	19030
375	19335	19641	19947	20255	20563
400	20872				

Tolerances (whichever is greater)

Class 1:	1.5°C or 0.4%	for -40°C to 350°C
Class 2:	1.0°C or 0.75%	for -40°C to 350°C
Class 3:	1.0°C or 1.5%	for -200°C to 40°C

Type T Reference Function Coefficients.

	−270°C to 0°C		0°C to 400°C
a_0	0	a_0	0
a_1	$3.8748106364E + 01$	a_1	$3.8748106364E + 01$
a_2	$4.4194434347E - 02$	a_2	$3.3292227880E - 02$
a_3	$1.1844323105E - 04$	a_3	$2.0618243404E - 04$
a_4	$2.0032973554E - 05$	a_4	$-2.1882256846E - 06$
a_5	$9.0138019559E - 07$	a_5	$1.0996880928E - 08$
a_6	$2.2651156593E - 08$	a_6	$-3.0815758772E - 11$
a_7	$3.6071154205E - 10$	a_7	$4.5479135290E - 14$
a_8	$3.8493939883E - 12$	a_8	$-2.7512901673E - 17$
a_9	$2.8213521925E - 14$		
a_{10}	$1.4251594779E - 16$		
a_{11}	$4.8768662286E - 19$		
a_{12}	$1.0795539270E - 21$		
a_{13}	$1.3945027062E - 24$		
a_{14}	$7.9795153927E - 28$		

Type T Inverse Function Coefficients.

	−200°C to 0°C −5603 μV to 0 μV		0°C to 400°C 0 μV to 20872 μV
d_0	0	d_0	0
d_1	$2.5949192E - 02$	d_1	$2.592800E - 02$
d_2	$-2.1316967E - 07$	d_2	$-7.602961E - 07$
d_3	$7.9018692E - 10$	d_3	$4.637791E - 12$
d_4	$4.2527777E - 13$	d_4	$-2.165394E - 17$
d_5	$1.3304473E - 16$	d_5	$6.048144E - 20$
d_6	$2.0241446E - 20$	d_6	$-7.293422E - 25$
d_7	$1.2668171E - 24$		

Properties

Nominal composition: Copper–constantan Copper vs. 55% Copper–45% Nickel.

Type T is a very useful low-temperature thermocouple having a high reproducibility and an ability to withstand reducing, inert, vacuum and mildly oxidising environments. It also has a moderate resistance to corrosion in the presence of moisture, making it suitable for use at sub-zero temperatures. If restricted to temperature ranges below 150°C its reproducibility is very good. The wire and instrumentation are readily available.

Index